Digital Electronics 2

Series Editor
Robert Baptiste

Digital Electronics 2

Sequential and Arithmetic Logic Circuits

Tertulien Ndjountche

iSTE

WILEY

First published 2016 in Great Britain and the United States by ISTE Ltd and John Wiley & Sons, Inc.

ISTE Ltd
27-37 St George's Road
London SW19 4EU
UK

www.iste.co.uk

John Wiley & Sons, Inc.
111 River Street
Hoboken, NJ 07030
USA

www.wiley.com

Library of Congress Control Number: 2016945589

British Library Cataloguing-in-Publication Data
A CIP record for this book is available from the British Library
ISBN 978-1-84821-985-4

Contents

Preface

The omnipresence of electronic devices in everyday life is accompanied by the size reduction and the ever-increasing complexity of digital circuits. This comprehensive and easy-to-understand work deals with basic principles of digital electronics and allows the reader to grasp the subtleties of digital circuits from logic gates to finite-state machines. It presents all the aspects related to combinational logic and sequential logic. It introduces techniques to establish in a simple and concise manner logic equations, as well as methods for the analysis and design of digital circuits. Emphasis has been especially laid on design approaches that can be used to ensure a reliable operation of finite-state machines. Various programmable logic circuit structures and their applications are also presented. Each chapter includes practical examples and well-designed exercises with worked solutions.

This series of books discusses all the different aspects of digital electronics, following a descriptive approach combined with a gradual, detailed, and comprehensive presentation of basic concepts. The principles of combinational and sequential logic are presented, as well as the underlying techniques for the analysis and design of digital circuits. The analysis and design of digital circuits with increasing complexity is facilitated by the use of abstractions at the circuit and architecture levels. This work consists of three volumes devoted to the following subjects:

1) combinational logic circuits;

2) sequential and arithmetic logic circuits;

3) finite state machines.

A progressive approach has been chosen and the chapters are relatively independent of each other. To help master the subject matter and put into practice the different concepts and techniques, topics are complemented by a selection of exercises with solutions.

P.1. Summary

Volume 2 deals with sequential circuits and arithmetic and logic circuits. The logic state of the output of a sequential logic circuit can depend, at any given time, on the inputs but also on the previous logic state of the outputs. Depending on whether a clock signal is used to synchronize the output state change or not, a sequential circuit is said to be synchronous or asynchronous. Arithmetic circuits can be used to perform addition, subtraction, multiplication and division operations on digital data. Volume 2 contains the following seven chapters:

1) Latch and Flip-flop;

2) Binary Counters;

3) Shift Registers;

4) Arithmetic and Logic Circuits;

5) Digital Integrated Circuit Technology;

6) Semiconductor Memory;

7) Programmable Logic Circuits.

P.2. The reader

This work is an indispensable tool for all engineering students on a bachelors or masters course who wish to acquire detailed and practical knowledge of digital electronics. It is detailed enough to serve as a reference for electronic, automation and computer engineers.

Tertulien NDJOUNTCHE
June 2016

1

Latch and Flip-Flop

1.1. Introduction

A latch or flip-flop is a bistable circuit that is most often used in applications that require data storage. Its chief characteristic is that the output is not dependent solely on the present state of the input but also on the preceding output state. A bistable circuit has two complementary outputs that can assume either of the two logic levels 0 or 1.

There are several common types of latches and flip-flops. Latches often have no dedicated input for the clock signal. They can be combined to implement level-triggered and edge-triggered flip-flops. Flip-flops can be triggered by one of the levels or one of the edges of a clock signal (or a digital signal).

1.2. General overview

A simple latch can be implemented using two NOR or two NAND logic gates.

A NOR gate based latch with initial conditions specified is represented in Figure 1.1(a). The characteristic equation for each of the outputs is determined by assuming that the logic gates have different propagation times[1] and this may be modeled as for a delay, Δ, between a signal that becomes available at the output and the feedback signal applied to the input. In this way, the logic circuit of the latch, as illustrated in Figure 1.1(b), may be transformed as shown in Figures 1.1(c) and 1.1(d).

1 Propagation delays in logic gates are assumed to take the form 1 and $1 + \Delta$, respectively.

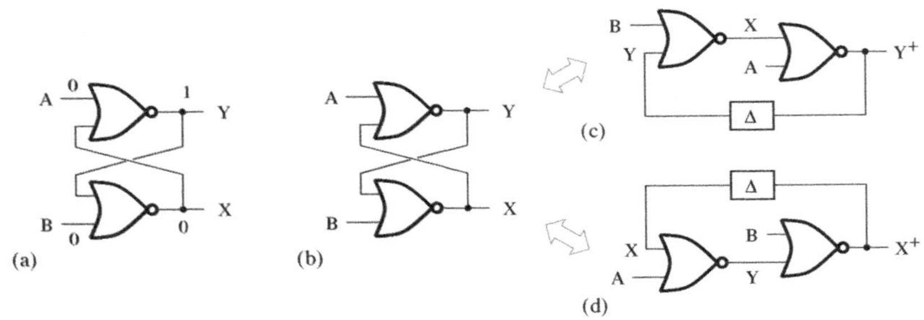

Figure 1.1. *a) NOR gate based latch with initial conditions specified; b) logic circuit for the latch and representations useful for the determination of c) Y^+ and d) X^+*

Referring to Figure 1.1(c), we can write:

$$X = \overline{B + Y} \quad [1.1]$$

$$Y^+ = \overline{A + X} \quad [1.2]$$

Substituting [1.1] into [1.2] yields:

$$Y^+ = \overline{A + \overline{B + Y}} \quad [1.3]$$

$$= \overline{A} \cdot \overline{\overline{B + Y}}$$

$$= \overline{A} \cdot (B + Y)$$

$$= \overline{A} \cdot B + \overline{A} \cdot Y \quad [1.4]$$

Similarly, the circuit shown in Figure 1.1(d) can be characterized using the following logic equations:

$$X^+ = \overline{B + Y} \quad [1.5]$$

$$Y = \overline{A + X} \quad [1.6]$$

By substituting [1.5] into [1.6], we have:

$$X^+ = \overline{B + \overline{A + X}} \quad [1.7]$$

$$= \overline{B} \cdot \overline{\overline{A + X}}$$

$$= \overline{B} \cdot (A + X)$$

$$= A \cdot \overline{B} + \overline{B} \cdot X \quad [1.8]$$

The characteristic equations of the NOR gate based latch are, thus, given by:

$$X^{+} = A \cdot \overline{B} + \overline{B} \cdot X \quad [1.9]$$

and

$$Y^{+} = \overline{A} \cdot B + \overline{A} \cdot Y \quad [1.10]$$

A	B	X	X^+	Y^+
0	0	0	0	1
0	0	1	1	0
0	1	0	1	0
0	1	1	1	0
1	0	0	0	1
1	0	1	0	1
1	1	0	0	0
1	1	1	0	0

Table 1.1. *State table of the NOR gate based latch*

For each output, the next state, X^+ or Y^+, depends on the present state, X or Y. In addition to the characteristic equations, the initial conditions must be specified to determine the operation of the latch. Table 1.1 gives the state table for the latch.

It must be noted that the two signals, X^+ and Y^+, are complementary except when both inputs, A and B, are set to 1.

Additionally, if the inputs A and B are simultaneously set to 0, the outputs can no longer be defined in a unique manner as the characteristic equations are verified by $(X, Y) = (1, 0)$ or by $(X, Y) = (0, 1)$. It is therefore impossible to predict the combination of the states held by the outputs.

In practice, sequential circuits are most often made to operate in the *fundamental mode*. This means that only one input can change states at any time. On the other hand, because of the difference in propagation delays between the logic gates, it is impossible to guarantee a simultaneous change in the state of two variables. Thus, the outputs of the latch are defined by $(X, Y) = (0, 1)$ when A is first set to 0 or by $(X, Y) = (1, 0)$ when B is first set to 0. In this case, the final state of the circuit is determined by the transient behavior, which depends on the order in which the state changes of the inputs take place. In general, if shifting from one state to another requires a change in at least two state variables, then a *race condition* will occur.

The race is said to be *non-critical* if the order in which the variables change state does not affect the final state of the circuit.

If, on the contrary, the circuit can assume two or more stable states depending on the order in which the variables change state, the race is said to be *critical.*

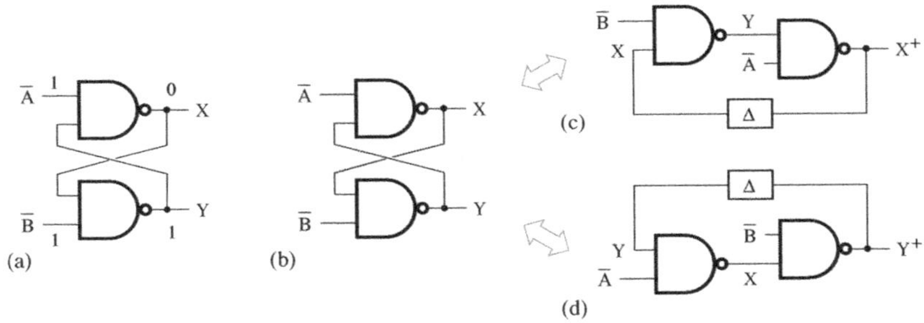

Figure 1.2. *a) NAND gate based latch with initial conditions specified; b) logic circuit of the latch and representations useful for the determination of c) X^+ and d) Y^+*

A NAND gate based latch with initial conditions specified is illustrated in Figure 1.2(a). Taking into account the fact that the differences in propagation delay of the two logic gates may translate into a delay, Δ, between an output and the feedback input, an equivalence may be established between the latch in Figure 1.2(b) and each representation shown in Figures 1.2(c) and 1.2(d).

The following logic equations may be derived based on the circuit shown in Figure 1.2(c):

$$X^+ = \overline{\overline{A} \cdot Y} \qquad [1.11]$$

$$Y = \overline{\overline{B} \cdot X} \qquad [1.12]$$

By substituting [1.12] into [1.11], we obtain:

$$X^+ = \overline{\overline{A} \cdot \overline{\overline{B} \cdot X}} \qquad [1.13]$$

$$= \overline{\overline{A}} + \overline{\overline{\overline{B} \cdot X}}$$

$$= A + \overline{B} \cdot X \qquad [1.14]$$

In the case of the circuit shown in Figure 1.2(d), the logic equations are written as follows:

$$X = \overline{\overline{A} \cdot Y} \qquad [1.15]$$

$$Y^{+} = \overline{\overline{B} \cdot X} \qquad [1.16]$$

Substituting [1.15] into [1.16], we obtain:

$$Y^{+} = \overline{\overline{B} \cdot \overline{\overline{A} \cdot Y}} \qquad [1.17]$$

$$= \overline{\overline{B}} + \overline{\overline{\overline{A} \cdot Y}}$$

$$= B + \overline{A} \cdot Y \qquad [1.18]$$

The characteristic equations of the NAND gate based latch are therefore in the following form:

$$X^{+} = A + \overline{B} \cdot X \qquad [1.19]$$

and

$$Y^{+} = B + \overline{A} \cdot Y \qquad [1.20]$$

$\overline{A}$	$\overline{B}$	X	X^{+}	Y^{+}
1	1	1	1	0
1	1	0	0	1
1	0	1	0	1
1	0	0	0	1
0	1	1	1	0
0	1	0	1	0
0	0	1	1	1
0	0	0	1	1

Table 1.2. *State table of the NAND gate based latch*

The state table of the NAND gate based latch may be constructed, as shown in Table 1.2, based on characteristic equations and initial conditions.

We can see that the signals X^{+} and Y^{+} are complementary except when the two inputs $\overline{A}$ and $\overline{B}$ are set at 0.

In addition, the signals X^{+} and Y^{+} are only defined uniquely when the inputs $\overline{A}$ and $\overline{B}$ cannot change states from 0 to 1 simultaneously. Thus, the outputs of the latch are defined by $(X,Y) = (0,1)$ if the input $\overline{A}$ is first set to 1 or by $(X,Y) = (1,0)$ if the input $\overline{B}$ is first set to 1. In this case, as the final state depends on the order in which the inputs change states, we have a critical race condition.

Among the combinations of states that the outputs of the latch can take, only those for which $X^+ = X$ and $Y^+ = Y$ are said to be *stable*.

1.2.1. *SR latch*

For the SR latch (S stands for *set*, and R for *reset*) represented in Figure 1.3, we can obtain the characteristic equations from equations [1.9] and [1.10], as follow:

$$Q^+ = \overline{R} \cdot S + \overline{R} \cdot Q = \overline{R} \cdot (S + Q) \qquad [1.21]$$

and:

$$\overline{Q^+} = \overline{S} \cdot R + \overline{S} \cdot \overline{Q} = \overline{S} \cdot (R + \overline{Q}) \qquad [1.22]$$

It must be noted that complementing Q^+ does not yield $\overline{Q^+}$. The state table is given in Table 1.3.

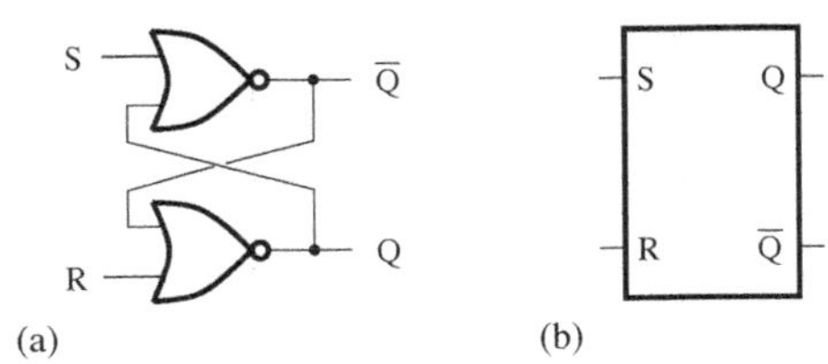

Figure 1.3. *SR latch: a) logic circuit; b) symbol*

S	R	Q	Q^+	$\overline{Q^+}$
0	0	0	0	1
0	0	1	1	0
0	1	0	0	1
0	1	1	0	1
1	0	0	1	0
1	0	1	1	0
1	1	0	*0*	*0*
1	1	1	*0*	*0*

Table 1.3. *State table of the SR latch*

S	R	Q^+	$\overline{Q^+}$	
0	0	Q	$\overline{Q}$	No change
0	1	0	1	Reset Q^+ to 0
1	0	1	0	Set Q^+ to 1
1	1	0	0	Forbidden state

Table 1.4. *Truth table of the SR latch*

An SR latch whose initial condition is specified can also be characterized using the truth table shown in Table 1.4. The SR latch is said to be *reset-dominant 0*, as setting both inputs to 1 causes the output Q to change to 0.

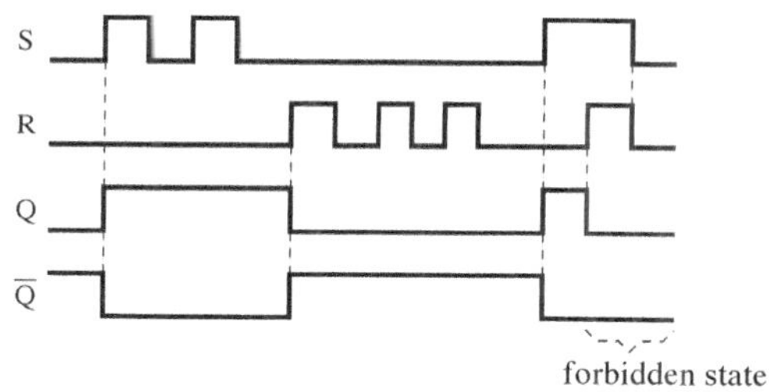

Figure 1.4. *Timing diagram for the SR latch*

Figure 1.4 shows the timing diagram of the SR latch where the different operating modes that appear in the truth table can be observed.

S	R	Q	Q^+
0	0	0	0
0	0	1	1
0	1	0	0
0	1	1	0
1	0	0	1
1	0	1	1
1	1	0	x
1	1	1	x

Table 1.5. *SR latch state table with do not care states*

However, if the forbidden state ($S = R = 1$) is considered as a do not care state, the state table takes the form given in Table 1.5. Constructing a Karnaugh map, as shown in Figure 1.5, we obtain another version of the characteristic equation given by:

$$Q^{+} = S + Q \cdot \overline{R} \qquad \text{and} \quad S \cdot R = 0 \qquad [1.23]$$

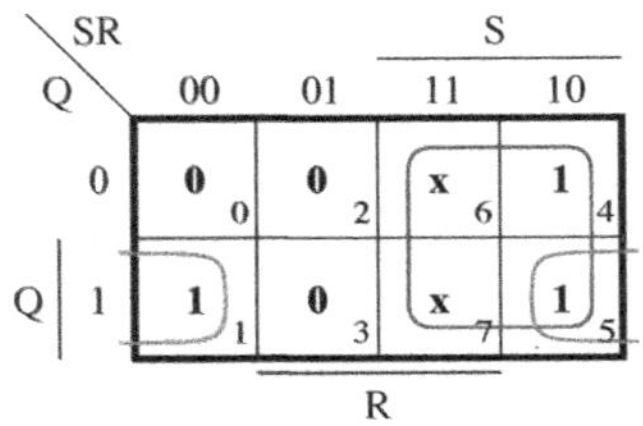

Figure 1.5. *Karnaugh map for the SR latch. For a color version of this figure, see www.iste.co.uk/ndjountche/electronics2.zip*

This last equation is used for applications where neither of the inputs S and R can take the state 1.

When a transition requires a change in state for at least two variables, an analysis based on Karnaugh maps, as shown in Figure 1.6, is necessary to detect the critical race conditions.

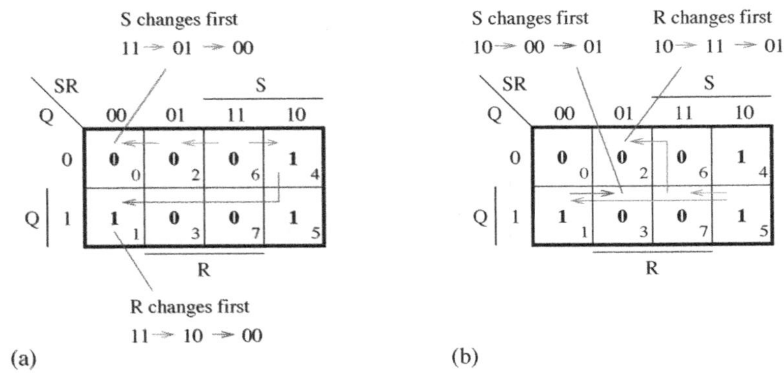

Figure 1.6. *Karnaugh map: a) critical race; b) non-critical race. For a color version of this figure, see www.iste.co.uk/ndjountche/electronics2.zip*

Let us consider that from the initial state, where $S = 1$, $R = 1$, and $Q = 0$, and which corresponds to the cell 6 in the Karnaugh map of Figure 1.6(a), both inputs S and R must be reset to zero.

The state of the input S can change before that of the input R, or *vice versa*.

The arrows entered in the Karnaugh map are used to illustrate the response of the latch in each case.

In SR terms, the transition $11 \rightarrow 01 \rightarrow 00$ is produced, and the output is maintained at the final state $Q^+ = 0$, corresponding to cell 0, if the input S changes first. However, if the input R changes first, the transition will be $11 \rightarrow 10 \rightarrow 00$, and the final state of the output is then $Q^+ = 1$, corresponding to cell 1.

In the case of Figure 1.6(b), the flip-flop is initially characterized by $S = 1$, $R = 0$ and $Q = 1$; this corresponds to the cell 5 in the Karnaugh map.

As a result of the possible transitions, $10 \rightarrow 00 \rightarrow 01$ when S changes first, or $10 \rightarrow 11 \rightarrow 01$ when R changes first, the output takes the same final state, $Q^+ = 0$, corresponding to cells 3 or 2. This corresponds to a non-critical race condition.

We can verify that the only critical race condition in an SR latch occurs when the inputs S and R that are initially set to 1 are reset to 0.

1.2.2. $\overline{S}\,\overline{R}$ latch

An $\overline{S}\,\overline{R}$ latch can be implemented using NAND gates, as shown in Figure 1.7(a). Its symbol is represented in Figure 1.7(b). Based on the truth table shown in Table 1.6, we can note that the inputs are activated by low-level signals. The $\overline{S}\,\overline{R}$ latch is said to be *set-dominant 1*, as setting both inputs to 1 changes the output Q to 1.

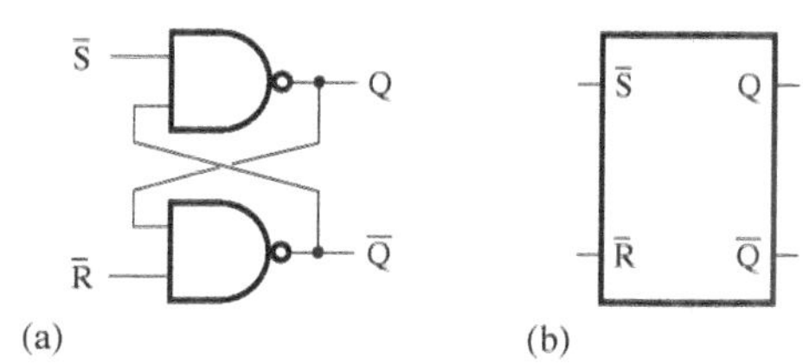

Figure 1.7. *$\overline{S}\,\overline{R}$ latch. a) logic circuit; b) symbol*

The effect of a race condition on the operation of the latch can be analyzed using a Karnaugh map.

$\overline{S}$	$\overline{R}$	Q^+	$\overline{Q^+}$	
1	1	Q	$\overline{Q}$	No change
1	0	0	1	Reset Q^+ to 0
0	1	1	0	Set Q^+ to 1
0	0	1	1	Forbidden state

Table 1.6. *Truth table of the $\overline{S}\,\overline{R}$ latch*

By referring to Figure 1.8(a), we can see that the flip-flop is initially characterized by $\overline{S} = 0$ and $\overline{R} = 0$, and $Q = 1$ (cell 1). The transition of the inputs $\overline{S}$ and $\overline{R}$ to 1 involves a change in two state variables. If, due to the difference in propagation delays, the input $\overline{S}$ changes first, this translates to the transitions, $00 \rightarrow 10 \rightarrow 11$, and the final state of the output is $Q^+ = 0$ (cell 6). If, on the other hand, the input $\overline{R}$ changes first, the latch follows the transitions, $00 \rightarrow 01 \rightarrow 11$, and the output takes the final state $Q^+ = 1$ (cell 7). This is a critical race condition because the final state of the outputs depends on the order in which the variables change.

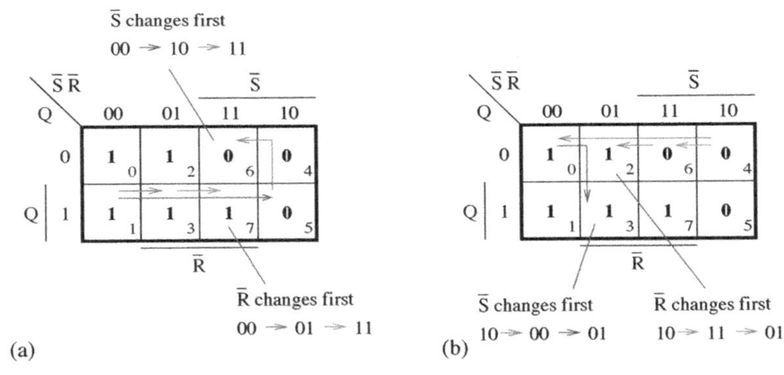

Figure 1.8. *Karnaugh map: a) critical race; b) non-critical race. For a color version of this figure, see www.iste.co.uk/ndjountche/electronics2.zip*

An example of a non-critical race condition is illustrated by the Karnaugh map, as shown in Figure 1.8(b). Starting from the state $\overline{S} = 1$ and $\overline{R} = 0$, and $Q = 0$ (cell 4), the inputs $\overline{S}$ and $\overline{R}$ must be set to 0 and 1, respectively. The two possible transitions $10 \rightarrow 00 \rightarrow 01$ (input $\overline{S}$ changes first) and $10 \rightarrow 11 \rightarrow 01$ (input $\overline{R}$ changes first) lead to the same final state for the output, $Q^+ = 1$ (cell 3 or 2).

For the $\overline{S}\,\overline{R}$ latch, the only critical race condition occurs when both inputs $\overline{S}$ and $\overline{R}$ move from 0 to 1.

1.2.3. *Application: switch debouncing*

Contact bounces of a push-button switch (see Figure 1.9) during its closing or opening can be eliminated using a $\overline{S}\ \overline{R}$ latch, as shown in Figure 1.10, where V_{CC} represents the supply voltage and R_P is the polarization resistor.

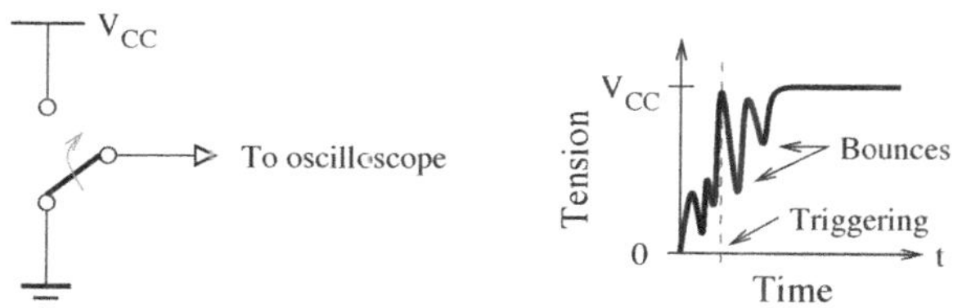

Figure 1.9. *Waveform illustrating switch contact bounces*

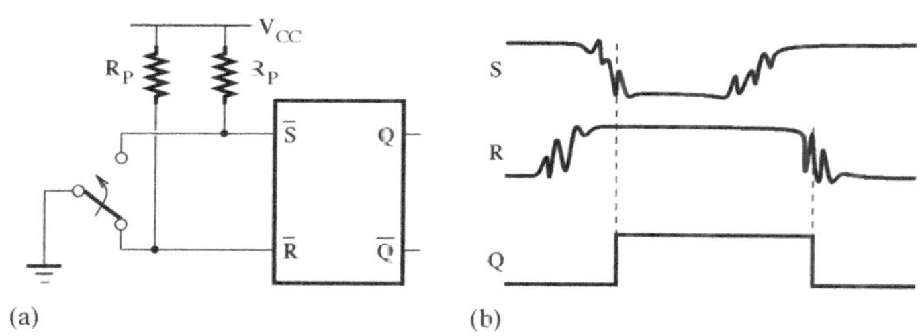

Figure 1.10. *Debouncing switch*

When $\overline{R} = 0$, the output Q of the latch is set to 1 as soon as the signal $\overline{S}$ reaches the logic level 1 for the first time. Subsequent fluctuations at the input $\overline{S}$ no longer affect the state of Q. Similarly, when $\overline{S}$ is at 0, the output Q is reset to 0 following the first transition attributing the logic level 1 to $\overline{R}$.

1.3. Gated SR latch

A gated or level-sensitive SR latch uses a control signal C that can be a clock signal. The signal C is used to enable (or inhibit) the latch at specific time intervals.

1.3.1. *Implementation based on an SR latch*

The gated SR latch in Figure 1.11(a) is made up of two AND gates and an SR latch. It is represented by the symbol shown in Figure 1.11(b). It can be characterized by equations of the form:

$$X^+ = A \cdot \overline{B} + \overline{B} \cdot X \quad [1.24]$$

and

$$Y^+ = \overline{A} \cdot B + \overline{A} \cdot Y \quad [1.25]$$

where:

$$A = RC, \quad B = SC, \quad X = Q, \quad X^+ = Q^+, \quad Y = \overline{Q},$$
$$\text{and} \quad Y^+ = \overline{Q^+} \quad [1.26]$$

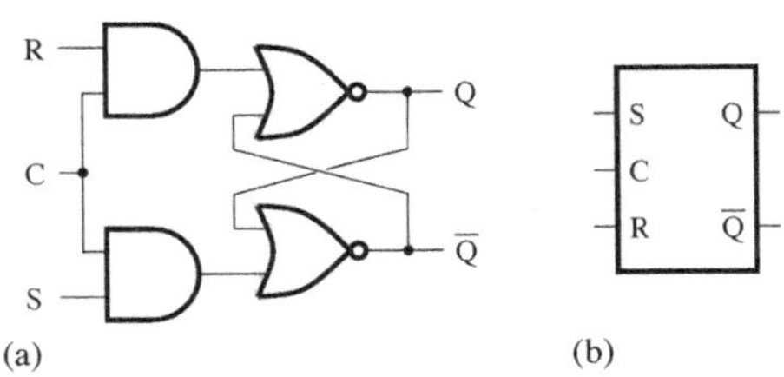

Figure 1.11. *Gated SR latch based on an SR latch: a) logic circuit; b) symbol*

The characteristic equations are, thus, given by:

$$\begin{aligned} Q^+ &= S \cdot C \cdot (\overline{R \cdot C}) + (\overline{R \cdot C}) \cdot Q \\ &= \overline{R} \cdot S \cdot C + (\overline{R} + \overline{C}) \cdot Q \end{aligned} \quad [1.27]$$

and

$$\begin{aligned} \overline{Q^+} &= (\overline{S \cdot C}) \cdot R \cdot C + (\overline{S \cdot C}) \cdot \overline{Q} \\ &= R \cdot \overline{S} \cdot C + (\overline{S} + \overline{C}) \cdot \overline{Q} \end{aligned} \quad [1.28]$$

– If $C = 0$, we have $Q^+ = Q$ and $\overline{Q^+} = \overline{Q}$.

– If $C = 1$, we have $Q^+ = \overline{R} \cdot (S + Q)$ and $\overline{Q^+} = \overline{S} \cdot (R + \overline{Q})$.

Table 1.7 presents the state table of the gated SR latch based on an SR latch. The truth table can be constructed as shown in Table 1.8. An example of the timing diagram is illustrated in Figure 1.12, for the case where $Q = 0$ and $\overline{Q} = 1$ initially.

C	S	R	Q	Q^+	$\overline{Q^+}$
0	x	x	0	0	1
0	x	x	1	1	0
1	0	0	0	0	1
1	0	0	1	1	0
1	0	1	x	0	1
1	1	0	x	1	0
1	1	1	x	0	0

Table 1.7. *State table of the gated SR latch based on an SR latch*

C	S	R	Q^+	$\overline{Q^+}$	
0	x	x	Q	$\overline{Q}$	No change
1	0	0	Q	$\overline{Q}$	
1	0	1	0	1	Reset
1	1	0	1	0	Set
1	1	1	0	0	Forbidden state

Table 1.8. *Truth table of the gated SR latch based on an SR latch*

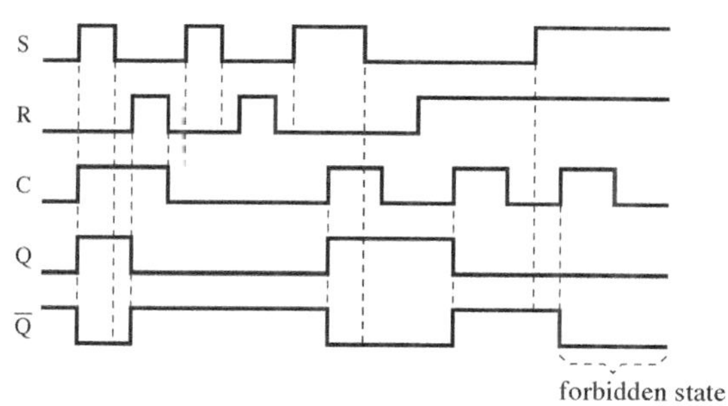

Figure 1.12. *Timing diagram of the gated SR latch*

1.3.2. *Implementation based on an $\overline{S}\,\overline{R}$ latch*

Another version of the gated SR latch, whose logic circuit and symbol are given in Figures 1.13(a) and 1.13(b), is implemented using two NAND gates and an $\overline{S}\ \overline{R}$ latch. By performing its analysis, the following equations can be derived:

$$X^+ = A + \overline{B} \cdot X \qquad [1.29]$$

and:

$$Y^+ = B + \overline{A} \cdot Y \qquad [1.30]$$

where:

$$\overline{A} = \overline{S \cdot C},\ \ \overline{B} = \overline{R \cdot C},\ X = Q,\ \ X^+ = Q^+,\ Y = \overline{Q},\ \text{and } Y^+ = \overline{Q^+} \qquad [1.31]$$

and finally we have:

$$Q^+ = S \cdot C + (\overline{R} + \overline{C}) \cdot Q \qquad [1.32]$$

and:

$$\overline{Q^+} = R \cdot C + (\overline{S} + \overline{C}) \cdot \overline{Q} \qquad [1.33]$$

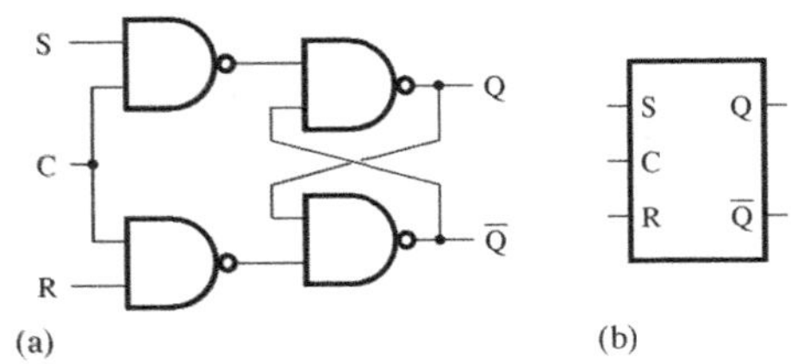

Figure 1.13. *Gated SR latch based on an $\overline{S}\ \overline{R}$ latch: a) logic circuit; b) symbol*

The truth table of the gated SR latch based on an $\overline{S}\ \overline{R}$ latch can, therefore, be constructed as shown in Table 1.9.

C	S	R	Q^+	$\overline{Q^+}$	
0	x	x	Q	$\overline{Q}$	No change
1	0	0	Q	$\overline{Q}$	
1	0	1	0	1	Reset
1	1	0	1	0	Set
1	1	1	1	1	Forbidden state

Table 1.9. *Truth table of the gated SR latch based on an $\overline{S}\,\overline{R}$ latch*

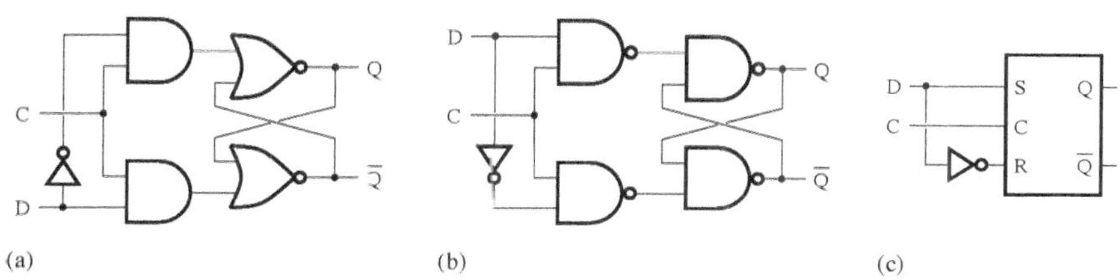

Figure 1.14. *Gated D latch: a) and b) logic circuits; c) symbol*

1.4. Gated D latch

A gated D latch (D stands for *data*) can be implemented from a gated SR latch, as shown in Figure 1.14. Connecting an inverter between the S and R inputs prevents the forbidden state from occurring. By inserting the expressions:

$$R = \overline{D} \quad \text{and} \quad S = D \qquad [1.34]$$

in any of the following two characteristic equations of the gated SR latches:

$$Q^+ = \overline{R} \cdot S \cdot C + (\overline{R} - \overline{C}) \cdot Q \qquad [1.35]$$

and

$$Q^+ = S \cdot C + (\overline{R} + \overline{C}) \cdot Q \qquad [1.36]$$

we obtain, for the gated D latch, the same characteristic equation, given by:

$$\begin{aligned} Q^+ &= D \cdot C + D \cdot Q + \overline{C} \cdot Q \\ &= D \cdot C \cdot (Q + \overline{Q}) + D \cdot (C + \overline{C}) \cdot Q + (D + \overline{D}) \cdot \overline{C} \cdot Q \\ &= D \cdot C \cdot (Q + \overline{Q}) + \overline{C} \cdot Q \cdot (D + \overline{D}) \\ &= D \cdot C + \overline{C} \cdot Q \end{aligned} \qquad [1.37]$$

– If $C = 1$, the characteristic equation becomes $Q^+ = D$.

– If $C = 0$, we have $Q^+ = Q$.

With a gated D latch, the state of the input D is transferred to the output when the control (or enable) input C is set to 1, while the state of the output does not change when the control input is reset to 0; this translates into a characteristic equation of the form:

$$Q^+ = D \cdot C + \overline{C} \cdot Q \quad [1.38]$$

The gated D latch is thus said to be *transparent* when $C = 1$. It is, therefore, sensitive to the high level of the signal applied at the input C.

Figure 1.15 shows the symbol of a gated D latch. The truth table of a gated D latch is represented in Table 1.10, where the outputs Q^+ and $\overline{Q^+}$ are complementary. An example of the timing diagram for the D latch is given in Figure 1.16, where the output Q is initially set to 0.

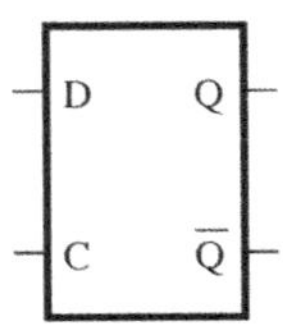

Figure 1.15. *Symbol of the gated D latch*

C	D	Q^+	$\overline{Q^+}$	
0	x	Q	$\overline{Q}$	No change
1	0	0	1	Reset
1	1	1	0	Set

Table 1.10. *Truth table of the gated D latch*

1.5. Basic JK flip-flop

The JK flip-flop (J as a set input, and K as a reset input) is the most versatile of the basic flip-flops. When it is activated, it permits the storage of a binary data based on the combination of states taken by the inputs J and K. A JK flip-flop can be implemented by using the logic circuit given in Figure 1.17(a). It is symbolically represented as shown in Figure 1.17(b). From the logic circuit of the JK flip-flop, we can obtain:

$$S = J \cdot C \cdot \overline{Q} \quad \text{and} \quad R = K \cdot C \cdot Q \qquad [1.39]$$

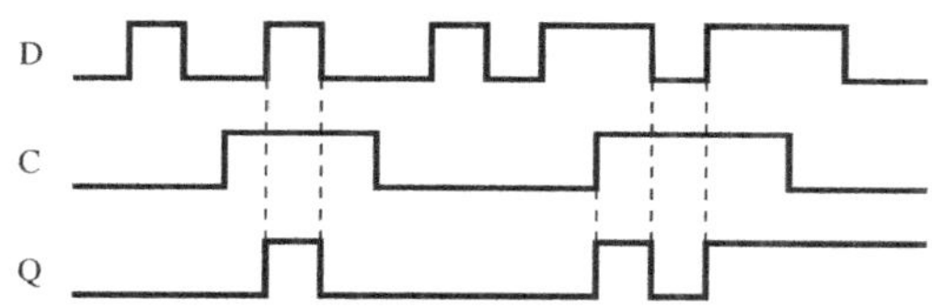

Figure 1.16. *Timing diagram for the gated D latch*

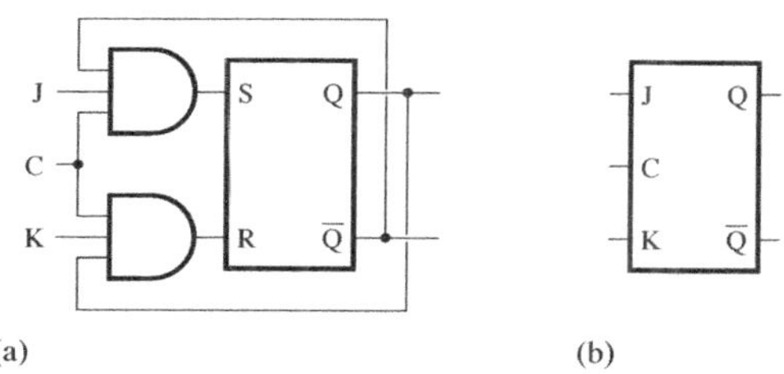

Figure 1.17. *Basic JK flip-flop: a) logic circuit; b) symbol*

By inserting these last expressions in the characteristic equation of the gated SR latch:

$$Q^+ = \overline{R} \cdot (S + Q) \qquad [1.40]$$

we get

$$\begin{aligned} Q^+ &= (\overline{K \cdot C \cdot Q}) \cdot (J \cdot C \cdot \overline{Q} + Q) \\ &= (\overline{K} + \overline{C} + \overline{Q}) \cdot (J \cdot C + Q) \\ &= J \cdot \overline{K} \cdot C + J \cdot \overline{Q} \cdot C + \overline{K} \cdot Q + Q \cdot \overline{C} \\ &= (1 + J \cdot C) \cdot \overline{K} \quad Q + (1 + \overline{K}) \cdot J \cdot \overline{Q} \cdot C + Q \cdot \overline{C} \\ &= J \cdot \overline{Q} \cdot C + (\overline{K} + \overline{C}) \cdot Q \end{aligned} \qquad [1.41]$$

– if $C = 1$, the characteristic equation takes the form $Q^+ = J \cdot \overline{Q} + \overline{K} \cdot Q$;

– if $C = 0$, we have $Q^+ = Q$.

The state table of the basic JK flip-flop can be constructed as shown in Table 1.11. The forbidden state, inherent to the SR latch, is eliminated by adding two feedback pathways in order to ensure that the output will be set to 1 only if $Q = 0$ and reset to 0 only if $Q = 1$. Table 1.12 presents the truth table of the basic JK flip-flop, where the outputs Q^+ and $\overline{Q^+}$ are complementary.

C	J	K	Q	Q^+
0	x	x	x	Q
1	0	0	0	0
1	0	0	1	1
1	0	1	0	0
1	0	1	1	0
1	1	0	0	1
1	1	0	1	1
1	1	1	0	1
1	1	1	1	0

Table 1.11. *State table for the JK flip-flop*

C	J	K	Q^+	$\overline{Q^+}$	
0	x	x	Q	$\overline{Q}$	No change
1	0	0	Q	$\overline{Q}$	
1	0	1	0	1	Reset
1	1	0	1	0	Set
1	1	1	$\overline{Q}$	Q	Toggle

Table 1.12. *Truth table of the basic JK flip-flop*

It must be noted that this JK flip-flop structure may be affected by undesirable oscillations. In fact, when the two inputs J and K are set at 1 and the clock signal changes to 1, the feedback of the values Q and $\overline{Q}$ taken by the outputs forces the flip-flop to toggle (or to switch from one state to its logical complement). And if the clock signal is still at the logic state 1, the process recommences and the flip-flop again changes state. To ensure smooth operation, the pulse width of the clock signal must be smaller than the propagation delay of the flip-flop.

1.6. T flip-flop

A JK flip-flop can be transformed into a T flip-flop (T stands for *toggle*), as shown in Figure 1.18. When the T flip-flop is activated, its outputs change state every time a pulse is applied to the input T. The characteristic equation of the JK flip-flop is given by:

$$Q^+ = J \cdot \overline{Q} \cdot C + (\overline{K} + \overline{C}) \cdot Q \qquad [1.42]$$

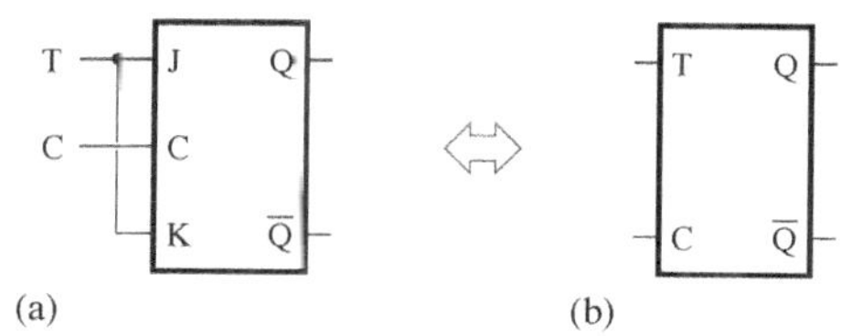

Figure 1.18. *T flip-flop: a) logic circuit; b) symbol*

Assuming that $J = K = T$, we obtain the characteristic equation of the T flip-flop:

$$Q^+ = T \cdot \overline{Q} \cdot C + (\overline{T} + \overline{C}) \cdot Q \qquad [1.43]$$

– if $C = 1$, the characteristic equation is reduced to $Q^+ = T \cdot \overline{Q} + \overline{T} \cdot Q = T \oplus Q$;

– if $C = 0$, we have $Q^+ = Q$.

Table 1.13 shows the state table of the T flip-flop. As the outputs Q^+ and $\overline{Q^+}$ are complementary, the truth table for the T flip-flop can be constructed as shown in Table 1.14.

C	T	Q	Q^+
0	x	x	Q
1	0	0	0
1	0	1	1
1	1	0	1
1	1	1	0

Table 1.13. *State table of the T flip-flop*

C	T	Q^+	$\overline{Q^+}$	
0	x	Q	$\overline{Q}$	No change
1	0	Q	$\overline{Q}$	
1	1	$\overline{Q}$	Q	Toggle

Table 1.14. *Truth table of the T flip-flop*

1.7. Master-slave and edge-triggered flip-flop

The operation of circuits implemented by coupling level-triggered flip-flops may become unpredictable, as the signal state can propagate from the output of one flip-flop to another as long as the clock signal is activated, thus preventing data storage.

One solution to this problem consists of using master-slave or edge-triggered flip-flops. This is implemented by memorizing only those state changes that occur on receiving one of the edges of the clock signal as illustrated in Figure 1.19.

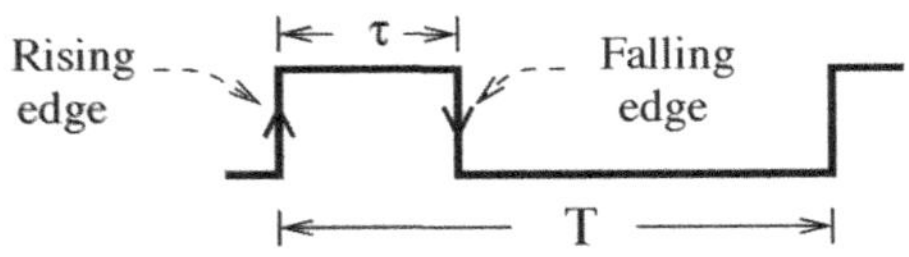

Figure 1.19. *Clock signal (τ: pulse width; T: signal period)*

1.7.1. *Master-slave flip-flop*

A Master-slave type flip-flop is implemented by connecting two flip-flops, called master and slave, whose clock signals are complementary.

1.7.1.1. *Master-slave D flip-flop*

An edge-triggered D flip-flop can be implemented using a master-slave structure that is composed of two gated D latches (see Figures 1.20 and 1.22).

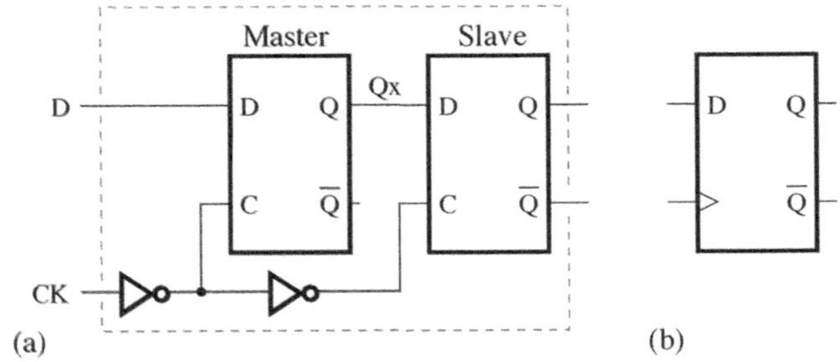

Figure 1.20. *Master-slave D flip-flop triggered by the clock signal rising edge: a) logic circuit; b) symbol*

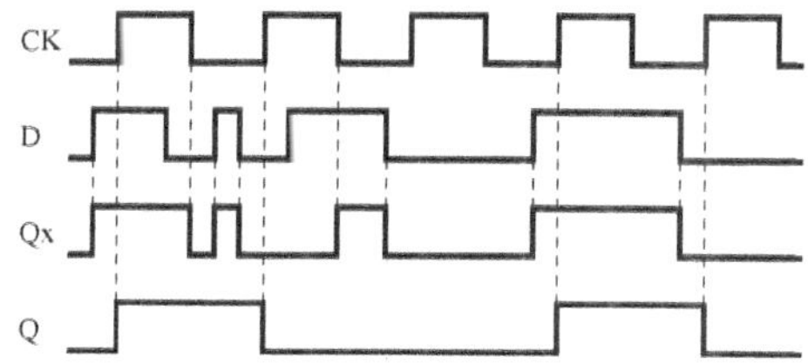

Figure 1.21. *Timing diagram of a master-slave D flip-flop triggered by the clock signal rising edge*

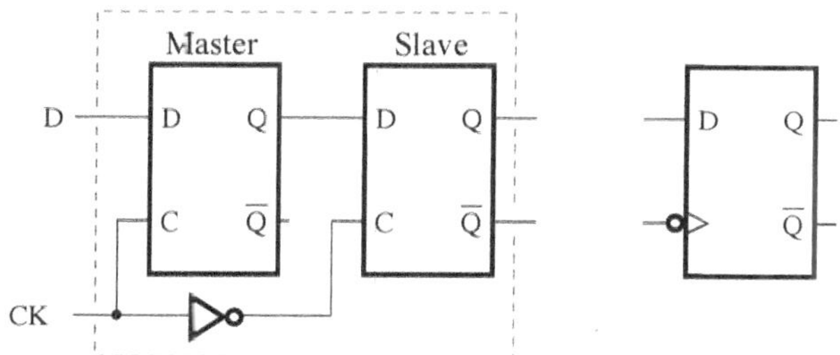

(a) (b)

Figure 1.22. *Master-slave D flip-flop triggered by the clock signal falling edge: a) logic circuit; b) symbol*

The master latch (or first latch) remains sensitive to changes in the input logic state as long as it is activated by the clock signal, but the output of the slave latch (or the second latch) only changes at the edges of the clock signal, when the master latch becomes deactivated and its state can no longer change. Thus, the output of the master-slave flip-flop only reflects the input logic state when the clock signal goes from high to low or vice versa.

Figures 1.20(a) and 1.20(b) show the logic circuit and symbol, respectively, for a D flip-flop triggered by the rising edge of the clock signal (or positive-edge-triggered D flip-flop). Table 1.15 gives the truth table. Figure 1.21 shows the timing diagram for a D flip-flop triggered by the rising edge of the clock signal.

In the case of the D flip-flop triggered by the falling edge of the clock signal (or negative-edge-triggered D flip-flop), the logic circuit and symbol are as represented in Figures 1.22(a) and 1.22(b), respectively. The truth table is given in Table 1.16.

D	CK	Q^+	$\overline{Q+}$
x	0	Q	$\overline{Q}$
x	1	Q	$\overline{Q}$
0	↑	0	1
1	↑	1	0

Table 1.15. *Truth table*

D	CK	Q^+	$\overline{Q+}$
x	0	Q	$\overline{Q}$
x	1	Q	$\overline{Q}$
0	↓	0	1
1	↓	1	0

Table 1.16. *Truth table*

1.7.1.2. *JK master-slave flip-flop*

A JK master-slave flip-flop can be described using the logic circuit and symbol represented in Figures 1.23(a) and 1.23(b), respectively, while its operation is characterized by the truth table given in Table 1.17.

J	K	CK	Q^+	$\overline{Q+}$
x	x	0	Q	$\overline{Q}$
0	0	⊓	Q	$\overline{Q}$
0	1	⊓	0	1
1	0	⊓	1	0
1	1	⊓	$\overline{Q}$	Q

Table 1.17. *Truth table*

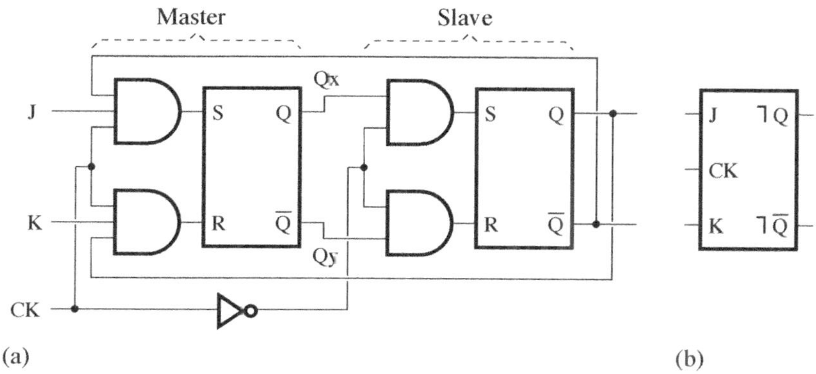

Figure 1.23. *JK master-slave flip-flop: a) logic circuit; b) symbol*

When the master flip-flop is activated, its output logic state is determined not only by the inputs J and K, but also by the outputs, Q and $\overline{Q}$, of the slave flip-flop. The master flip-flop state is then transferred to the slave flip-flop only when the clock signal transitions from high to low (falling edge).

Thus, to ensure the normal operation of the JK master-slave flip-flop, the logic state taken by each input, J and K, must not change when the master flip-flop is activated (or the clock signal CK is set to 1). If this condition is not satisfied, the outputs of the JK master-slave flip-flop may be affected by the undesirable catching of a logic state 1 or 0 by the master flip-flop.

– when the output Q of the slave flip-flop is at 0, the transition from 0 to 1 of the input J when $CK = 1$ results in the master flip-flop output being set at 1, and the slave flip-flop output can then be set to 1 when CK goes from 1 to 0. Once the master flip-flop is set to 1 following a change to 1 in the input J, a subsequent assignment of 1 to the input K when $CK = 1$ cannot bring the master flip-flop output back to 0. This is because the slave flip-flop remains in the same state until the clock signal, CK, again changes to 0 and the feedback signal $Q = 0$ keeps the input K deactivated. This behavior is known as *1s catching*;

– in the case where the slave flip-flop output is at 1 and a transient disturbance forces the input K to change to 1 while $CK = 1$, the master flip-flop acquires this reset condition, which is then transferred to the slave flip-flop when the clock signal CK goes from 1 to 0. It must be noted that K subsequently changing to 1 while $CK = 1$ has no effect on the master flip-flop that can only be set to 1 by a high-going pulse at the input J, which is actually deactivated by the feedback signal $\overline{Q} = 0$. This phenomenon is called *0s catching*.

The JK master-slave flip-flop can be considered to be level triggered. The symbol ⌉ is used in Figure 1.23(b) to indicate that the outputs of the JK master-slave flip-flop only reflect the state of the J and K inputs at the end of the pulse of the clock signal CK.

The truth table (see Table 1.17) is constructed assuming that the input signals J and K remain constant while the clock signal is set to 1 and, thus, does not take into account the 1s catching and 0s catching problem.

Figure 1.24 shows a timing diagram showing a 1s catching and 0s catching in a JK master-slave flip-flop.

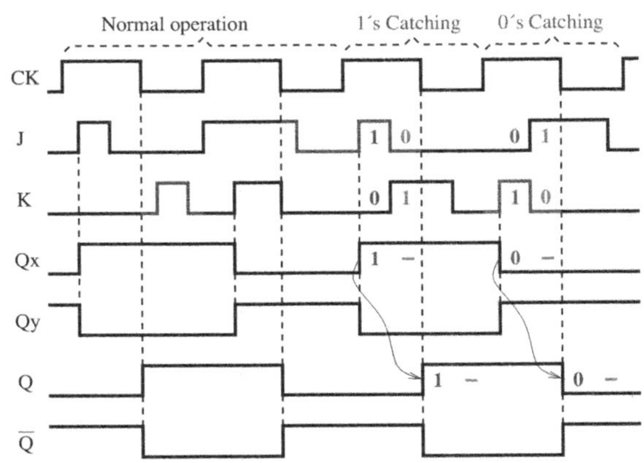

Figure 1.24. *Timing diagram for the JK master-slave flip-flop (illustration of 1s and 0s catching). For a color version of this figure, see www.iste.co.uk/ndjountche/electronics2.zip*

1.7.2. *Edge-triggered flip-flop*

An edge-triggered flip-flop is designed so as to ensure that the output can only change at the rising or falling edge of the clock signal and remains constant between two consecutive edges.

1.7.2.1. *Principle of edge detection*

Even though the circuits shown in Figure 1.25 are not exactly the same as those found in integrated flip-flops, they clearly demonstrate the detection principle of the edge of a signal.

The propagation delay caused by an inverter is exploited to generate a signal with a very small width during any transition of the clock signal.

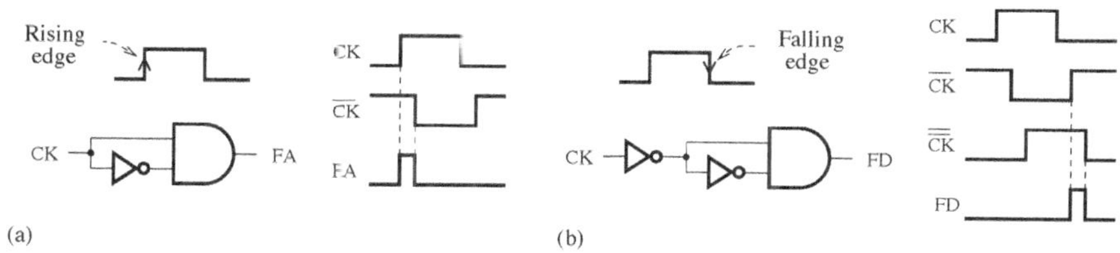

Figure 1.25. *Principle for detecting a) the rising edge and b) falling edge*

1.7.2.2. *Edge-triggered D flip-flop*

In an edge-triggered D flip-flop, the detection of the clock signal transition in a given direction can be carried out by making use of the fact that the change in state (set, reset) of an SR or $\overline{S}\ \overline{R}$ latch occurs only when the logic states of both inputs change. Thus, the state acquired by a latch after a clock signal transition occurs at one of the inputs while the other input is set to 1 or reset to 0; it cannot change only because of subsequent changes in the logic state of the clock signal.

Flip-flops can be triggered by the rising edge or the falling edge of the clock signal.

D flip-flop triggered by the clock signal rising edge

A D flip-flop triggered by the rising edge of the clock signal can be implemented using $\overline{S}\ \overline{R}$ latches, as illustrated in Figure 1.26(a). The input signal D and the clock signal CK are applied to the input stage that generates the signal required by the output stage to determine the outputs Q and $\overline{Q}$. When the clock signal goes from 0 to 1, the state (0 or 1) of the input D is converted by the input stage in a $(\overline{S}\ \overline{R}) = (10)$ or (01) combination that results in the output stage being reset to 0 or set to 1. For other states that can be taken by the clock signal, the combination $(\overline{S}\ \overline{R}) = (11)$ is generated by the input stage regardless of the value on the D input. This forces the output stage to maintain its logic level unchanged.

A triangle is placed at the clock signal input, as shown by the symbol in Figure 1.26(b), to indicate that the flip-flop is active on the rising edge of the clock signal. Table 1.18 gives the truth table.

Flip-flop triggered by the clock signal falling edge

Similarly, a D flip-flop triggered by the falling edge of the clock signal can be implemented using SR latches, as shown in Figure 1.27(a). Depending on whether the D input state is 0 or 1 the input stage generates, in response to a falling edge of the clock signal CK, the combination $(SR) = (01)$ or (10) that sets the output stage to 1 or resets the output stage to 0. For the other states that can be taken by the clock signal, the input stage produces the combination $(SR) = (00)$ and the output stage holds its previous state.

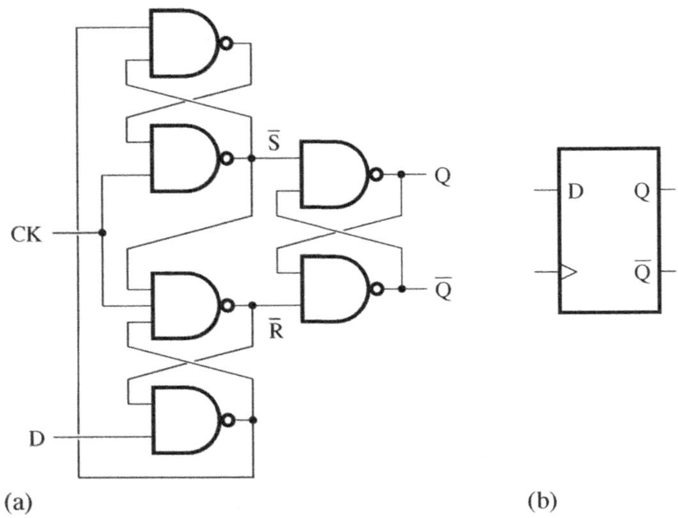

Figure 1.26. *D flip-flop triggered by the clock signal rising edge: a) logic circuit; b) symbol*

D	CK	Q^+	$\overline{Q+}$
x	0	Q	$\overline{Q}$
x	1	Q	$\overline{Q}$
0	↑	0	1
1	↑	1	0

Table 1.18. *Truth table of the flip-flop*

Referring to the symbol shown in Figure 1.27(b), a circle is placed before the triangle at the clock signal input to indicate that the flip-flop is activated by the clock signal falling edge. The truth table is represented in Table 1.19.

APPLICATION.– Implementation of a JK and T flip-flops using a D flip-flop

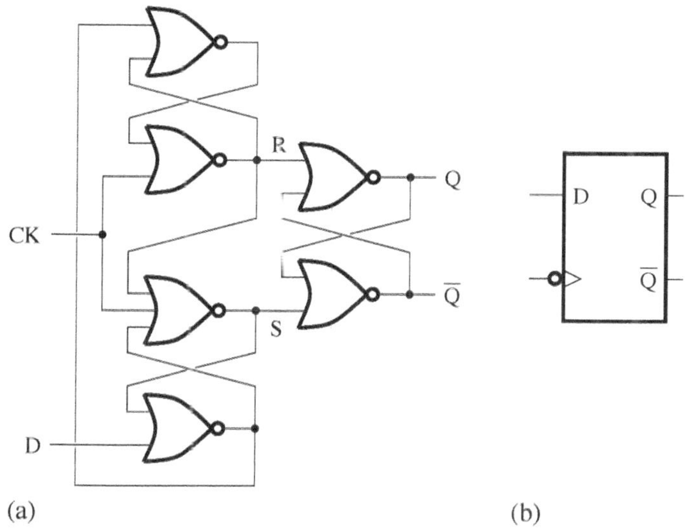

Figure 1.27. *D flip-flop triggered by the clock signal falling edge: a) logic circuit; b) symbol*

D	CK	Q^+	$\overline{Q^+}$
x	0	Q	$\overline{Q}$
x	1	Q	$\overline{Q}$
0	↓	0	1
1	↓	1	0

Table 1.19. *Truth table of the flip-flop*

A JK flip-flop triggered by the rising edge of the clock signal (or positive-edge-triggered JK flip-flop), as shown in Figure 1.28, can be implemented by adding a combinational circuit to a D flip-flop. The characteristic equation takes the following form:

$$Q^+ = D = J \cdot \overline{Q} + \overline{K} \cdot Q \qquad [1.44]$$

where Q is the present state and Q^+ represents the next state. Table 1.20 shows the truth table. The timing diagram of the JK flip-flop triggered by rising edge is given in Figure 1.29.

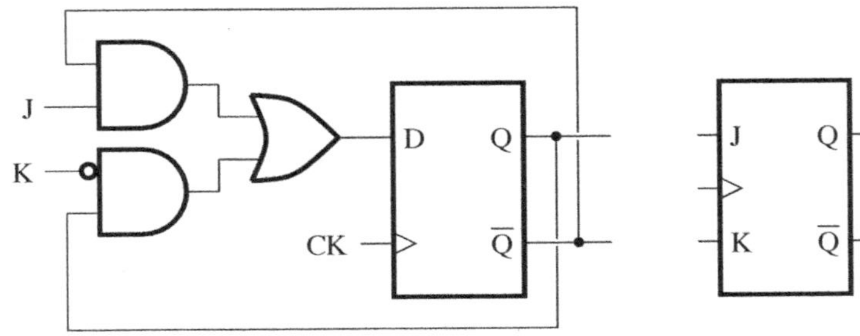

Figure 1.28. *JK flip-flop triggered by the clock signal rising edge: a) logic circuit; b) symbol*

J	K	CK	Q^+	$\overline{Q+}$
x	x	0	Q	$\overline{Q}$
x	x	1	Q	$\overline{Q}$
0	0	↑	Q	$\overline{Q}$
0	1	↑	0	1
1	0	↑	1	0
1	1	↑	$\overline{Q}$	Q

Table 1.20. *Truth table*

We can also implement a T flip-flop by connecting a combinational circuit to a D flip-flop as illustrated in Figure 1.30. If Q is the present state and Q^+ denotes the next state, the characteristic equation for the edge-triggered flip-flop is given by:

$$Q^+ = D = T \oplus Q \qquad [1.45]$$

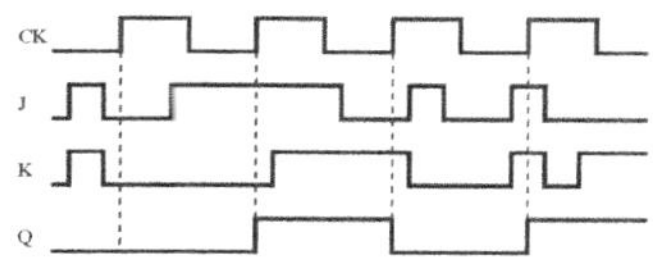

Figure 1.29. *Timing diagram of the JK flip-flop triggered by the clock signal rising edge*

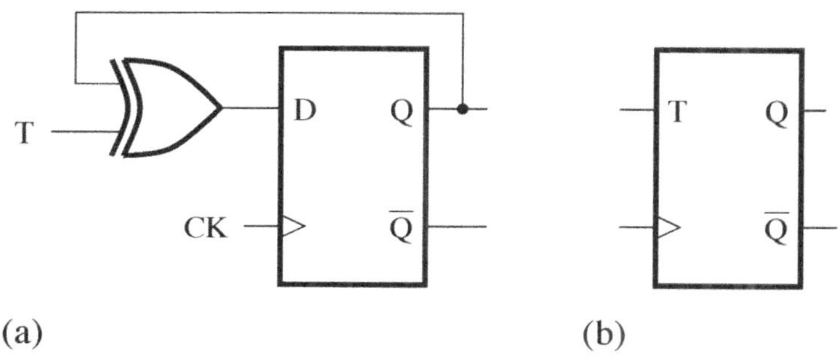

Figure 1.30. *T flip-flop triggered by the clock signal rising edge: a) logic circuit; b) symbol*

T	CK	Q^+	$\overline{Q+}$
x	0	Q	$\overline{Q}$
x	1	Q	$\overline{Q}$
0	↑	Q	$\overline{Q}$
1	↑	$\overline{Q}$	Q

Table 1.21. *Truth table*

Table 1.21 gives the corresponding truth table. The timing diagram of the T flip-flop triggered by the clock signal rising edge is represented in Figure 1.29.

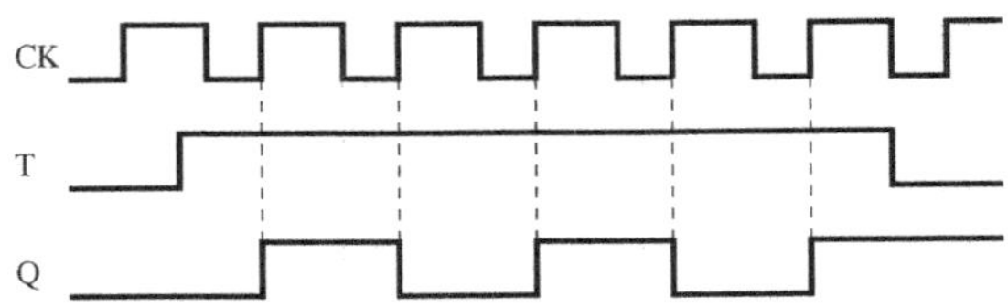

Figure 1.31. *Timing diagram for the T flip-flop triggered by the clock signal rising edge*

1.8. Flip-flops with asynchronous inputs

Just after power-up, for instance, asynchronous inputs can be used to define initial conditions of a flip-flop, regardless of the states of synchronous inputs and the clock signal in order to prevent any possible hazards. They are generally low active.

– The D flip-flop shown in Figure 1.32(a) has two asynchronous inputs, $\overline{PR}$ and $\overline{CLR}$, that can be used to determine the output state, regardless of the clock signal. Its symbol is given in Figure 1.32(b). Based on the truth table, shown in Table 1.22, the input $\overline{PR}$ sets the output to 1 (asynchronous preset), and $\overline{CLR}$ resets the output to 0 (asynchronous clear). For operation in the synchronous mode, the inputs $\overline{PR}$ and $\overline{CLR}$ must be kept in the high logic state.

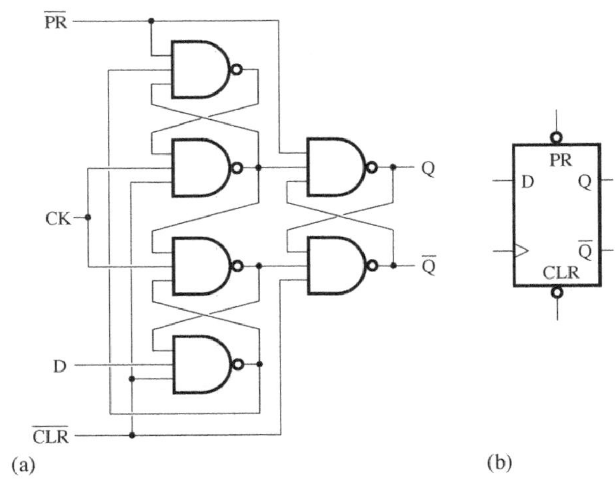

Figure 1.32. *Structure of a D flip-flop with asynchronous inputs (integrated circuit 74LS74): a) logic circuit; b) symbol*

– A JK flip-flop triggered by the clock signal falling edge (or negative edge triggered JK flip-flop), as depicted in Figure 1.34(a), consists of a synchronous SR latch connected to NAND gates. It can be set to 1 or reset to 0 using the asynchronous inputs $\overline{PR}$ and $\overline{CLR}$, respectively. Its symbol is shown in Figure 1.34(b).

$\overline{PR}$	$\overline{CLR}$	D	CK	Q^+	$\overline{Q^+}$	
0	1	x	x	1	0	Asynchronous preset
1	0	x	x	0	1	Asynchronous clear
0	0	x	x	1	1	Forbidden state
1	1	x	0	Q	$\overline{Q}$	Normal operation
1	1	x	1	Q	$\overline{Q}$	Normal operation
1	1	0	↑	0	1	Normal operation
1	1	1	↑	1	0	Normal operation

Table 1.22. *Truth table of the D flip-flop with asynchronous inputs*

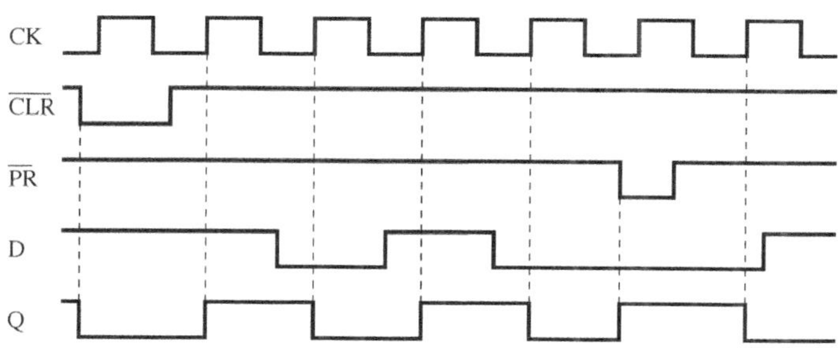

Figure 1.33. *Timing diagram of the D flip-flop with asynchronous inputs*

During a normal (or synchronous) operation of the flip-flop, the asynchronous inputs are set to 1. When the clock signal changes from 1 to 0, the 0 logic state is directly transferred to the output gated latch of the type SR, which is then activated so that the state of the inputs J and K can be taken into account. Because the NAND gates are sized to have a propagation delay in the order of the time required by the flip-flop outputs to change states, just enough time passed before the clock signal propagating through the NAND gates can affect the flip-flop, thereby preventing any other change in logic state. When the clock signal CK takes the 0 logic state, each NAND gate is then set to 1 and the state of the AND gate connected to the output is now only dependent on the feedback signal. This allows the flip-flop to preserve its earlier state. When the clock signal CK changes from 0 to 1, or takes the logic state 1, the output of each AND gate directly connected to the clock signal is dependent only on the feedback signal. This prevents the flip-flop from changing state.

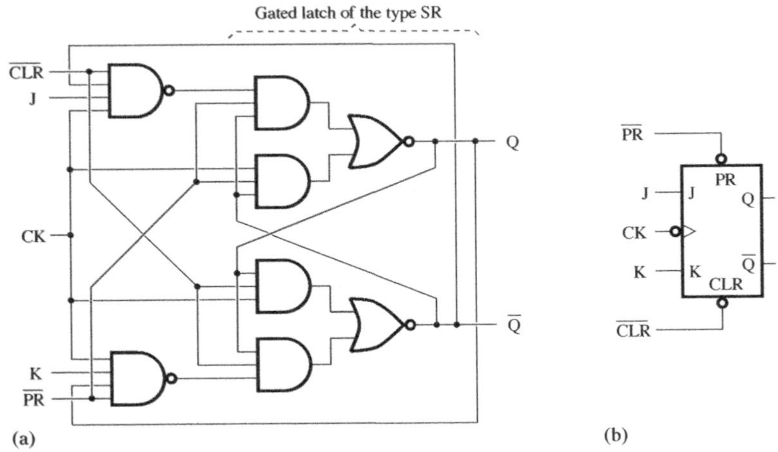

Figure 1.34. *Logic circuit and symbol of the JK flip-flop with asynchronous inputs*

Edge triggering is implemented by exploiting the difference in propagation delays associated with the clock signal CK, that is applied directly and *via* the NAND gates to the SR latch.

Table 1.23 shows the truth table of a JK flip-flop with asynchronous inputs.

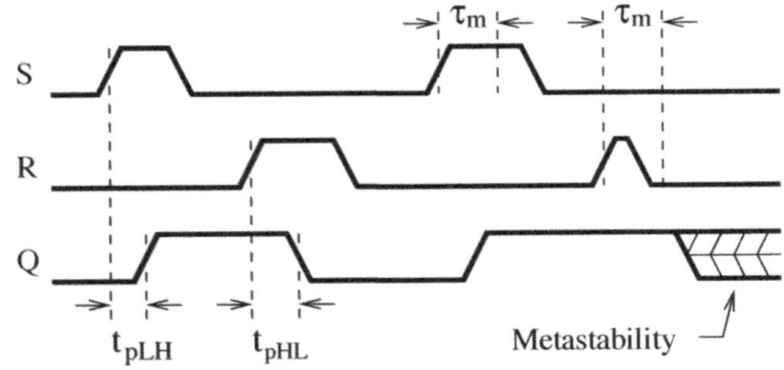

Figure 1.35. *Operational characteristics of an SR latch*

NOTE.– By simultaneously applying a data D to the input J and its complement, $\overline{D}$, to the input K, the JK flip-flop operates as a D flip-flop.

$\overline{PR}$	$\overline{CLR}$	CK	J	K	Q^+		
0	0	x	x	x	1	Forbidden state	
0	1	x	x	x	1	Asynchronous preset	
1	0	x	x	x	0	Asynchronous clear	
1	1	↓	0	0	Q	No change	Normal operation
1	1	↓	0	1	0	Reset	Normal operation
1	1	↓	1	0	1	Set	Normal operation
1	1	↓	1	1	$\overline{Q}$	Toggle	Normal operation
1	1	1	x	x	Q	No change	Normal operation

Table 1.23. *Truth table of the JK flip-flop with asynchronous inputs*

1.9. Operational characteristics of flip-flops

A flip-flop only acquires a signal whose level can remain stable for a certain time. Thus, it can operate normally only when the setup time requirements are met.

The timing diagram shown in Figure 1.35 illustrates the effect of the following characteristics on the state of the Q output of an SR latch:

– propagation delay t_p: this is the interval of time between the application of an input signal and the appearance of the resulting signal at the output. The delay t_{pLH} is measured on the rising edge of the output, while t_{pHL} is measured on the falling edge;

– minimum pulse width τ_m: in order for the flip-flop to operate reliably, the width of each pulse must be greater than τ_m, otherwise the state of the output may become metastable.

Flip-flops available in the form of integrated circuits have propagation delays of the order of a few nanoseconds.

In addition, with reference to waveforms of a synchronous D flip-flop shown in Figure 1.36, we can define:

– the *setup time* as the minimum time during which the input logic levels must be kept constant before the transition of the clock signal in order to ensure a reliable triggering;

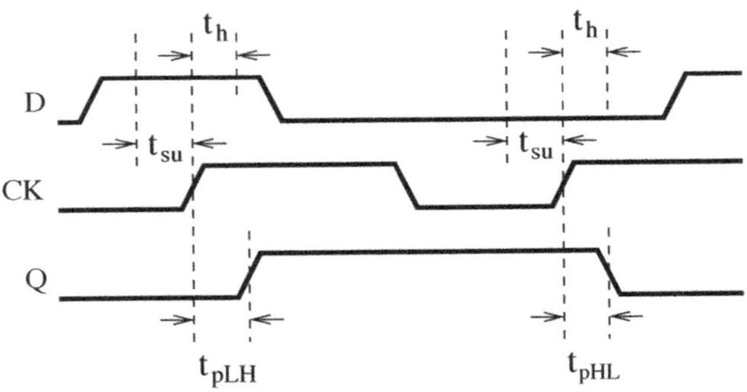

Figure 1.36. *Operational characteristics of an edge-triggered D flip-flop*

– the *hold time*, which is the minimum time interval during which the logic levels of the input must be kept constant after the transition of the clock signal in order to guarantee a reliable triggering.

The set-up time and hold time for integrated-circuit flip-flops are of the order of a few nanoseconds. When the set-up and hold conditions are not satisfied, the output state of the flip-flop may become unpredictable (either 0 or 1). In some cases, we can observe an oscillation of the output signal or a metastable state situated between the high and low logic levels.

1.10. Exercises

EXERCISE 1.1.– Propose an equivalent switch-based circuit for each of the circuits in Figure 1.37.

What is the function of these circuits?

EXERCISE 1.2.– Consider the T latch whose logic circuit and symbol are given in Figure 1.38.

Determine the characteristic equations of this latch.

EXERCISE 1.3.– Analyze and construct the truth table for the flip-flop shown in Figure 1.39.

EXERCISE 1.4.– Consider the positive edge-triggered D flip-flop shown in Figure 1.40(a). Complete the timing-diagram in Figure 1.40(b).

EXERCISE 1.5.– Figure 1.41(a) shows a positive edge-triggered JK flip-flop. Complete the timing diagram in Figure 1.41(b).

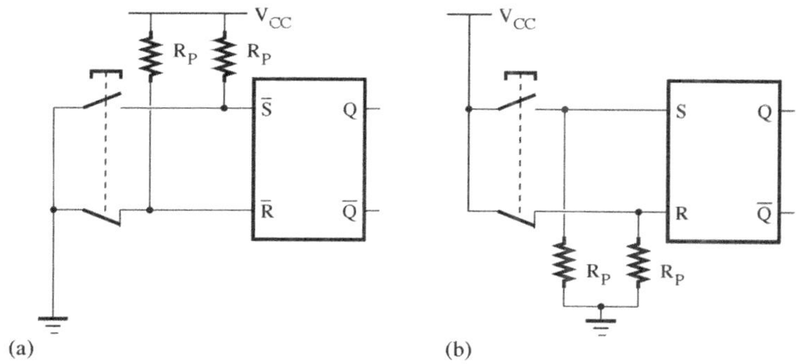

Figure 1.37. *Logic circuits*

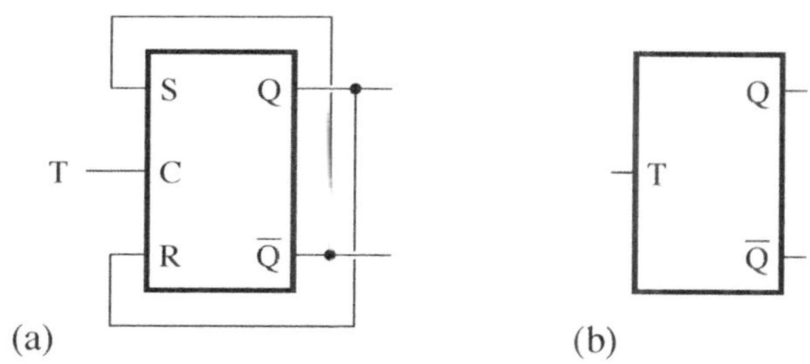

Figure 1.38. *T latch: a) logic circuit; b) symbol*

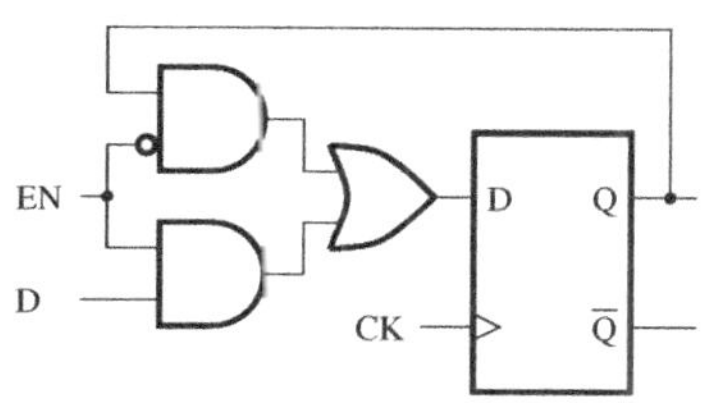

Figure 1.39. *Logic circuit for the flip-flop*

EXERCISE 1.6.– Consider the master-slave JK flip-flop in Figure 1.42(a). Complete the timing diagram shown in Figure 1.42(b).

EXERCISE 1.7.– Figure 1.43(a) shows a JK flip-flop with asynchronous inputs. Complete the timing diagram in Figure 1.43(b).

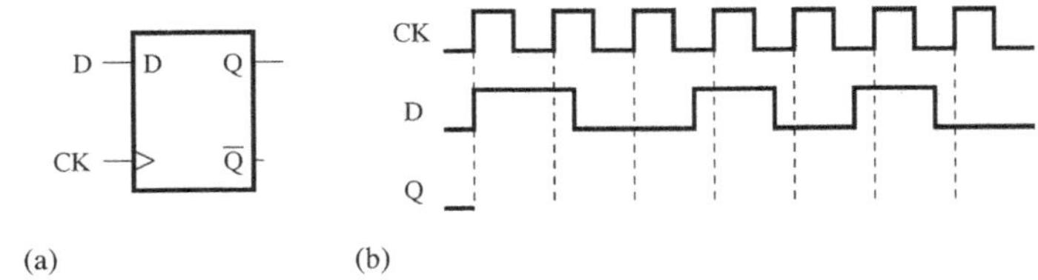

Figure 1.40. *a) D flip-flop; b) timing diagram*

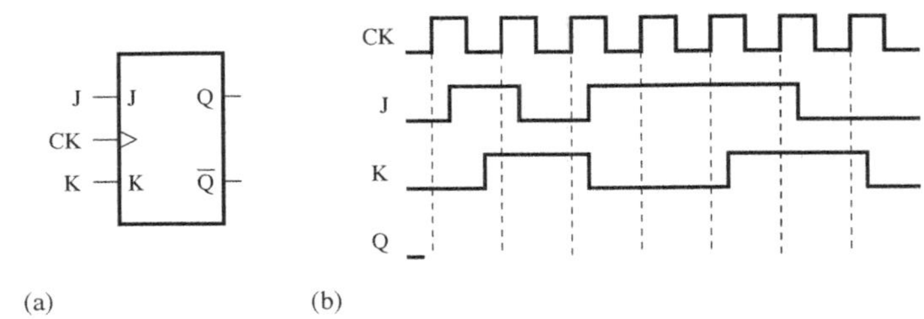

Figure 1.41. *a) JK flip-flop; b) timing diagram*

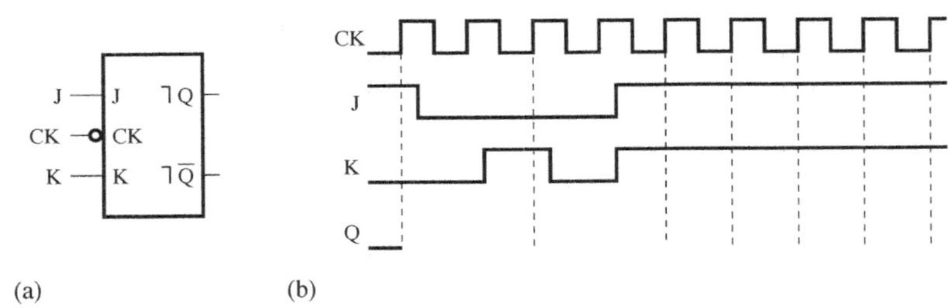

Figure 1.42. *a) Master-slave JK flip-flop; b) timing diagram*

EXERCISE 1.8.– The logic circuit for a D flip-flop with asynchronous inputs is represented in Figure 1.44(a). Complete the timing diagram in Figure 1.44(b).

EXERCISE 1.9.– For each circuit using two D flip-flops, as represented in Figures 1.45–1.47, complete the corresponding timing diagram.

EXERCISE 1.10.– Complete the timing diagram corresponding to each of the circuits using two JK flip-flops, as represented in Figures 1.48 and 1.49.

EXERCISE 1.11.– Consider the logic circuit shown in Figure 1.50(a), which is made up of two D flips-flops and a combinational logic circuit section F to be determined.

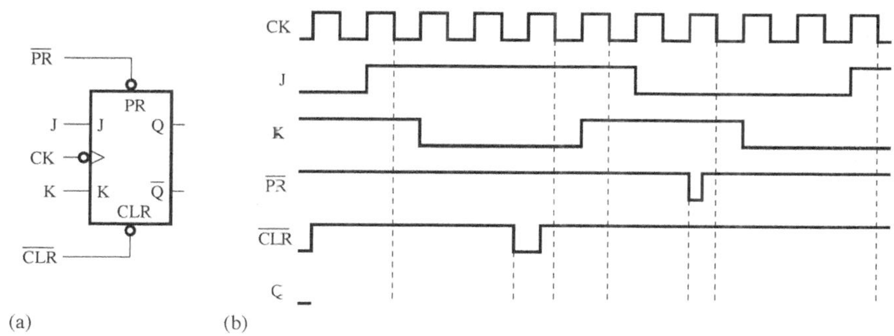

Figure 1.43. *a) JK flip-flop; b) timing diagram*

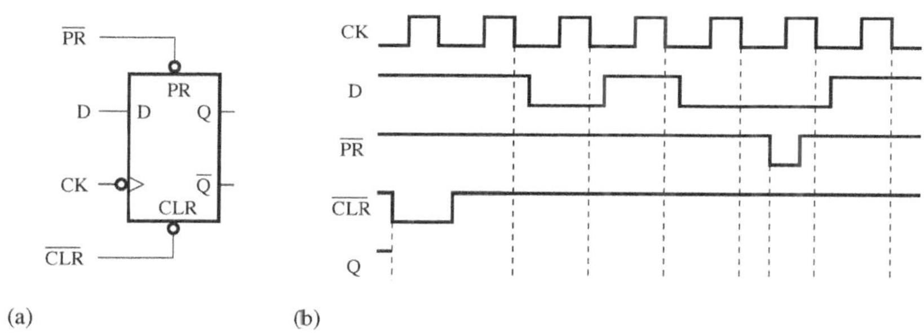

Figure 1.44. *a) D flip-flop; b) timing diagram*

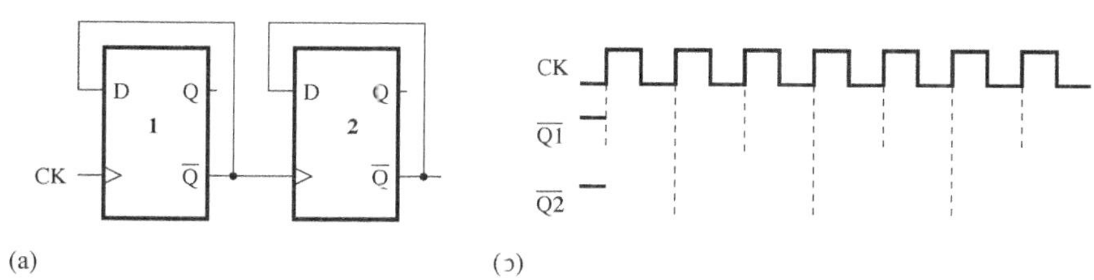

Figure 1.45. *a) Logic circuit 1; b) timing diagram*

Complete the timing diagram (signals $Q1$ and $Q2$) in Figure 1.50(b).

Determine the logic function F and suggest how it can be implemented.

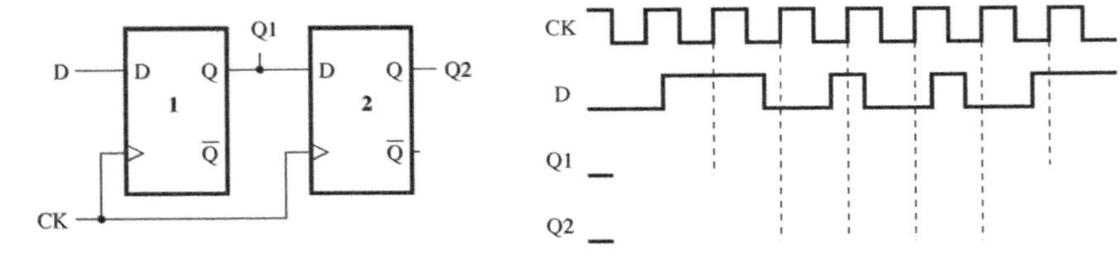

(a) (b)

Figure 1.46. *a) Logic circuit 2; b) timing diagram*

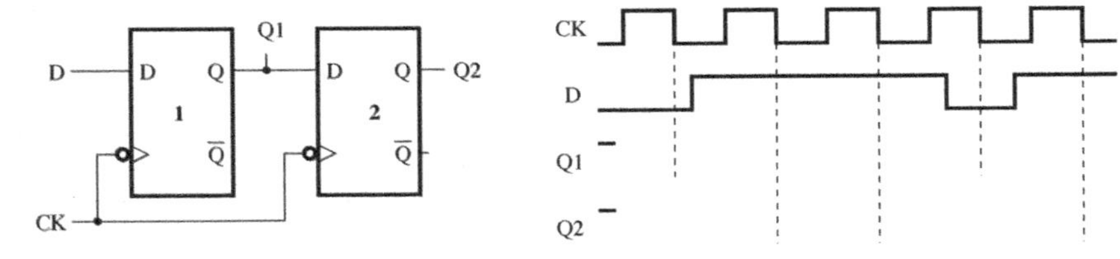

(a) (b)

Figure 1.47. *a) Logic circuit 3; b) timing diagram*

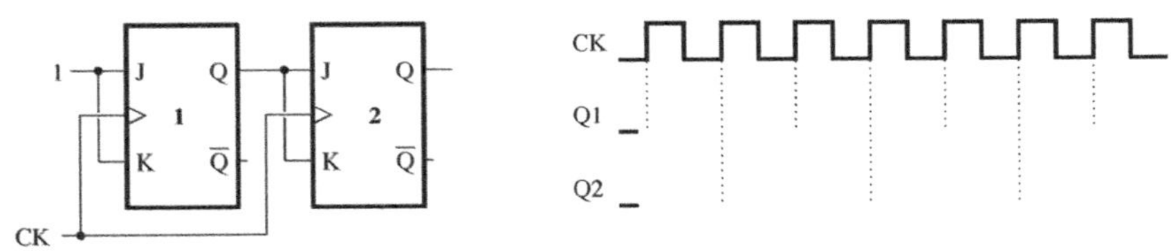

(a) (b)

Figure 1.48. *a) Logic circuit 1; b) timing diagram*

EXERCISE 1.12.– Determine the characteristic equation for each of the synchronous D flip-flops in Figure 1.51. To compare these two D flip-flops, we use the set-up shown in Figure 1.52(a) and assume that the propagation delay of the inverter is not equal to zero.

Complete the timing diagram in Figure 1.52(b).

Which of the two flip-flops operates correctly? Why? Justify your response using Karnaugh maps.

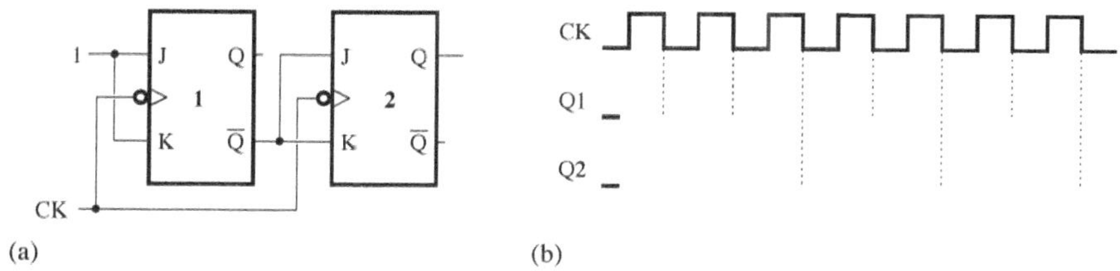

Figure 1.49. *a) Logic circuit 2; b) timing diagram*

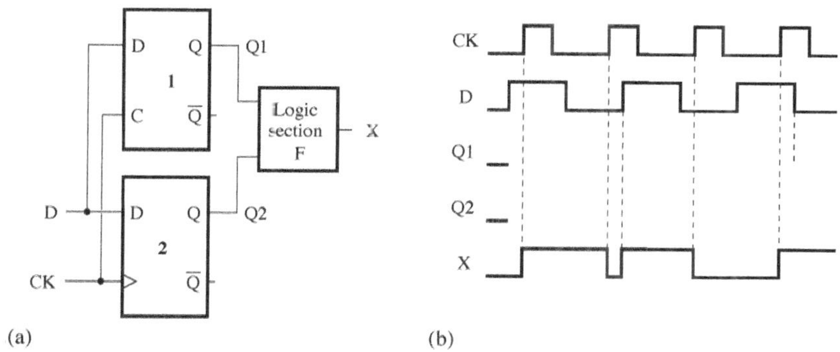

Figure 1.50. *a) Logic circuit; b) timing diagram*

EXERCISE 1.13.– Converting between different types of flip-flops. Verify the equivalence between the flip-flops represented on each of the lines a, b, c and d in Figure 1.53.

1.11. Solutions

SOLUTION 1.1.– The equivalent circuit for each of the proposed circuits is represented in Figure 1.54.

It is a switch debouncer.

SOLUTION 1.2.– T latch.

An SR latch is characterized by:

$$Q^+ = S \cdot C + (\overline{R} + \overline{C}) \cdot Q \qquad [1.46]$$

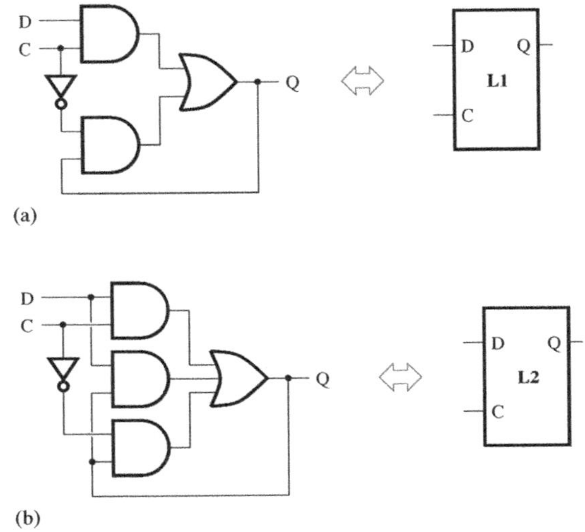

Figure 1.51. *Logic circuits and symbols for the synchronous D flip-flops*

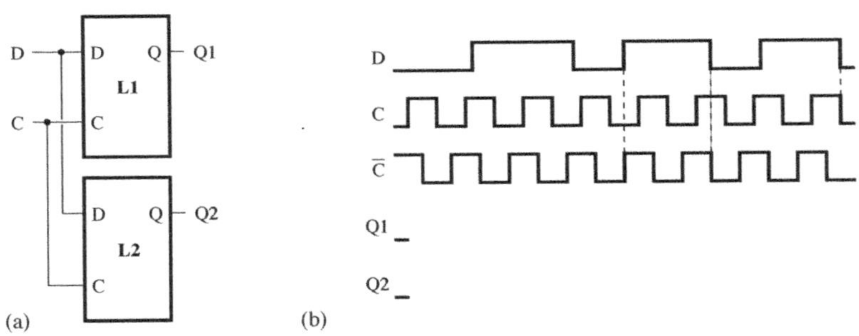

Figure 1.52. *a) Logic circuit; b) timing diagram*

or

$$Q^{+} = \overline{R} \cdot S \cdot C + (\overline{R} + \overline{C}) \cdot Q \tag{1.47}$$

Assuming that for the T latch, $S = \overline{Q}$, $R = Q$ and $C = T$, we obtain the same characteristic equation in both cases, which can be written as follows:

$$Q^{+} = \overline{Q} \cdot T + \overline{T} \cdot Q = T \oplus Q \tag{1.48}$$

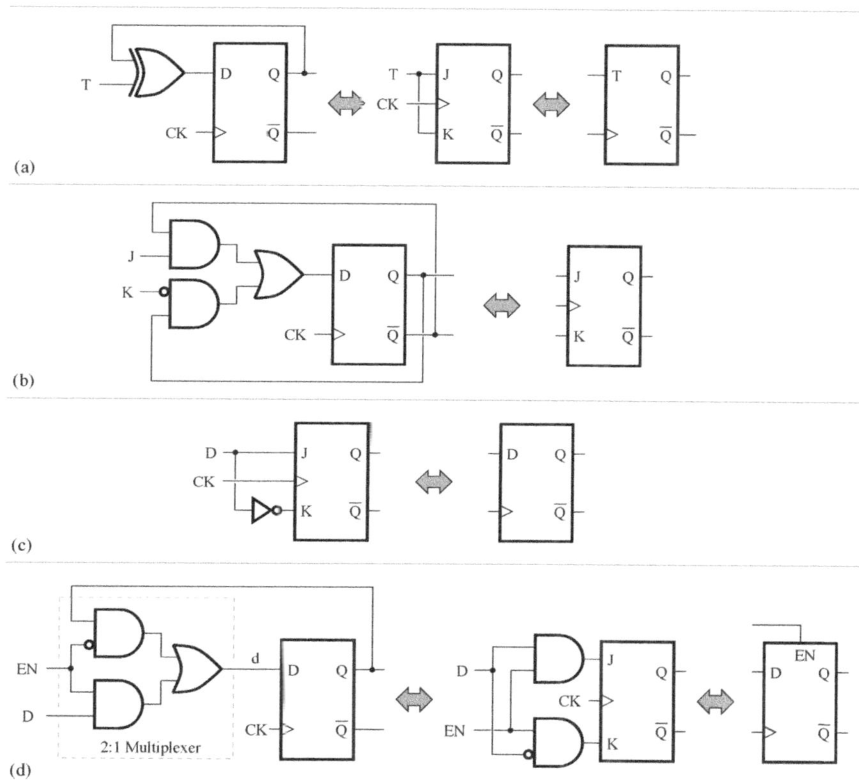

Figure 1.53. *Flip-flops*

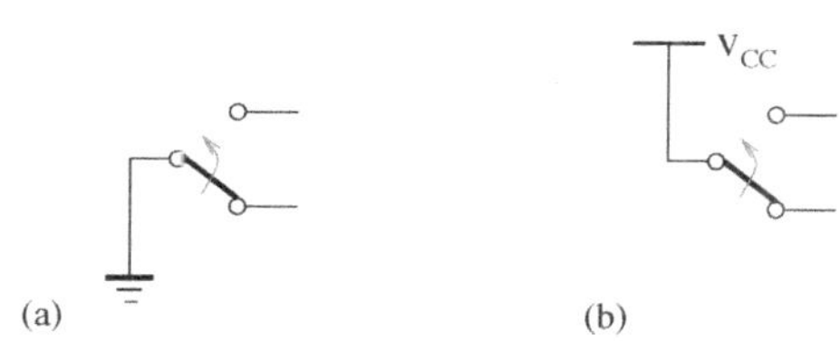

Figure 1.54. *Equivalent circuits*

SOLUTION 1.3.– D flip-flop with enable input.

The characteristic equation for the D flip-flop with enable input is given by:

$$Q^+ = D \cdot EN + \overline{EN} \cdot Q \qquad [1.49]$$

Figure 1.55 shows the logic circuit and the truth table for the D flip-flop with enable input is represented in Table 1.24.

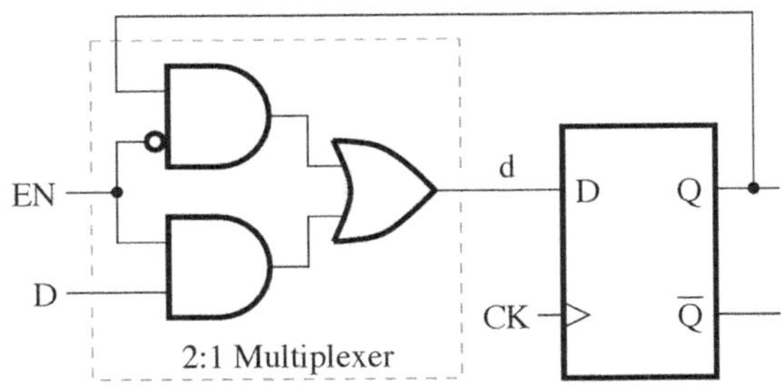

Figure 1.55. *Logic circuit for D flip-flop with enable input*

EN	D	CK	Q^+	$\overline{Q^+}$
x	x	0	Q	$\overline{Q}$
x	x	1	Q	$\overline{Q}$
0	x	↑	Q	$\overline{Q}$
1	0	↑	0	1
1	1	↑	1	0

Table 1.24. *Truth table of the flip-flop*

SOLUTION 1.4.– Positive edge-triggered DD flip-flop.

For the positive edge-triggered D flip-flop, Figure 1.56 shows the logic circuit and the timing diagram that can be obtained from the truth table.

Figure 1.57 shows the logic circuit and the timing diagram for the level-triggered D flip-flop.

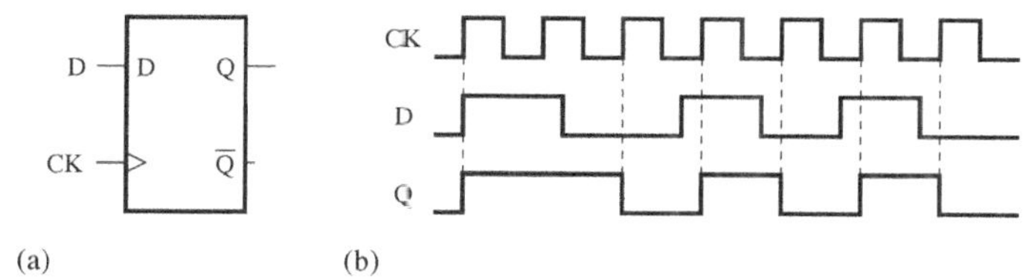

Figure 1.56. *a) Positive edge-triggered D flip-flop; b) timing diagram*

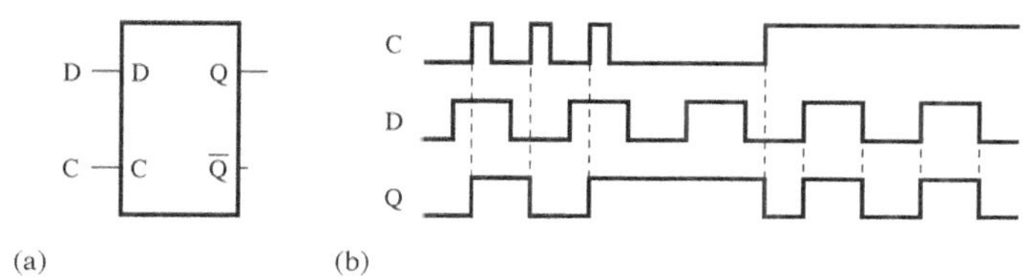

Figure 1.57. *a) Level-triggered D flip-flop; b) timing diagram*

Figure 1.58 shows the logic circuit and the timing diagram that can be used to compare these two types of D flip-flops.

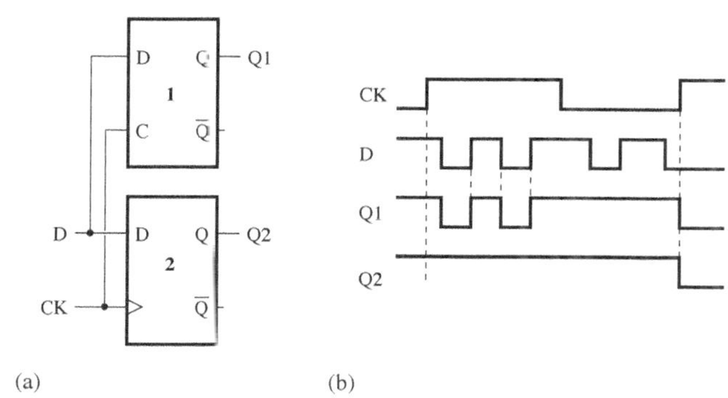

Figure 1.58. *Comparison of two D flip-flops: a) logic circuit; b) timing diagram*

SOLUTION 1.5.– Positive edge-triggered JK flip-flop.

Figure 1.59 shows the logic circuit and timing diagram for the positive edge-triggered JK flip-flop.

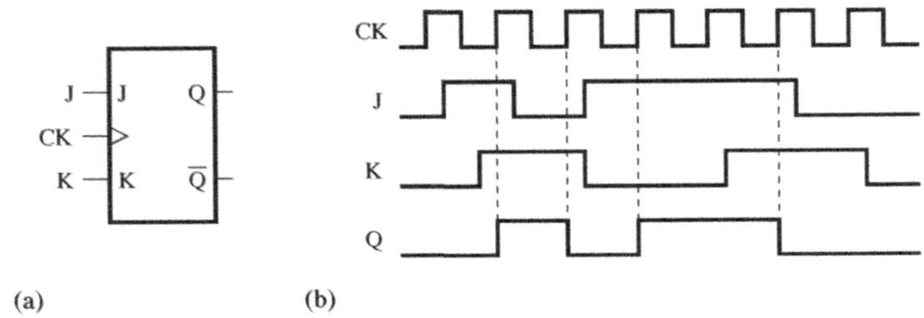

Figure 1.59. *a) JK flip-flop; b) timing diagram*

SOLUTION 1.6.– Master-slave JK flip-flop.

Figure 1.60 shows the logic circuit and timing diagram for the master-slave JK flip-flop.

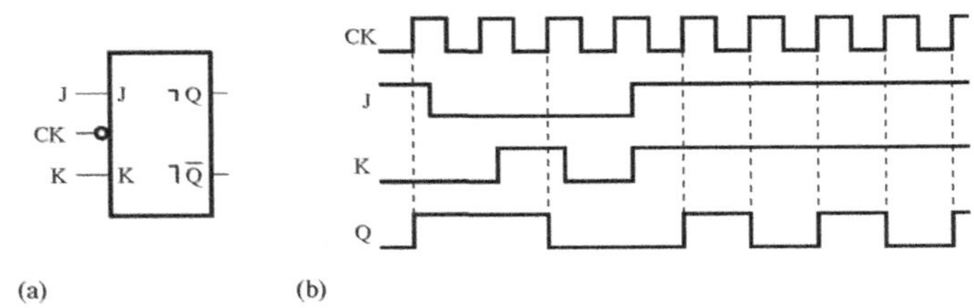

Figure 1.60. *a) Master-slave JK flip-flop; b) timing diagram*

SOLUTION 1.7.– JK flip-flop with asynchronous inputs.

Figure 1.61 shows the logic circuit and the timing diagram for the JK flip-flop with asynchronous inputs.

SOLUTION 1.8.– D flip-flop with asynchronous inputs.

Figure 1.62 shows the logic circuit and timing diagram for the D flip-flop with asynchronous inputs.

SOLUTION 1.9.– Connection of two D flip-flops.

Figure 1.63 shows logic circuit 1 and the corresponding timing diagram.

Logic circuit 2 and its timing diagram are represented in Figure 1.64.

Figure 1.65 shows logic circuit 3 and the corresponding timing diagram that can be obtained based on the truth table.

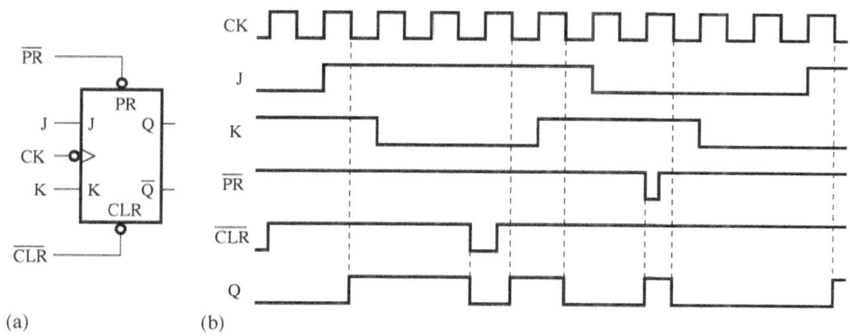

Figure 1.61. *a) JK flip-flop; b) timing diagram*

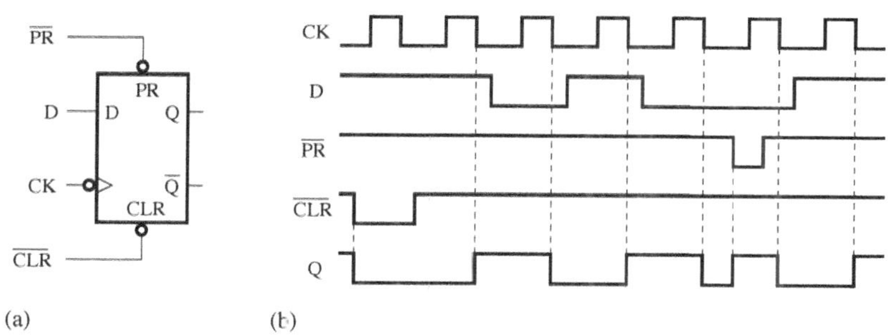

Figure 1.62. *a) D flip-flop; b) timing diagram*

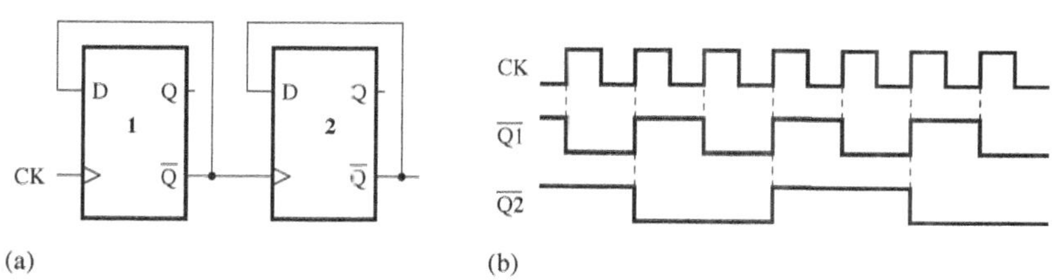

Figure 1.63. *a) Logic circuit 1; b) timing diagram*

SOLUTION 1.10.– Connection of two JK flip-flops.

Figure 1.66 shows logic circuit 1 and the timing diagram that can be obtained based on the truth table.

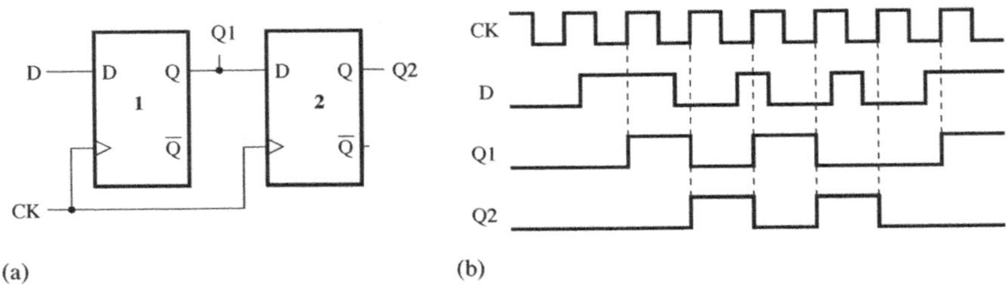

Figure 1.64. *a) Logic circuit 2; b) timing diagram*

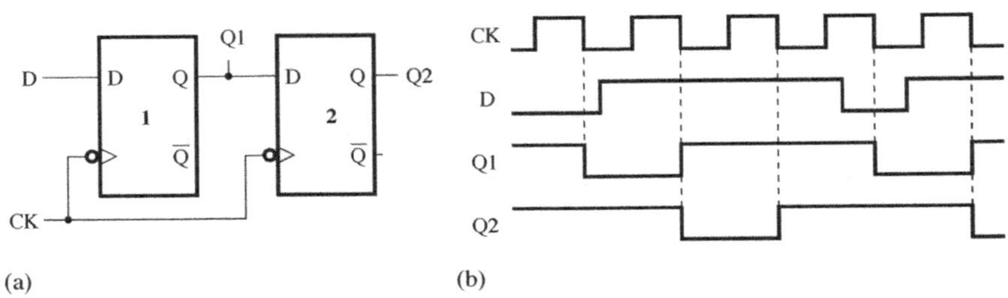

Figure 1.65. *a) Logic circuit 3; b) timing diagram*

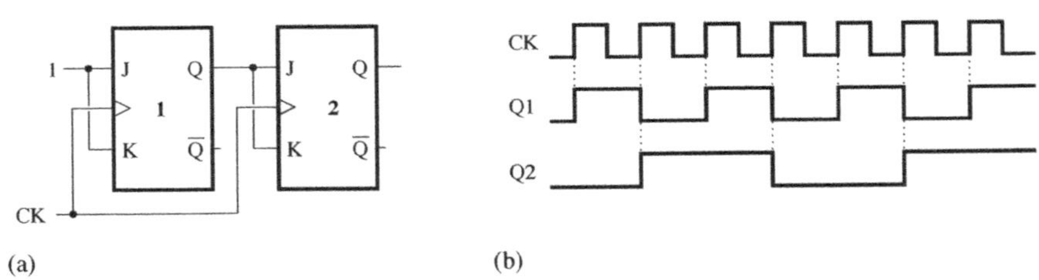

Figure 1.66. *a) Logic circuit 1; b) timing diagram*

Figure 1.67 shows logic circuit 2 and the corresponding timing diagram.

SOLUTION 1.11.– Circuit using D flip-flops.

The truth table for the level-triggered D flip-flop and the truth table for the positive edge-triggered D flip-flop can be used to complete the timing diagram (for the outputs $Q1$ and $Q2$) for the circuit shown in Figure 1.68(a), as illustrated in Figure 1.68(b).

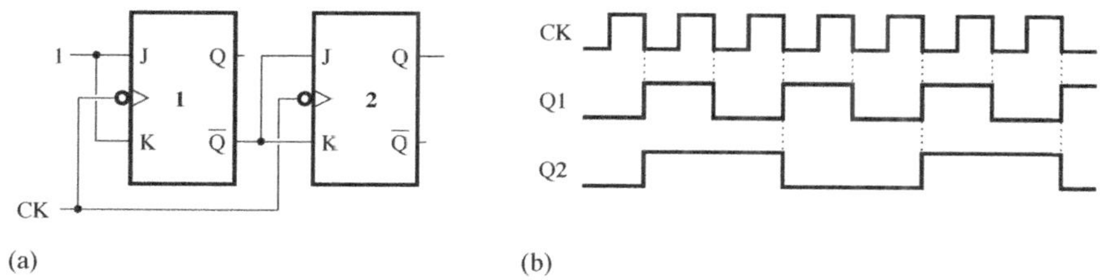

Figure 1.67. *a) Logic circuit 2; b) timing diagram*

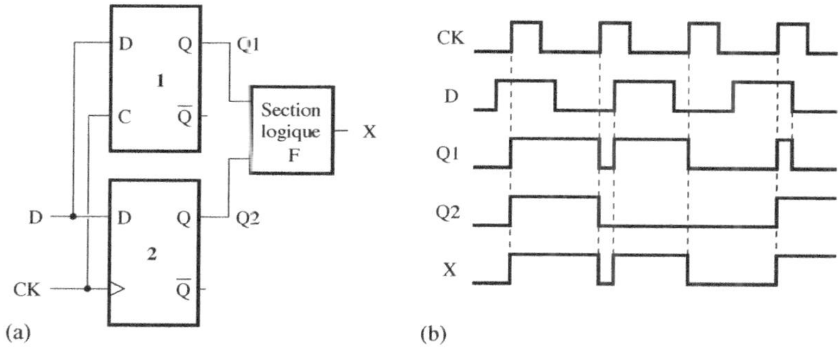

Figure 1.68. *a) Logic circuit; b) timing diagram*

Considering $Q1$ and $Q2$ as the inputs and X as the output, the truth table (see Table 1.25) obtained based on the timing diagram helps define the logic relationship that exists between $Q1$, $Q2$ and X.

Because the resulting logic equation is of the form, $X = Q1 + Q2$, the function F can be implemented by an OR gate (see Table 1.25).

Q1	Q2	X
0	0	0
0	1	1
1	0	1
1	1	1

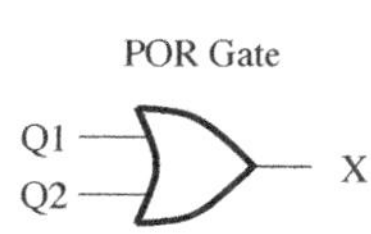

Table 1.25. *Truth table (OR gate)*

SOLUTION 1.12.– Gated D latches.

By analyzing each latch, we obtain a characteristic equation of the following form:

– latch L1:

$$Q^+ = D \cdot C + Q \cdot \overline{C} \quad [1.50]$$

– latch L2:

$$Q^+ = D \cdot C + Q \cdot \overline{C} + D \cdot Q \quad [1.51]$$

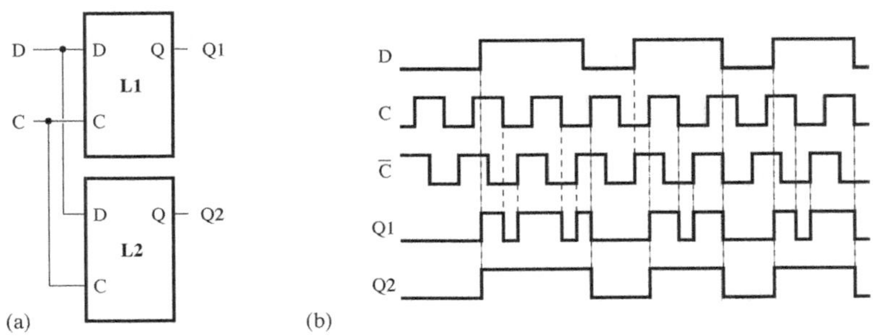

Figure 1.69. *a) Logic circuit; b) timing diagram*

Figure 1.69 shows the timing diagram that can be used to compare the latches L1 and L2.

The operation of the latch L1 is affected by the propagation delay of the inverter used to generate the signal $\overline{C}$. Hence, if $D = 1$ and $Q = 1$, we have:

– latch L1:

$$Q^+ = C + \overline{C} \quad [1.52]$$

– latch L2:

$$Q^+ = 1 + C + \overline{C} = 1 \quad [1.53]$$

Adding the redundant term $D \cdot Q$ corresponding to cells 5 and 7 of the Karnaugh map (see Figure 1.70) is useful for the elimination of the aforementioned functional hazard in the case of the latch L2.

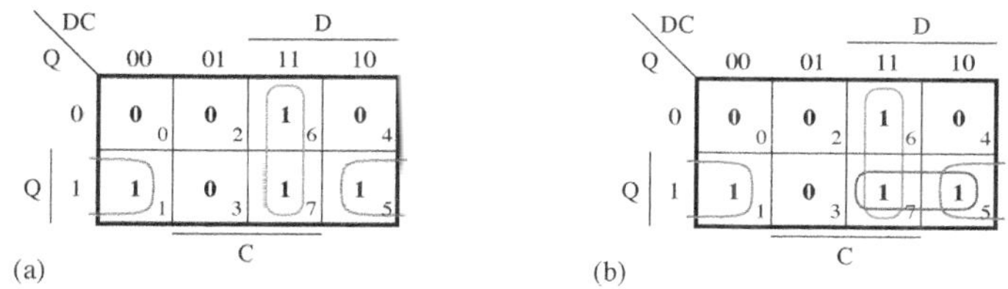

Figure 1.70. *Karnaugh maps: a) latch L1; b) latch L2. For a color version of this figure, see www.iste.co.uk/ndjountche/electronics2.zip*

SOLUTION 1.13.– Conversion of one type of flip-flop to another.

– T flip-flop

For the circuit based on the D flip-flop, we get:

$$Q^+ = D = T \oplus Q \qquad [1.54]$$

Considering the circuit based on the JK flip-flop, we have:

$$Q^+ = D = J \cdot \overline{Q} + \overline{K} \cdot Q = T \cdot \overline{Q} + \overline{T} \cdot Q = T \oplus Q \qquad [1.55]$$

In both cases, we have the characteristic equation for the T flip-flop.

– JK flip-flop

By analyzing the circuit based on the D flip-flop, we can write:

$$Q^+ = D = J \cdot \overline{Q} + \overline{K} \cdot Q \qquad [1.56]$$

this is the characteristic equation of the JK flip-flop.

– D flip-flop

For the circuit based on the JK flip-flop, we have:

$$J = D, \quad \overline{K} = D, \quad \text{and} \quad Q^+ = J \cdot \overline{Q} + \overline{K} \cdot Q = D \qquad [1.57]$$

this is the characteristic equation of the D flip-flop.

The logic expression obtained for the circuit based on the D flip-flop is of the form:

$$Q^+ = d = D \cdot EN + Q \cdot \overline{EN} \qquad [1.58]$$

The equation associated with the circuit based on the JK flip-flop is given by:

$$Q^+ = J \cdot \overline{Q} + \overline{K} \cdot Q \qquad [1.59]$$

where $J = D \cdot EN$ and $K = \overline{D} \cdot EN$. By applying Boolean algebra theorems, we can successively find that:

$$\begin{aligned} Q^+ &= D \cdot EN \cdot \overline{Q} + \overline{\overline{D} \cdot EN} \cdot Q \qquad [1.60] \\ &= D \cdot EN \cdot \overline{Q} + (D + \overline{EN})Q \\ &= D(EN + Q) + Q \cdot \overline{EN} \\ &= D \cdot EN + D \cdot Q(EN + \overline{EN}) + Q \cdot \overline{EN} \end{aligned}$$

$$\begin{aligned} &= D \cdot EN(1 + Q) + Q \cdot \overline{EN}(1 + D) \\ &= D \cdot EN + Q \cdot \overline{EN} \qquad [1.61] \end{aligned}$$

In both cases, the characteristic equation obtained is that of a D flip-flop with enable input.

2

Binary Counters

2.1. Introduction

Binary counters are circuits that generate binary sequences that can be associated with the number of clock signal pulses applied to the input. They are used in applications such as event synchronization and frequency measurement, estimation of angular position and the duration of an event.

An asynchronous counter is often called a *ripple counter*. The clock signal is only directly applied to the first flip-flop and it is subsequently transmitted, with a propagation delay, from one flip-flop to another.

In a synchronous counter, all the flip-flops are triggered by the same clock signal. Thus, the outputs of the counter change state at the same time and there is no time lag between the different outputs.

The state of a counter is defined by a specific combination formed by all the outputs together.

A state diagram, which shows the states and possible transitions, is most often used to illustrate the counter operation. It is made up of circles with labels representing the states and with arrows symbolizing the transitions.

The following parameters can be used to characterize a counter:

– the number of different states (also called *modulo*);

– the direction of counting (up or down);

– the operating mode (asynchronous or synchronous).

To analyze counters it is always supposed, unless indicated otherwise, that the initial state is 0.

2.2. Modulo 4 counter

A modulo 4 (or two-bit) counter has four different states ($2^2 = 4$). The output Q_0 represents the least significant bit (LSB) and Q_1 corresponds to the most significant bit (MSB). Initially, we have: $Q_0 = 0$ and $Q_1 = 0$. The counting is cyclic and once the state $Q_0 = 1$ and $Q_1 = 1$ is achieved, the next pulse of the clock signal allows the counter to reset to its initial state.

The implementation of a modulo 4 counter requires at least two flip-flops that can be configured for synchronous or asynchronous operation.

Figures 2.1(a) and 2.1(b) show the logic circuit and the timing diagram for an asynchronous modulo 4 counter.

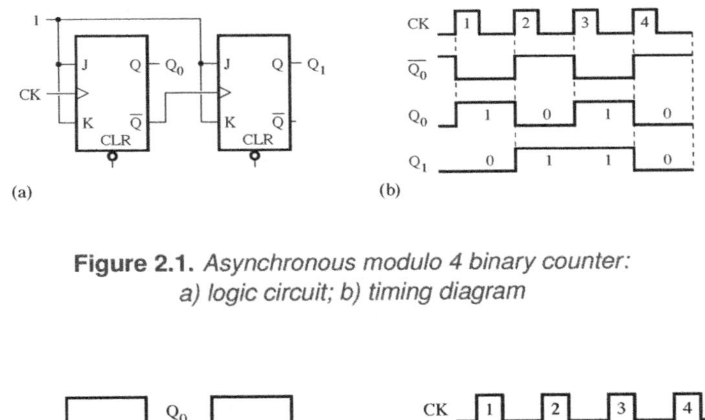

Figure 2.1. *Asynchronous modulo 4 binary counter: a) logic circuit; b) timing diagram*

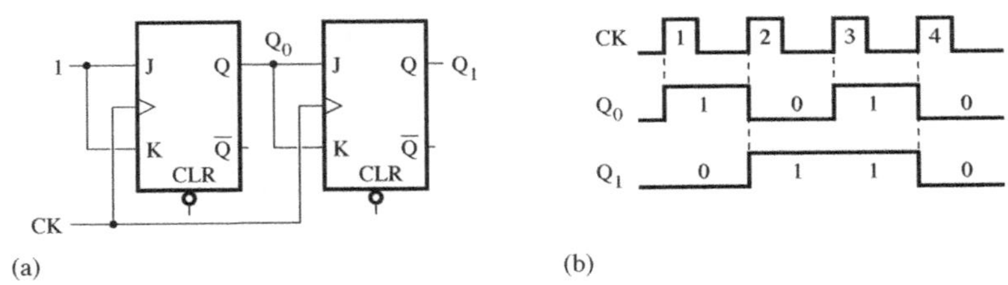

Figure 2.2. *Synchronous modulo 4 binary counter: a) logic circuit; b) timing diagram*

The logic circuit and the timing diagram for a synchronous modulo 4 counter are illustrated in Figures 2.2(a) and 2.2(b).

Table 2.1 describes the count sequence and Figure 2.3 gives the state diagram, where the initial state is $Q_1Q_0 = 00$.

Clock signal	Outputs	
pulse	Q_1	Q_0
Initial state	0	0
1	0	1
2	1	0
3	1	1
4	0	0

Table 2.1. *Table describing the count sequence*

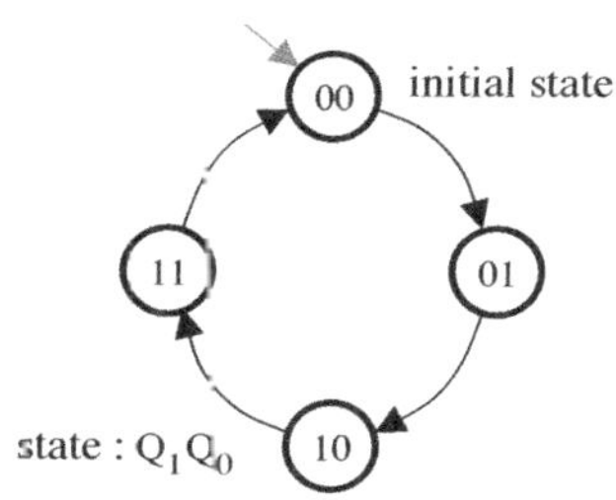

Figure 2.3. *State diagram for the modulo 4 counter*

2.3. Modulo 8 counter

A modulo 8 (or three-bit) counter has eight different states and is made up of at least three flip-flops. The output Q_0 represents the LSB and Q_2 is the MSB.

Initially, we have $Q_0 = 0$, $Q_1 = 0$ and $Q_2 = 0$. The counter follows an ascending sequence from 0 to 7. At the eighth pulse of the clock signal, the counter is reset to its initial state.

Figures 2.1(a) and 2.1(b) show the logic circuit and the timing diagram for an asynchronous modulo 8 counter.

The logic circuit and the timing diagram for a synchronous 8 modulo counter are illustrated in Figures 2.2(a) and 2.2(b). The logic equations for the inputs of each flip-flop are given by:

– flip-flop 0: $J_0 = K_0 = 1$;

– flip-flop 1: $J_1 = K_1 = Q_0$;

– flip-flop 2: $J_2 = K_2 = Q_1 \cdot Q_0$.

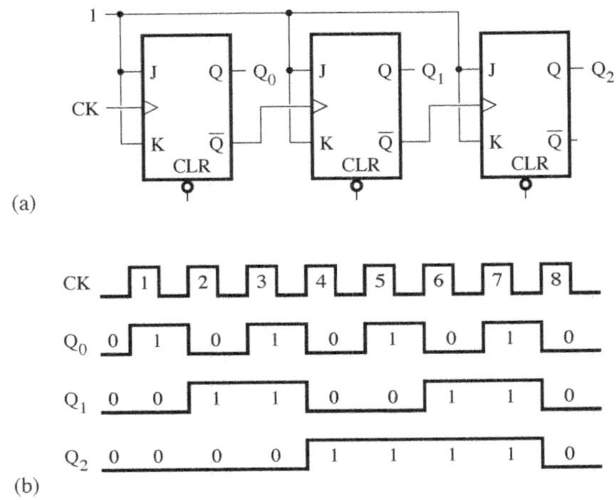

Figure 2.4. *Asynchronous modulo 8 binary counter: a) logic circuit; b) timing diagram*

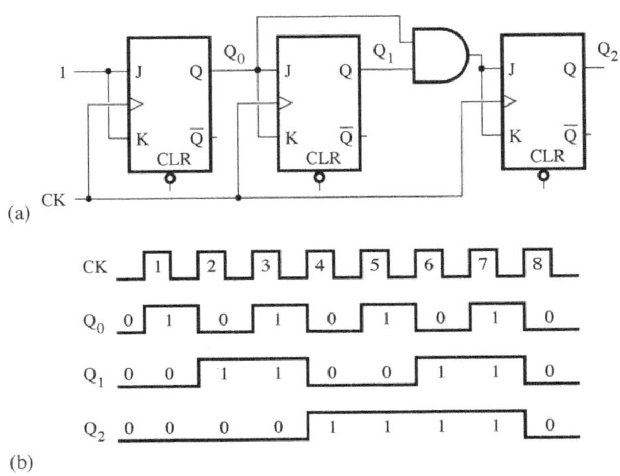

Figure 2.5. *Synchronous modulo 8 binary counter: a) logic circuit; b) timing diagram*

Table 2.2 gives the count sequence, assuming that the initial state of the synchronous and asynchronous counters is 0. The state diagram is represented in Figure 2.6.

Clock signal	Outputs		
pulse	Q_2	Q_1	Q_0
Initial state	0	0	0
1	0	0	1
2	0	1	0
3	0	1	1
4	1	0	0
5	1	0	1
6	1	1	0
7	1	1	1
8	0	0	0

Table 2.2. *Table describing the count sequence*

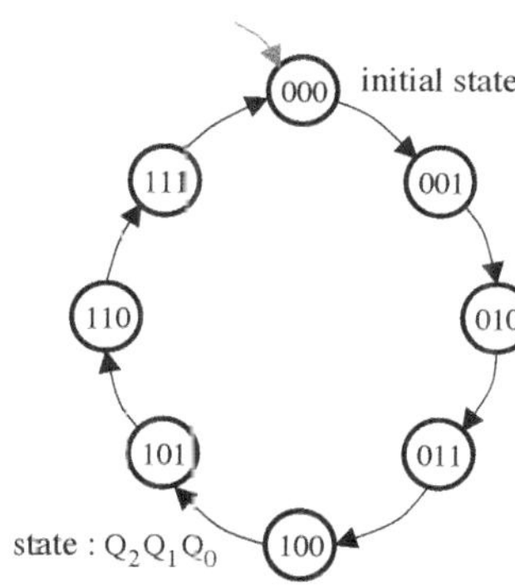

Figure 2.6. *State diagram of the modulo 8 counter*

2.4. Modulo 16 counter

A modulo 16 (or four-bit) counter has 16 (($2^4 = 16$)) states and at least four flip-flops are required for its implementation. It can generate binary sequences corresponding to the numbers from 1 to 15.

At each clock signal pulse, the counter moves from one number to the next. The counter is reset to 0 upon reaching the sequence 1111, that is the number 15.

The logic circuit for an asynchronous modulo 16 counter is given in Figure 2.7(a). Asynchronous counters present the advantage of being easy to implement. However, as the timing diagram of Figure 2.7b shows, the output signals are affected by different propagation delays that can become too high as the flip-flop number increases. This limits the maximum operating frequency of the counter.

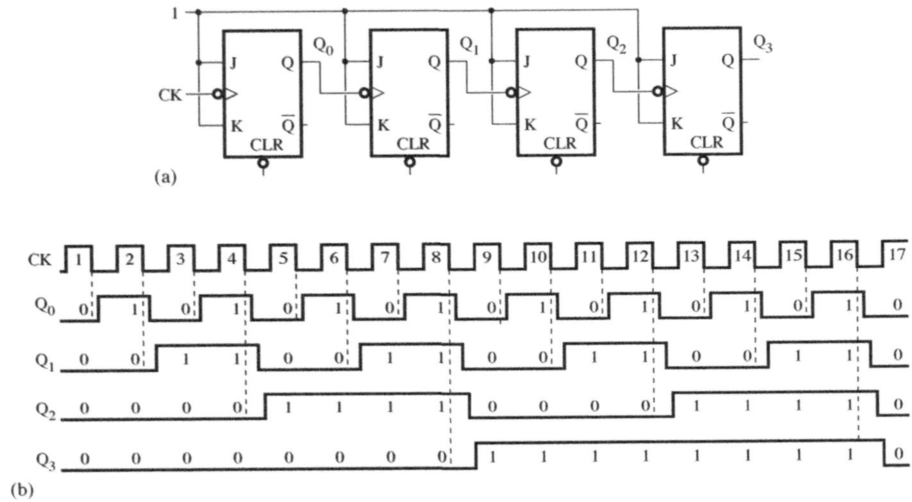

Figure 2.7. *Asynchronous modulo 16 binary counter: a) logic circuit; b) timing diagram*

Synchronous counters can be implemented, with all outputs having identical propagation delays. Figures 2.8(a) and 2.8(b) show the logic circuit and the timing diagram of the synchronous modulo 16 counter when $EN = 1$. For each flip-flop, the logic equations of the inputs are written as follows:

– flip-flop 0: $J_0 = K_0 = EN$;

– flip-flop 1: $J_1 = K_1 = Q_0 \cdot EN$;

– flip-flop 2: $J_2 = K_2 = Q_1 \cdot Q_0 \cdot EN$;

– flip-flop 3: $J_3 = K_3 = Q_2 \cdot Q_1 \cdot Q_0 \cdot EN$.

The counter is activated when the signal EN is set to 1, and when the signal EN assumes the logic state 0 the counter remains in its previous state. Furthermore, we have:

$$RCO = Q_3 \cdot Q_2 \cdot Q_1 \cdot Q_0 \cdot EN$$

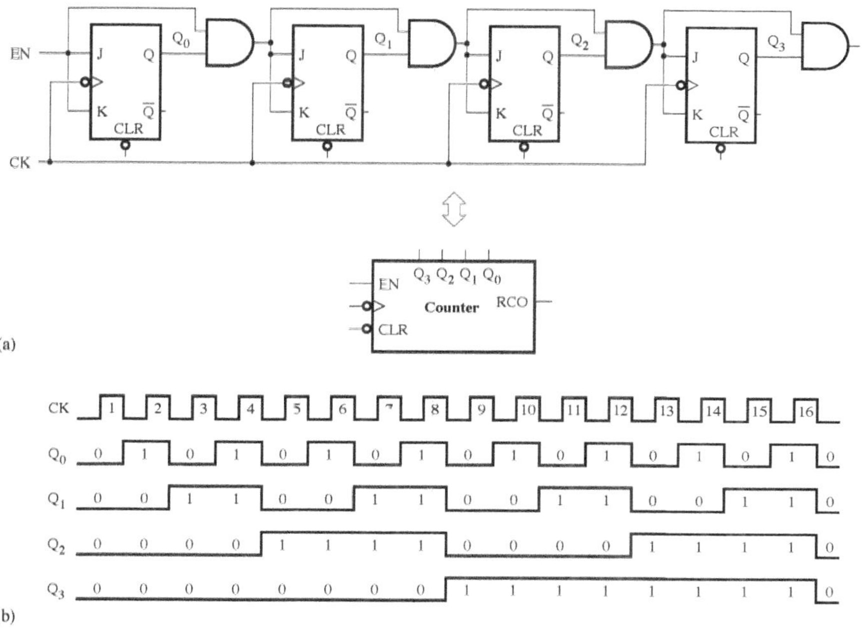

Figure 2.8. *Synchronous modulo 16 binary counter: a) logic circuit; b) timing diagram*

The ripple-carry output, RCO, is set to 1 at the end of the counting cycle. To increase the count range, two counters can be cascaded by connecting the output, RCO, of the first to the input, EN, of the second.

The output of one flip-flop only changes states if the outputs of all the flip-flops of lower numbers are set to 1. The flip-flops used by the counter have identical J and K inputs and can thus only function in hold or toggle modes such as T flip-flops.

The operation of asynchronous and synchronous modulo 16 counters is described by the state diagram given in Figure 2.9, where the initial state is 0.

Figure 2.10 shows the logic circuit of a modulo 256 counter implemented by cascading two modulo 16 counters.

2.4.1. *Modulo 10 counter*

In general, a counter using N flip-flops has a maximum number of states equal to 2^N. It is thus said to have completed a cycle if it can generate binary sequences corresponding to the numbers from 0 to $2^N - 1$.

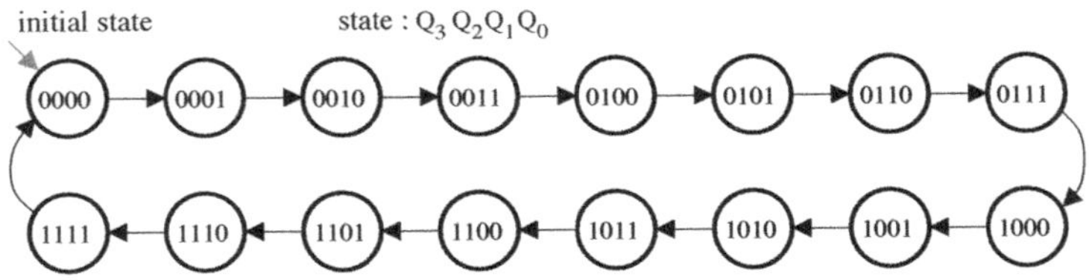

Figure 2.9. *State diagram of the modulo 16 counter*

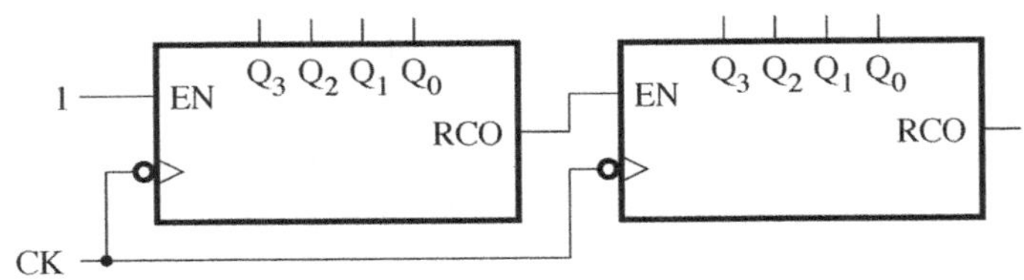

Figure 2.10. *Modulo 256 counter*

The decade counter or the modulo 10 counter is used in applications where a digital display is required.

The asynchronous modulo 10 counter in Figure 2.11(a) is implemented by reducing the number of states of a four-bit asynchronous counter. It produces an ascending sequence going from (0000) to 9 (1001), corresponding to BCD codes. The counter must be reset at the 10th pulse of the clock signal. A NAND gate is used to detect the state $Q_1 = 1$ and $Q_3 = 1$ corresponding to the count of 10 (1010) in order to reset the counter. This can result in unwanted transients during the state transition, as shown in the timing diagram in Figure 2.11(b).

The logic circuit and the timing diagram of a synchronous modulo 10 counter are given in Figures 2.12(a) and 2.12(b). The logic equations for the inputs J and K are written as follows:

– flip-flop 0: $J_0 = K_0 = 1$;

– flip-flop 1: $J_1 = K_1 = Q_0 \cdot \overline{Q_3}$;

– flip-flop 2: $J_2 = K_2 = Q_0 \cdot Q_1$;

– flip-flop 3: $J_3 = K_3 = Q_0 \cdot Q_1 \cdot Q_2 + Q_0 \cdot Q_3$.

NOTE.– Three flip-flops are not enough to implement a modulo 10 counter.

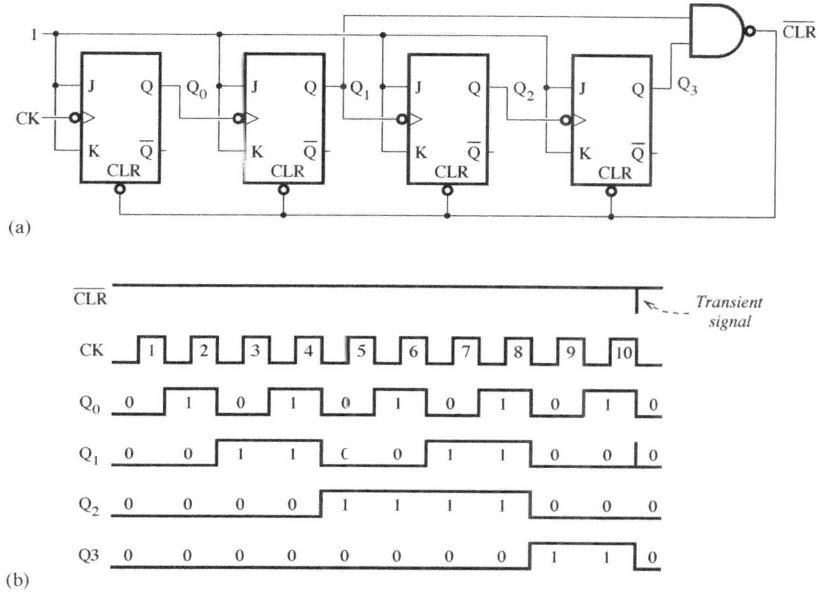

Figure 2.11. *Asynchronous modulo 10 binary counter: a) logic circuit; b) timing diagram*

Considering that the counter assumes one of the states corresponding to the numbers from 10 to 15 (or unused states) after it is powered on, the input equations for the flip-flops can be used to determine the next states. Table 2.3 shows the transition table for unused states. The state diagram of the modulo 10 counter is represented in Figure 2.13. If the counter takes one of the unused states, it can reenter the main counting cycle after, at most, two clock pluses.

PS	Inputs				NS
$Q_3Q_2Q_1Q_0$	J_3K_3	J_2K_2	J_1K_1	J_0K_0	$Q_3^+Q_2^+Q_1^+Q_0^+$
1 0 1 0	0 0	0 0	0 0	1 1	1 0 1 1
1 0 1 1	1 1	1 1	0 0	1 1	0 1 1 0
1 1 0 0	0 0	0 0	0 0	1 1	1 1 0 1
1 1 0 1	1 1	0 0	0 0	1 1	0 1 0 0
1 1 1 0	0 0	0 0	0 0	1 1	1 1 1 1
1 1 1 1	1 1	1 1	0 0	1 1	0 0 1 0

Table 2.3. *Transition table for unused states (PS, present state; NS, next state)*

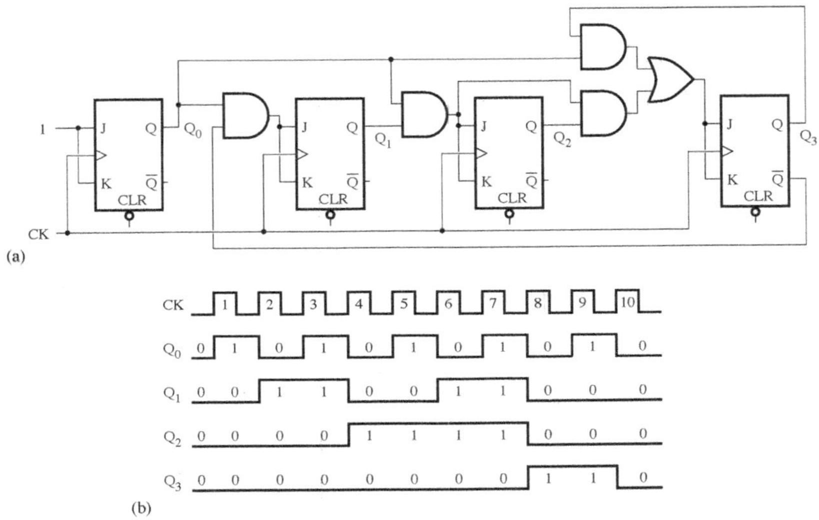

Figure 2.12. *Synchronous modulo 10 binary counter: a) logic circuit; b) timing diagram*

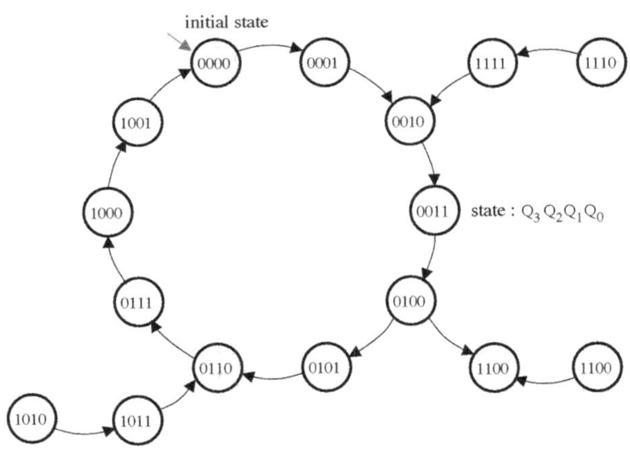

Figure 2.13. *State diagram for the synchronous modulo 10 binary counter*

2.5. Counter with parallel load

A counter with parallel load offers more flexibility in the selection of the count sequence.

Figure 2.14 depicts a four-bit counter with parallel load. It consists of logic gates (AND, XOR), 2-to-1 multiplexers and D flip-flops whose input logic equations can be written as follows:

$$D_0 = P_0 \cdot Load + (Q_0 \oplus En)\overline{Load} \quad [2.1]$$

$$D_1 = P_1 \cdot Load + [Q_1 \oplus (Q_0 \cdot En)]\overline{Load} \quad [2.2]$$

$$D_2 = P_2 \cdot Load + [Q_2 \oplus (Q_1 \cdot Q_0 \cdot En)]\overline{Load} \quad [2.3]$$

$$D_3 = P_3 \cdot Load + [Q_3 \oplus (Q_2 \cdot Q_1 \cdot Q_0 \cdot En)]\overline{Load} \quad [2.4]$$

and:

$$RCO = Q_3 \cdot Q_2 \cdot Q_1 \cdot Q_0 \cdot En \quad [2.5]$$

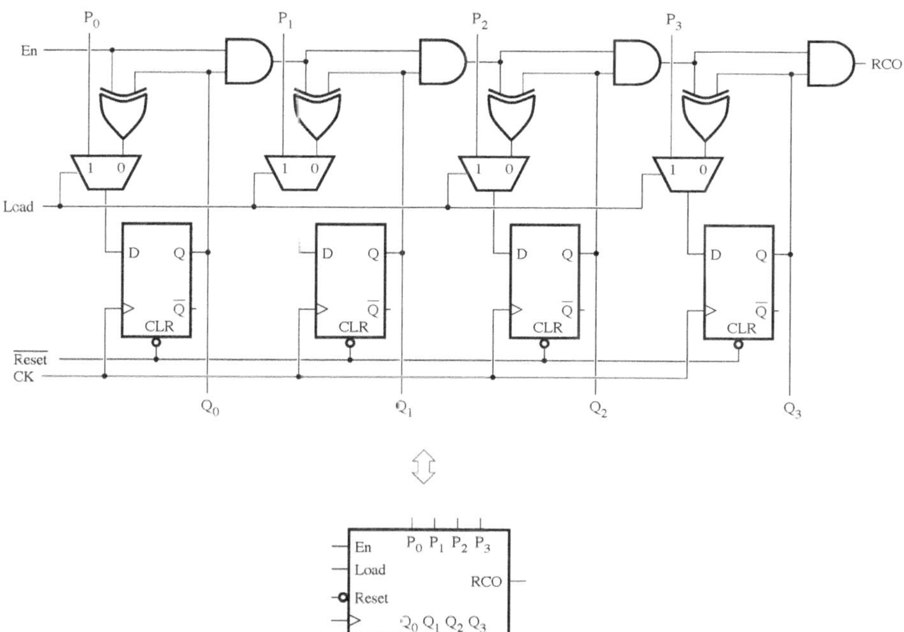

Figure 2.14. *Logic circuit of a four-bit counter with parallel load*

The counter can be reset synchronously and initialized using parallel loads. Table 2.4 presents the function table.

Inputs			
Reset	En	Load	Function
0	x	x	Reset
1	0	0	Hold
1	x	1	Load
1	1	0	Count

Table 2.4. *Function table*

For applications requiring a modulo 7 or modulo 10 counter, the four-bit counter with parallel load can be configured as shown in Figure 2.15. The AND gate detects the state 0110 (6) (modulo 7 counter) or 1001 (9) (modulo 10 counter) to initiate the counter reset through the parallel load.

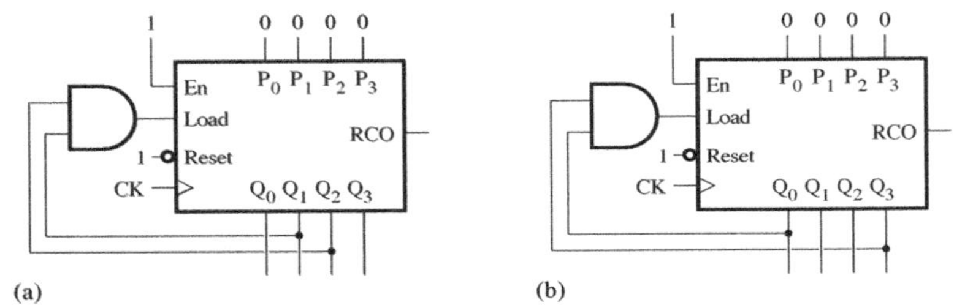

Figure 2.15. *a) Modulo 7 counter; b) modulo 10 counter*

Using two counter stages, the count range can be expanded by connecting the output, RCO, of the first stage to the input, EN, of the second stage.

2.6. Down counter

A down counter generates, from a given initial state, a sequence of numbers in decreasing order.

The logic circuit of an asynchronous modulo 16 down counter is given in Figure 2.16. It consists of JK flip-flops whose inputs are connected to an enable signal EN. The down counter is activated or deactivated by setting the signal EN to 1 or to 0.

Figure 2.17 shows the logic circuit and the symbol for a synchronous modulo 16 down counter. The logic equations for the inputs J and K are given by:

– flip-flop 0: $J_0 = K_0 = EN$;

– flip-flop 1: $J_1 = K_1 = \overline{Q_0} \cdot EN$;

– flip-flop 2: $J_2 = K_2 = \overline{Q_1} \cdot \overline{Q_0} \cdot EN$;

– flip-flop 3: $J_3 = K_3 = \overline{Q_2} \cdot \overline{Q_1} \cdot \overline{Q_0} \cdot EN$.

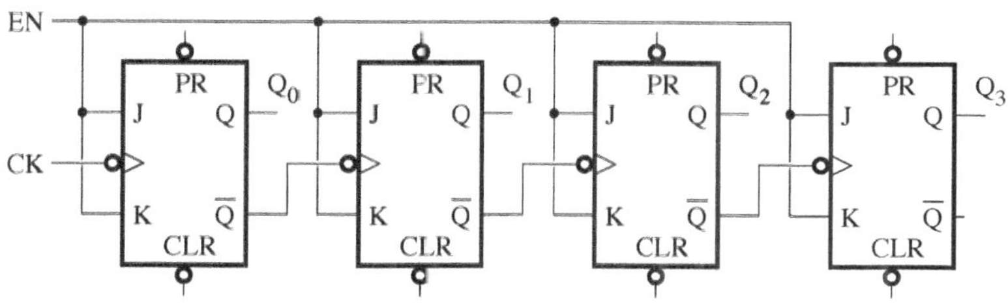

Figure 2.16. *Logic circuit for an asynchronous modulo 16 down counter*

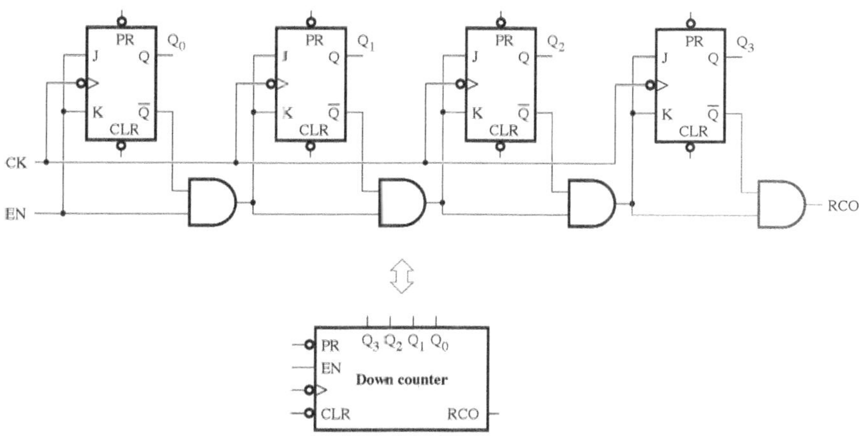

Figure 2.17. *Logic circuit and symbol for a synchronous modulo 16 down counter*

During the normal operation, the enable signal EN must be set to 1.

The state diagram given in Figure 2.18 can be used to describe the operation of an asynchronous down counter, as well as a synchronous down counter. However, the maximum frequency of the asynchronous down counter is limited by the different propagation delays that can affect the clock signal.

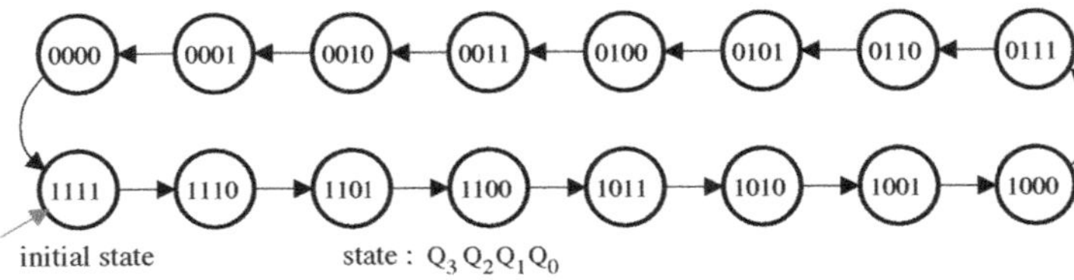

Figure 2.18. *State diagram for a modulo 16 down counter*

2.7. Synchronous reversible counter

A reversible counter is also called a bidirectional counter or an up/down counter. It can generate a sequence of numbers in increasing or decreasing order. In general, it is possible to change the count direction from any state.

Figures 2.19(a) and 2.19(b) show the logic circuit and the timing diagram of a reversible synchronous modulo 8 counter.

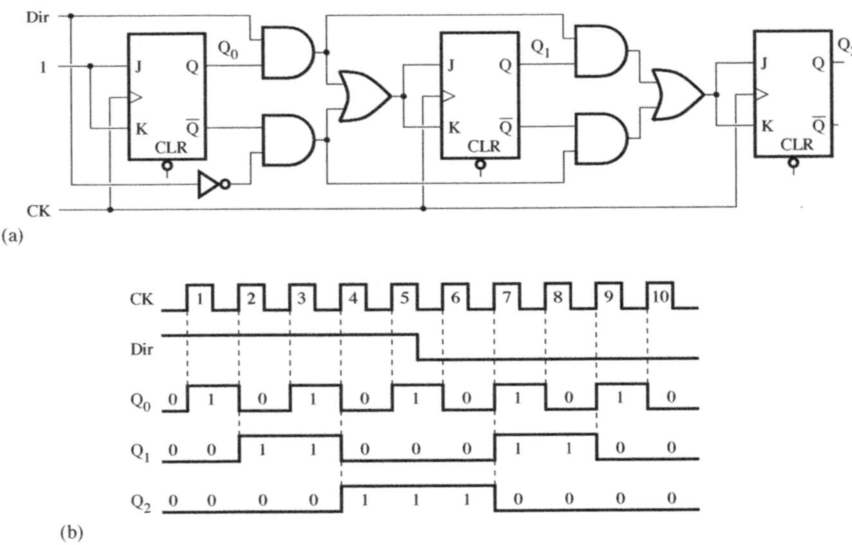

Figure 2.19. *Reversible synchronous modulo 8 counter: a) logic circuit; b) timing diagram*

For each flip-flop, the input logic equations are given by:

– flip-flop 0: $J_0 = K_0 = 1$;

– flip-flop 1: $J_1 = K_1 = Q_0 \cdot \text{Dir} + \overline{Q_0} \cdot \overline{\text{Dir}}$;

– flip-flop 2: $J_2 = K_2 = Q_0 \cdot Q_1 \cdot \text{Dir} + \overline{Q_0} \cdot \overline{Q_1} \cdot \overline{\text{Dir}}$.

The Dir signal is used to choose the type of counting:

– Dir = 0 for counting in decreasing order;

– Dir = 1 for counting in increasing order.

Table 2.5 describes the count sequence.

Clock signal pulse	Q_2	Q_1	Q_0	Number
Initial state	0	0	0	0
1	0	0	1	1
2	0	1	0	2
3	0	1	1	3
4	1	0	0	4
5	1	0	1	5
6	1	0	0	4
7	0	1	1	3
8	0	1	0	2
9	0	0	1	1
10	0	0	0	0

Table 2.5. *Table describing the counter sequence*

2.8. Decoding a down counter

To control the sequence of operation or the display of the numbers represented by flip-flop states, a decoder must be associated with the counter.

An asynchronous three-bit counter with a decoder is depicted in Figure 2.20. Each decoder output Y_i $(i = 0, 1, 2, 3, 4, 5, 6, 7)$ is only in the high state when the count result is equal to i. We thus have:

$$Y_0 = \overline{Q_2} \cdot \overline{Q_1} \cdot \overline{Q_0} \quad [2.6]$$

$$Y_1 = \overline{Q_2} \cdot \overline{Q_1} \cdot Q_0 \quad [2.7]$$

$$Y_2 = \overline{Q_2} \cdot Q_1 \cdot \overline{Q_0} \quad [2.8]$$

$$Y_3 = \overline{Q_2} \cdot Q_1 \cdot Q_0 \quad [2.9]$$

$$Y_4 = Q_2 \cdot \overline{Q_1} \cdot \overline{Q_0} \quad [2.10]$$

$$Y_5 = Q_2 \cdot \overline{Q_1} \cdot Q_0 \quad [2.11]$$

$$Y_6 = Q_2 \cdot Q_1 \cdot \overline{Q_0} \quad [2.12]$$

and:

$$Y_7 = Q_2 \cdot Q_1 \cdot Q_0 \quad [2.13]$$

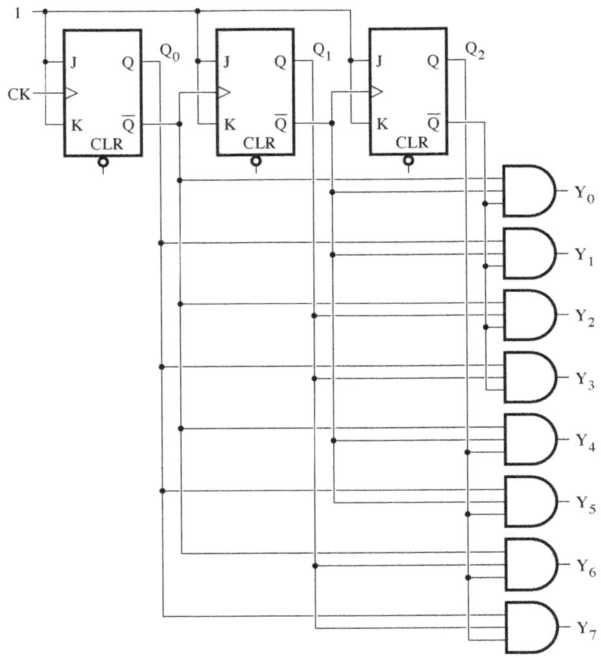

Figure 2.20. *Asynchronous three-bit counter with a decoder*

In general, for binary counters the number of inputs for the AND gates that make up the decoder increases with the modulo.

2.9. Exercises

EXERCISE 2.1.– Design an asynchronous modulo 6 counter.

EXERCISE 2.2.– Asynchronous counter using D flip-flop:

a) verify that the logic circuits shown in Figures 2.21(a) and 2.21(b), where the clock signal input, CK, is used as the T input, are equivalent to a T flip-flop;

b) complete the timing diagram shown in Figure 2.21(c);

c) for the asynchronous counter shown in Figure 2.22(a), complete the timing diagram in Figure 2.22(b) and deduce the modulo. We will assume that the counter is initially set to 0.

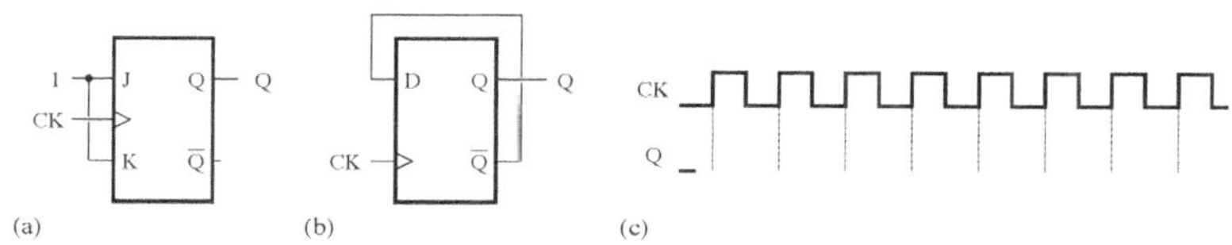

Figure 2.21. *a) and b) Logic circuit; c) timing diagram*

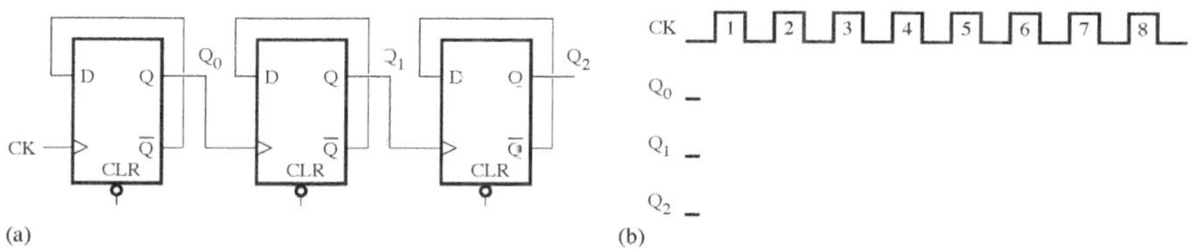

Figure 2.22. *a) Logic circuit of the counter; b) timing diagram*

EXERCISE 2.3.– The counter shown in Figure 2.23 is initially set to 1 (thus, $Q_1Q_0 = 11$).

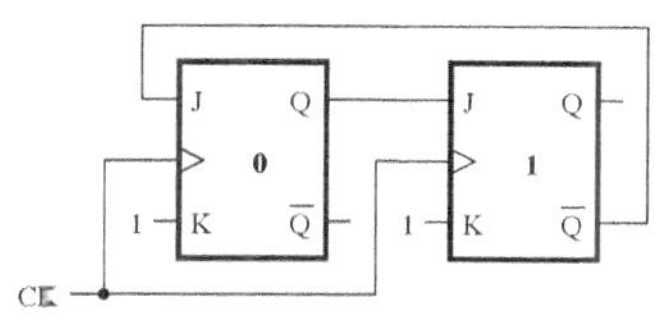

Figure 2.23. *Counter using two JK flip-flops*

Neglecting the propagation delays, determine the count sequence after five pulses of the clock signal.

Represent the state diagram of the counter.

EXERCISE 2.4.– Asynchronous circuit using two flip-flops.

The asynchronous circuit shown in Figure 2.24(a) consists of two D flip-flops and a NAND gate.

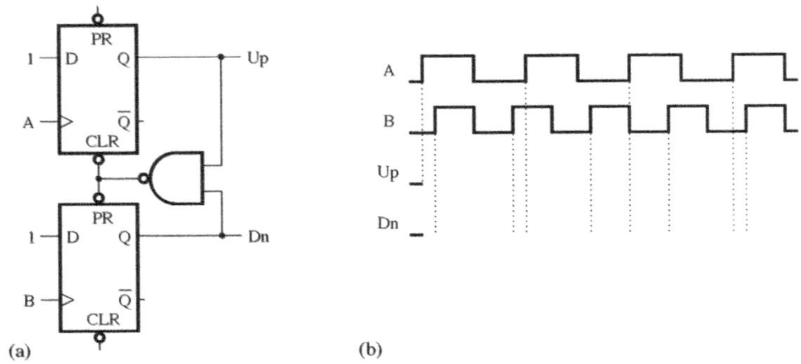

Figure 2.24. *a) Asynchronous circuit using two flip-flops; b) timing diagram*

Complete the timing diagram as shown in Figure 2.24(b).

What function does this circuit serve?

EXERCISE 2.5.– Counter using three flip-flops.

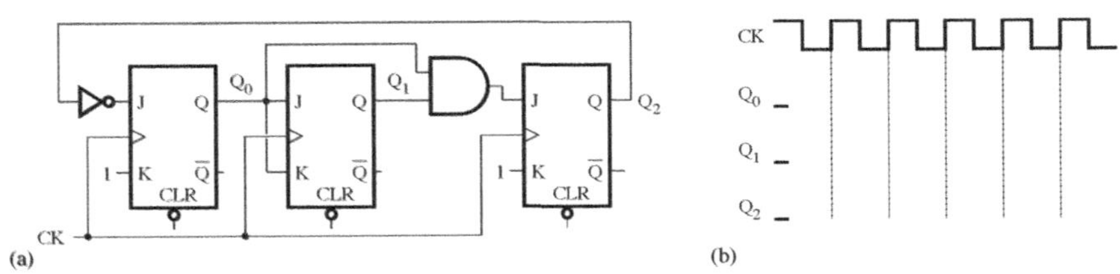

Figure 2.25. *a) Counter using three flip-flops; b) timing diagram*

Consider the counter shown in Figure 2.25(a), whose flip-flops are initially set to 0:

– determine the logic equations for the J and K inputs of each flip-flop;

– complete the timing diagram in Figure 2.25(b);

– construct the state diagram of the counter;

– deduce the counter modulo.

EXERCISE 2.6.– Counter using four flip-flops.

Give the timing diagram for the counter shown in Figure 2.26, assuming that all the flip-flops are initially set to 0.

Modify the counter shown in Figure 2.26 by adding a set input ($\overline{PR}$) and a reset input ($\overline{CLR}$), which, when activated, determine the state: $Q_3Q_2Q_1Q_0 = 1010$.

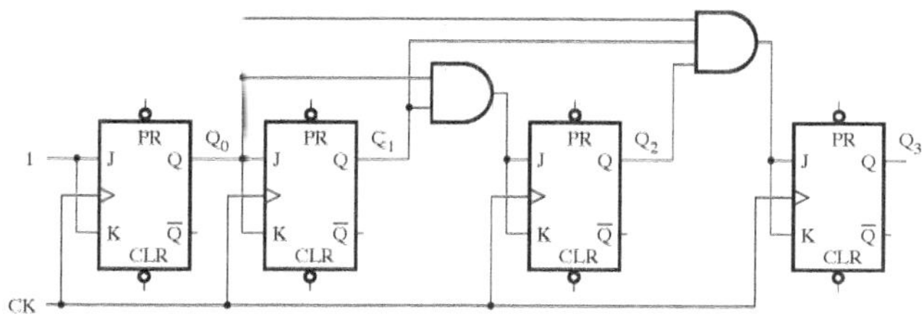

Figure 2.26. *Logic circuit for a four-bit counter*

EXERCISE 2.7.– Counter using three JK flip-flops.

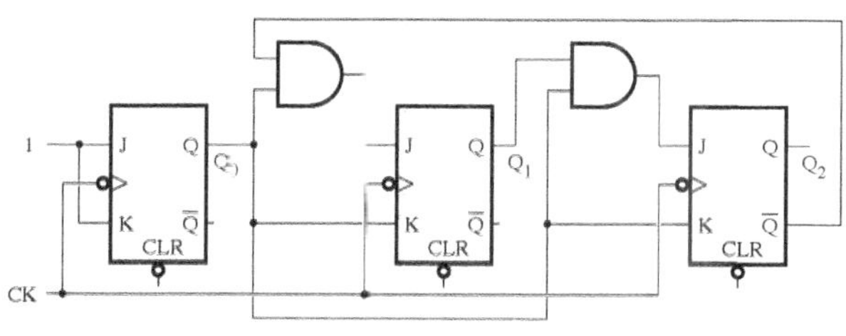

Figure 2.27. *Logic circuit for a counter using three JK flip-flops*

Consider the logic circuit of the counter shown in Figure 2.27:

– determine the logic equations for the inputs of each flip-flop;

– give the timing diagram for the counter assuming that all the flip-fops are initially reset to 0;

– give the state diagram for the counter;

– deduce the counter modulo.

EXERCISE 2.8.– Synchronous D flip-flop counter.

For the counter shown in Figure 2.28, determine the logic equation of the input for each flip-flop and construct the transition table. We will assume that the initial state of each counter is 0.

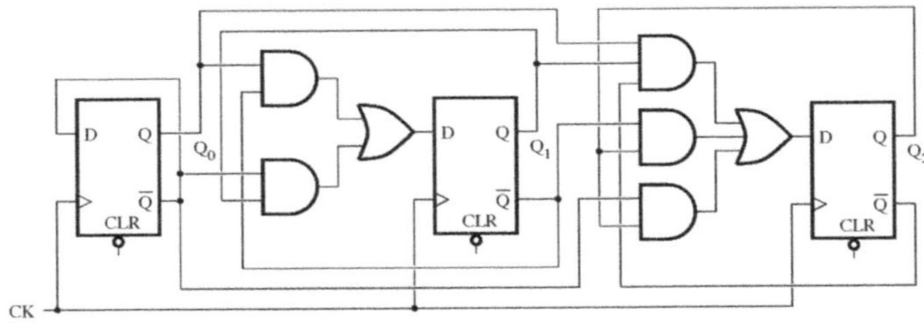

Figure 2.28. *Logic circuit for the synchronous D flip-flop counter*

Deduce the counter modulo.

EXERCISE 2.9.– The counter shown in Figure 2.29 is initially set to 0 (that is $Q_2Q_1Q_0 = 000$).

a) Determine the logic expressions for the inputs J_0, K_0, J_1, K_1, J_2 and K_2.

b) Derive the state diagram for this counter.

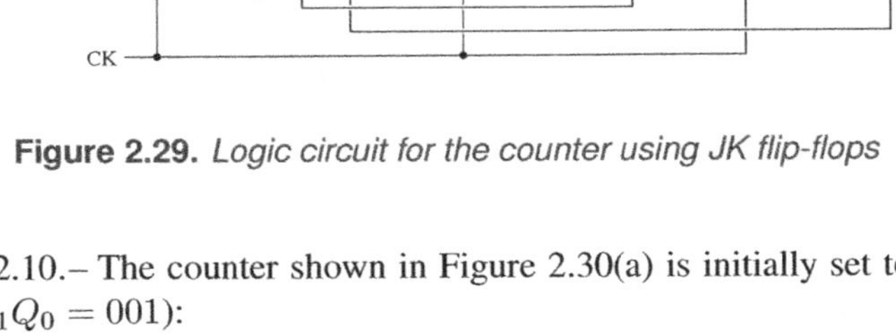

Figure 2.29. *Logic circuit for the counter using JK flip-flops*

EXERCISE 2.10.– The counter shown in Figure 2.30(a) is initially set to 0 (in other words $Q_2Q_1Q_0 = 001$):

– determine the logic expressions for the inputs J_0, K_0, J_1, K_1, J_2 and K_2;

– complete the timing diagram shown in Figure 2.30(b);

– determiner Q_2 Q_1 Q_0 when the diode D_7 is switched on;

– can the counter simultaneously switch on both the diodes, D_2 and D_7?

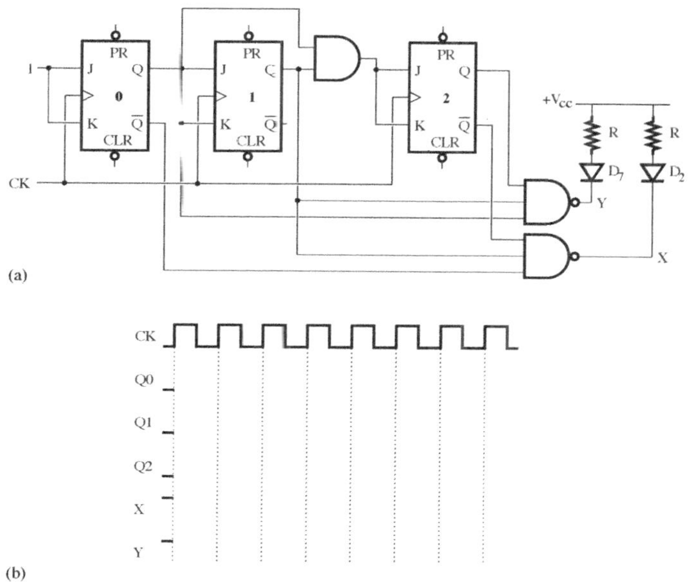

Figure 2.30. *a) Counter connected to D_1 and D_2 diodes; b) timing diagram*

EXERCISE 2.11.– Synchronous counter.

The synchronous counter shown in Figure 2.31 consists of a debouncing switch (S), a D flip-flop for data synchronization on the falling edge of the clock signal and a counter circuit with a reset push button.

Determine the logic equations for the inputs, D_1 and D_0, and the outputs, Y_1 and Y_0, as functions of X, Q_1 and Q_0.

Complete the timing diagram given in Figure 2.32 and 2.33.

What is the function of the switch S?

Complete the output decoding table given in Table 2.6.

EXERCISE 2.12.– Temporization circuit for a quartz clock.

An m modulo counter can be implemented, as shown in Figure 2.34, with a reset input, $\overline{CLR}$, an enable input, EN, and a ripple-carry output, RCO.

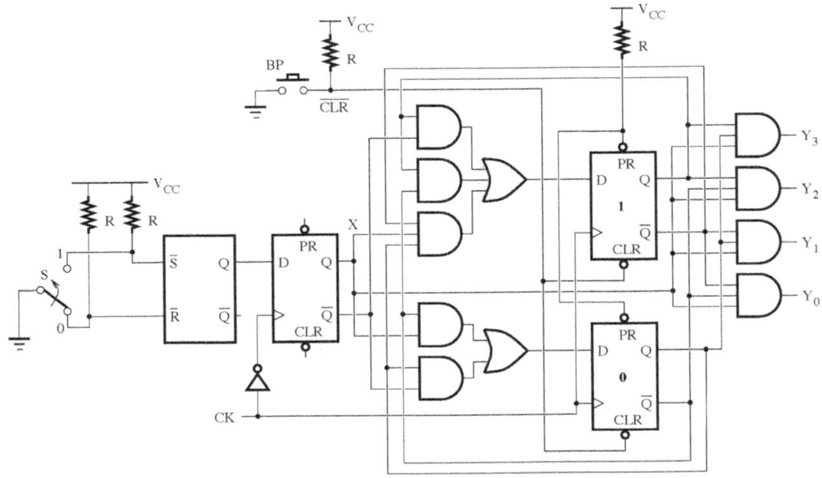

Figure 2.31. *Synchronous circuit*

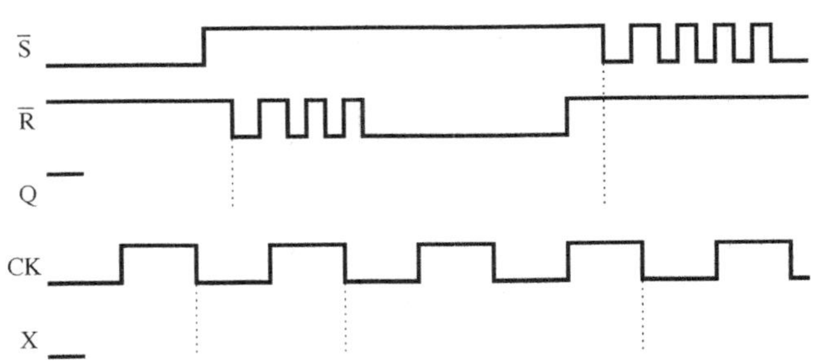

Figure 2.32. *Timing diagram 1*

X	Q_1	Q_0	Y_3	Y_2	Y_1	Y_0
0	x	x				
1	0	0				
1	0	1				
1	1	0				
1	1	1				

Table 2.6. *Output decoding table*

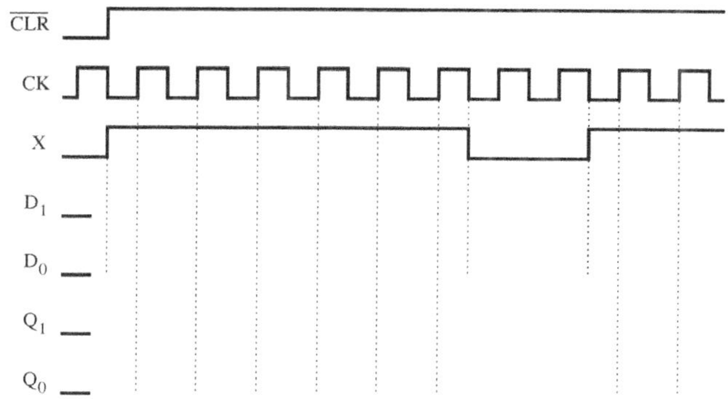

Figure 2.33. *Timing diagram 2*

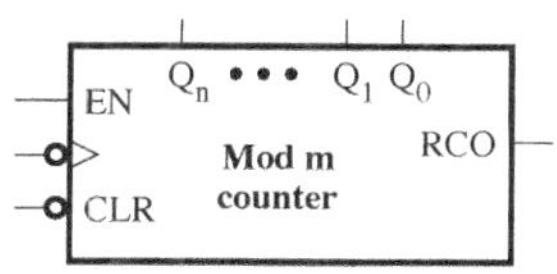

Figure 2.34. *Modulo m counter*

Determine the relationship that exists between the modulo, m, and the number of bits, n.

Using a quartz oscillator, which yields a clock signal, CK, whose frequency is $65,536$ Hz and using modulo 16, 10, 6 and 4 counters, implement a temporization circuit that can generate a signal for each day (D), each hour (H), each minute (M) and each second (S).

2.10. Solutions

SOLUTION 2.1.– Asynchronous modulo 6 counter.

Figure 2.35 shows the logic circuit for an asynchronous modulo 6 counter. The NAND gate detects the combinations $Q_2Q_1Q_0$ taking the form 110 (6) or 111 (7) to initiate the flip-flops reset.

SOLUTION 2.2.– Asynchronous D flip-flop.

For the logic circuit shown in Figure 2.36(a), we have:

$$J = K = 1 \quad \text{and} \quad Q^+ = J \cdot \overline{Q} + \overline{K} \cdot Q = \overline{Q}$$

and for the logic circuit shown in Figure 2.36(b), we have:

$$D = \overline{Q} \quad \text{and} \quad Q^+ = D = \overline{Q}$$

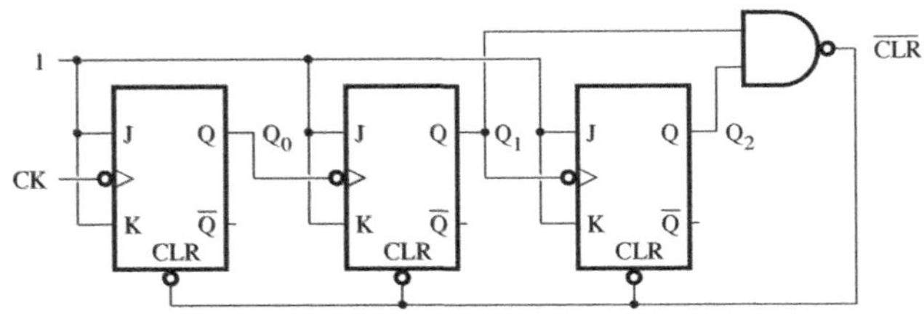

Figure 2.35. *Asynchronous modulo 6 counter*

They then operate in toggle mode or like a T flip-flop, as illustrated by the timing diagram shown in Figure 2.36(c).

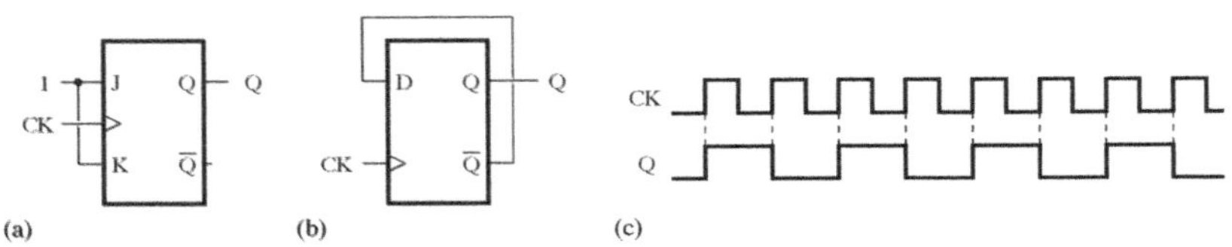

Figure 2.36. *a) and b) Logic circuit; c) timing diagram*

The timing diagram for the counter in Figure 2.37(a) is represented in Figure 2.37(b). As the count cycle has eight distinct states, the counter modulo is equal to 8.

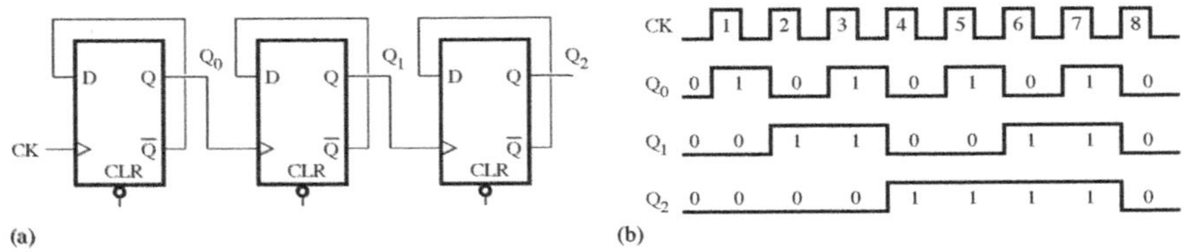

Figure 2.37. *a) Logic circuit of the counter; b) timing diagram*

SOLUTION 2.3.– Counter using two flip-flops.

For each of the counter's flip-flops, the input equations are given by:

– flip-flop 0: $J_0 = \overline{Q_1}$ and $K_0 = 1$;

– flip-flop 1: $J_1 = Q_0$ and $K_1 = 1$.

Table 2.7 shows the transition table. The state diagram is given in Figure 2.38.

Q_1	Q_0	$J_1 K_1$	$J_0 K_0$	Q_1^+	Q_0^+
1	1	1 1	0 1	0	0
0	0	0 1	1 1	0	1
0	1	1 1	1 1	1	0
1	0	0 1	0 1	0	0

Table 2.7. *Transition table for the counter*

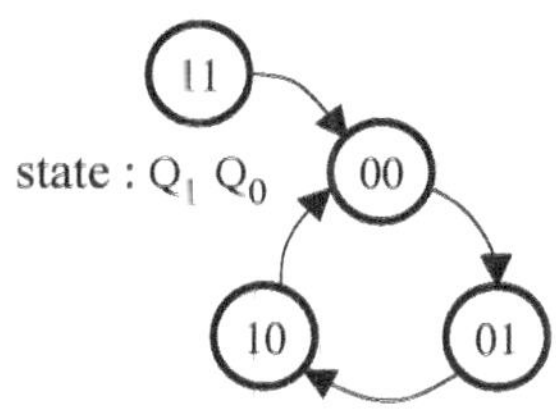

Figure 2.38. *State diagram for the counter*

SOLUTION 2.4.– Asynchronous circuit using two flip-flops.

Figure 2.39 depicts an asynchronous circuit using two flip-flops and the corresponding timing diagram.

By analyzing the timing diagram, we can determine which of the two signals, A or B, is in advance or delayed. This is, thus, a phase detector.

SOLUTION 2.5.– Counter using three flip-flops.

By referring to the counter's logic circuit, we obtain the following logic equations:

– flip-flop 0: $J_0 = \overline{Q_2}$ and $K_0 = 1$;

– flip-flop 1: $J_1 = Q_0$ and $K_1 = Q_0$;

– flip-flop 2: $J_2 = Q_1 \cdot Q_0$ and $K_2 = 1$.

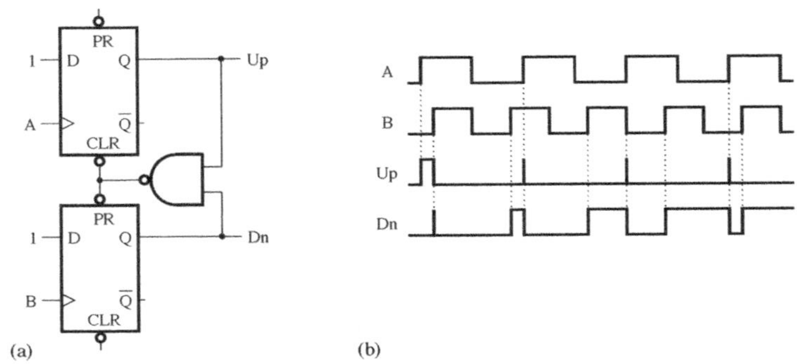

Figure 2.39. *a) Asynchronous circuit using two flip-flops; b) timing diagram*

Figure 2.40(a) depicts a timing diagram, where five different states can be differentiated. As the counter uses three flip-flops, the next states corresponding to the three unused states (101, 110, and 111) must be determined.

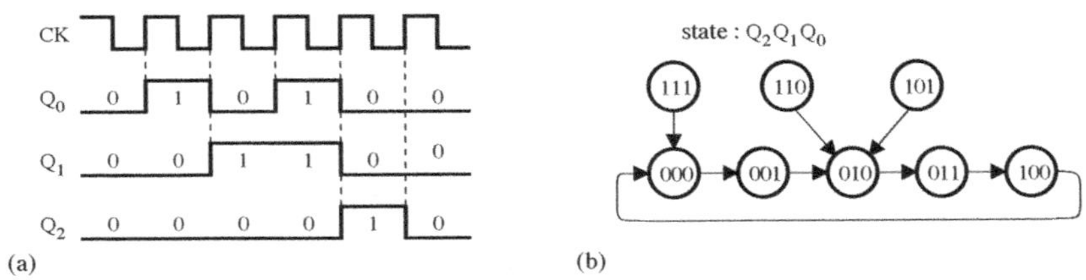

Figure 2.40. *a) Timing diagram; b) state diagram*

Table 2.8 depicts the transition table for the unused states.

PS	Inputs			NS
$Q_2Q_1Q_0$	J_2K_2	J_1K_1	J_0K_0	$Q_2^+Q_1^+Q_0^+$
1 0 1	0 1	1 1	0 1	0 1 0
1 1 0	0 1	0 0	0 1	0 1 0
1 1 1	1 1	1 1	0 1	0 0 0

Table 2.8. *Transition table for the unused states (PS, present state; NS, next state)*

The state diagram for the counter is given in Figure 2.40(b).

This is a 5 modulo counter.

SOLUTION 2.6.– Counter using four flip-flops.

The logic equations for the inputs of the flip-flops are given by:

– flip-flop 0: $J_0 = K_0 = 1$;

– flip-flop 1: $J_1 = K_1 = Q_0$;

– flip-flop 2: $J_2 = K_2 = Q_1 \cdot Q_0$;

– flip-flop 3: $J_2 = K_2 = Q_2 \cdot Q_1 \cdot Q_0$.

Based on the timing diagram in Figure 2.41, it is possible to deduce that this is a 16 modulo counter.

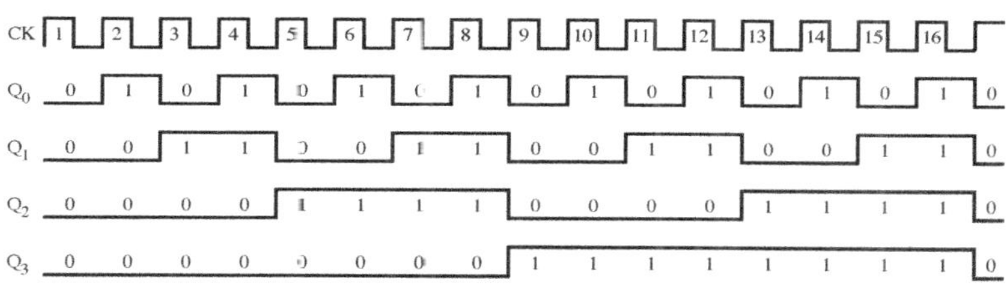

Figure 2.41. *Timing diagram*

Figure 2.42 depicts the logic circuit for the modified counter as well as a representation of the $\overline{PR}$ and $\overline{CLR}$ signals.

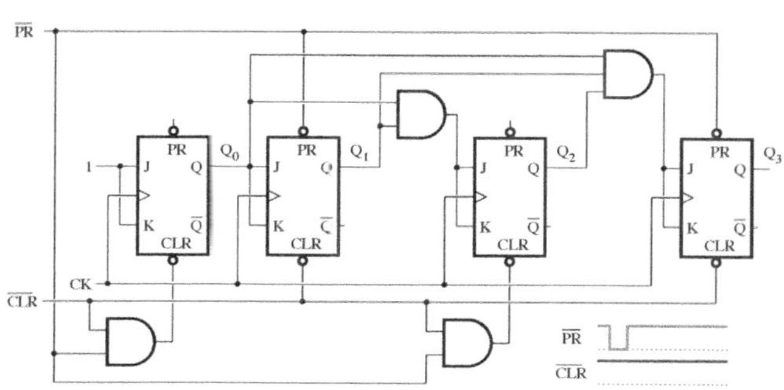

Figure 2.42. *Logic circuit for the modified counter. For a color version of this figure, see www.iste.co.uk/ndjountche/electronics2.zip*

SOLUTION 2.7.– Counter using three flip-flops.

By analyzing the logic circuit for the counter, we obtain the following logic equations:

– flip-flop 0: $J_0 = K_0 = 1$;

– flip-flop 1: $J_1 = \overline{Q_2} \cdot Q_0$ and $K_1 = Q_0$;

– flip-flop 2: $J_2 = Q_1 \cdot Q_0$ and $K_2 = Q_0$.

Figure 2.43(a) depicts the timing diagram of the counter. The states 110 and 111 are unused. The transition table for the unused states is given in Table 2.9. Figure 2.43(b) shows the state diagram of the counter. Because the count cycle consists of six different states, it is a modulo 6 counter.

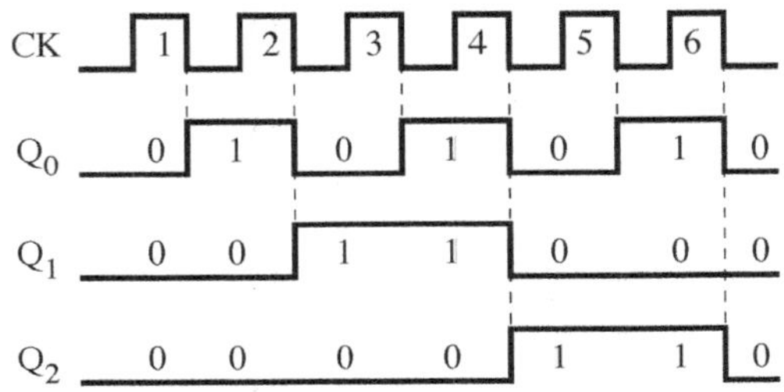

Figure 2.43. *Timing diagram of the counter*

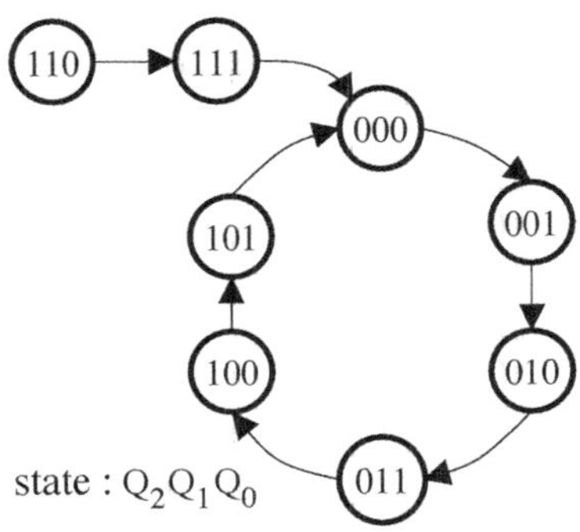

Figure 2.44. *State diagram of the counter*

SOLUTION 2.8.– Synchronous D flip-flop counter.

The analysis of the logic circuit of the counter yields the following equations:

– flip-flop 0: $D_0 = \overline{Q_0}$;

– flip-flop 1: $D_1 = \overline{Q_1} \cdot Q_0 + Q_1 \cdot \overline{Q_0}$;

– flip-flop 2: .

PS	Inputs			NS
$Q_2Q_1Q_0$	J_2K_2	J_1K_1	J_0K_0	$Q_2^+Q_1^+Q_0^+$
1 1 0	0 0	0 0	1 1	1 1 1
1 1 1	1 1	0 1	1 1	0 0 0

Table 2.9. *Transition table for the unused states*

Because we have $Q^+ = D$ for a D flip-flop, we can construct the transition table given in Table 2.10.

PS	Inputs			NS
$Q_2Q_1Q_0$	D_2	D_1	D_0	$Q_2^+Q_1^+Q_0^+$
0 0 0	0	0	1	0 0 1
0 0 1	0	1	0	0 1 0
0 1 0	0	1	1	0 1 1
0 1 1	1	0	0	1 0 0
1 0 0	1	0	1	1 0 1
1 0 1	1	1	0	1 1 0
1 1 0	1	1	1	1 1 1
1 1 1	0	0	0	0 0 0

Table 2.10. *Transition table (PS, present state; NS: next state)*

The count cycle is composed of eight different states, corresponding to the numbers from 0 to 7. The modulo for the counter is, thus, equal to 8.

SOLUTION 2.9.– The input equations for the flip-flops are written as follows:

– flip-flop 0: $J_0 = \overline{Q_2 \oplus Q_1}$ and $K_0 = Q_2 \oplus Q_1$;

– flip-flop 1: $J_1 = \overline{Q_2} \cdot Q_0$ and $K_1 = Q_2 \cdot Q_0$;

– flip-flop 2: $J_2 = Q_1 \cdot \overline{Q_0}$ and $K_2 = \overline{Q_1} \cdot \overline{Q_0}$.

The transition table of the counter is represented in Table 2.11. Figure 2.45 depicts the state diagram. The count cycle consists of eight different states.

Furthermore, because the successive states differ by only one bit, this is a modulo 8 Gray counter.

SOLUTION 2.10.– The logic equations for the inputs of the flip-flops are given by:

– flip-flop 0: $J_0 = K_0 = 1$;

– flip-flop 1: $J_1 = K_1 = Q_0$;

– flip-flop 2.

PS	Inputs			NS
$Q_2\ Q_1\ Q_0$	J_2K_2	J_1K_1	J_0K_0	$Q_2^+\ Q_1^+\ Q_0^+$
0 0 0	0 1	0 0	1 0	0 0 1
0 0 1	0 0	1 0	1 0	0 1 1
0 1 1	0 0	1 0	0 1	0 1 0
0 1 0	1 0	0 0	0 1	1 1 0
1 1 0	1 0	0 0	1 0	1 1 1
1 1 1	0 0	0 1	1 0	1 0 1
1 0 1	0 0	0 1	0 1	1 0 0
1 0 0	0 1	0 0	0 1	0 0 0

Table 2.11. *Transition table (PS, present state; NS, next state)*

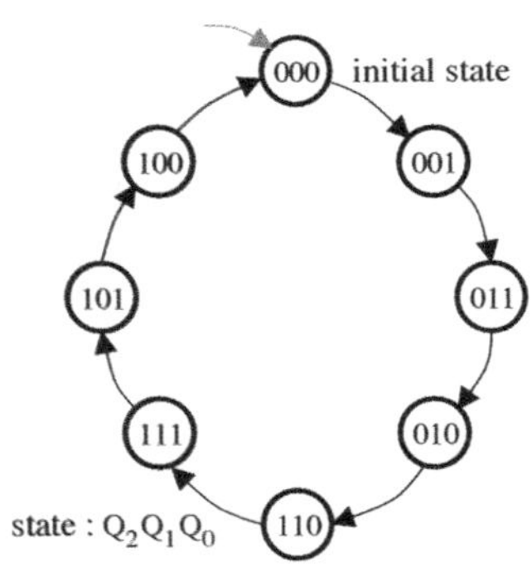

Figure 2.45. *State diagram*

For the outputs connected to the diodes, we have:

$$X = \overline{\overline{Q_2} \cdot Q_1 \cdot \overline{Q_0}} \qquad [2.14]$$

$$Y = \overline{Q_2 \cdot Q_1 \cdot Q_0} \qquad [2.15]$$

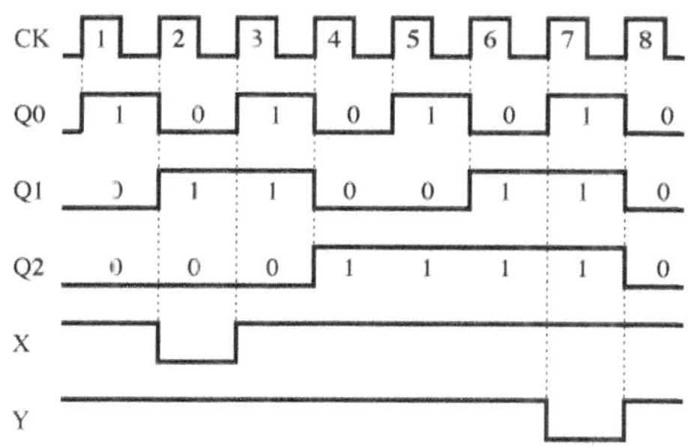

Figure 2.46. *Timing diagram*

Figure 2.46 depicts the timing diagram of the counter.

D_7 is switched on when $Q_2Q_1Q_0 = 111$.

Based on the timing diagram, the counter cannot simultaneously switch on both diodes.

SOLUTION 2.11.– Synchronous counter.

By analyzing the logic circuit of the counter, we obtain the following equations:

$$D_0 = X \cdot \overline{Q_0} + \overline{X} \cdot Q_0 \qquad [2.16]$$

$$D_1 = X \cdot \overline{Q_1} \cdot Q_0 + Q_1 \cdot \overline{Q_0} + \overline{X} \cdot Q_1 \qquad [2.17]$$

$$Y_3 = X \cdot Q_1 \cdot Q_0 \qquad [2.18]$$

$$Y_2 = X \cdot Q_1 \cdot \overline{Q_0} \qquad [2.19]$$

$$Y_1 = X \cdot \overline{Q_1} \cdot Q_0 \qquad [2.20]$$

$$Y_0 = X \cdot \overline{Q_1} \cdot \overline{Q_0} \qquad [2.21]$$

Figure 2.47 depicts timing diagram 1. Bounces of the switch S are eliminated by the flip-flop $\overline{S}\ \overline{R}$, while the signal X is synchronized to the falling edge of the clock signal by the D flip-flop.

Figure 2.48 depicts timing diagram 2. The counting commences when the signal X takes the logic state 1 and stops when the signal X is reset.

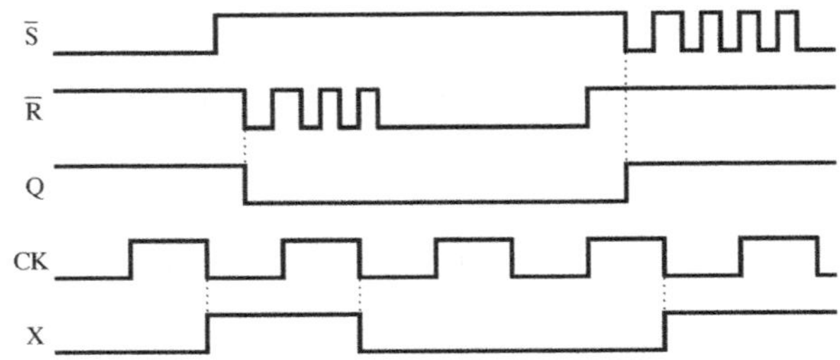

Figure 2.47. *Timing diagram 1*

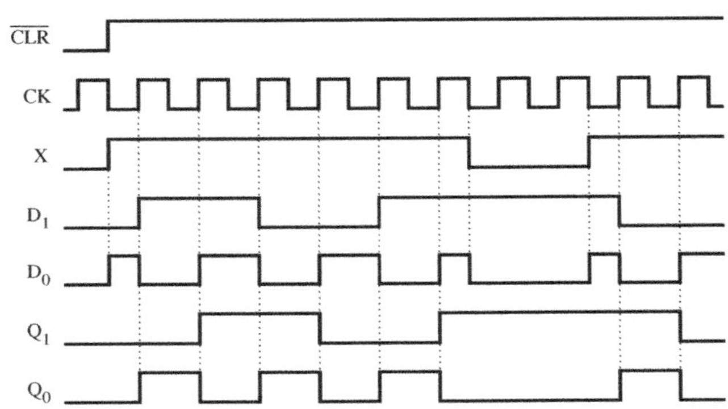

Figure 2.48. *Timing diagram 2*

The switch S is, thus, used to turn on and off the counter.

The table for output decoding is given in Table 2.12.

X	Q_1	Q_0	Y_3	Y_2	Y_1	Y_0
0	x	x	0	0	0	0
1	0	0	0	0	0	1
1	0	1	0	0	1	0
1	1	0	0	1	0	0
1	1	1	1	0	0	0

Table 2.12. *Table for output decoding*

SOLUTION 2.12.– Temporization of a quartz clock.

It is possible to establish the following relationship between the modulo, m, and the number of bits, n:

$$n = \lceil \log(m)/\log(2) \rceil \quad [2.22]$$

where $\lceil x \rceil$ represents the smallest integer greater than or equal to x.

Figure 2.49 depicts the temporization circuit for the quartz clock.

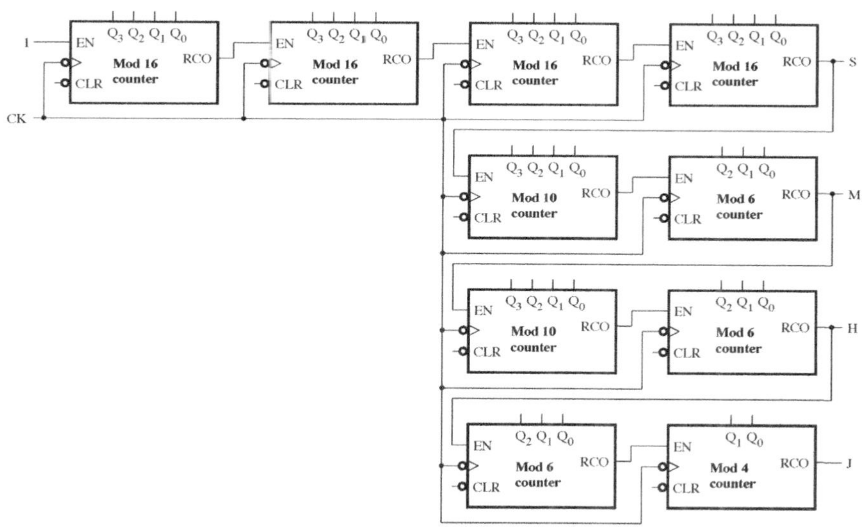

Figure 2.49. *Temporization circuit*

3

Shift Register

3.1. Introduction

Shift registers are used to shift bit positions of data words either to the left or to the right. They are used in data storage and transfer applications. Unlike binary counters, shift registers do not have a specific sequence of states.

A register composed of n flip-flops possesses the storage capacity for a binary word of n bits. There are different types of registers depending on the mode of access (series or parallel) to the register (for read and write operations).

3.2. Serial-in shift register

A shift register is implemented by serially connecting D flip-flops activated by the same clock signal. Figure 3.1 depicts a 5-bit shift register. Data bits applied to the serial input (SI) are transferred from one flip-flop to another at the edge of each clock pulse. Signals that appear at the parallel outputs or at the outputs of different flip-flops are identical but delayed with respect to each other, while data are available as a serial sequence at the output of the last flip-flop. The timing diagram shown in Figure 3.2 illustrates the temporization function carried out by a shift register.

3.3. Parallel-in shift register

A shift register can be implemented with synchronous inputs, which, if necessary, can be used to modify its content as shown in Figure 3.3. In this case, the operating mode of the shift register depends on the state of the enable signal En:

– when En is low, the gates 1, 2 and 3 are activated and the gates 4, 5 and 6 are deactivated. Each D_i $(i = 1, 2, 3)$ bit can thus be applied to the data input of the corresponding flip-flop;

– when En is high, gates 1, 2 and 3 are deactivated and gates 4, 5 and 6 are activated, thereby enabling the shift-right operation. D_0 can be considered as a SI and Q_3 can be regarded as a serial output (SO).

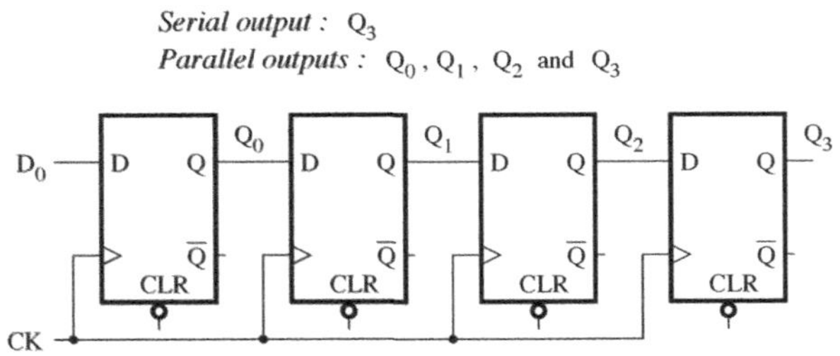

Figure 3.1. *Shift register*

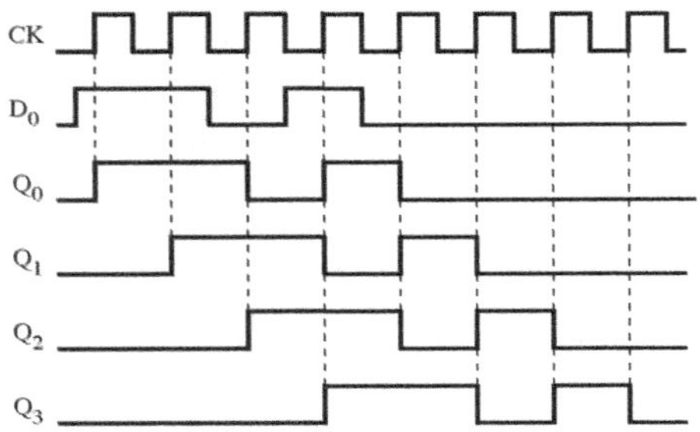

Figure 3.2. *Timing diagram*

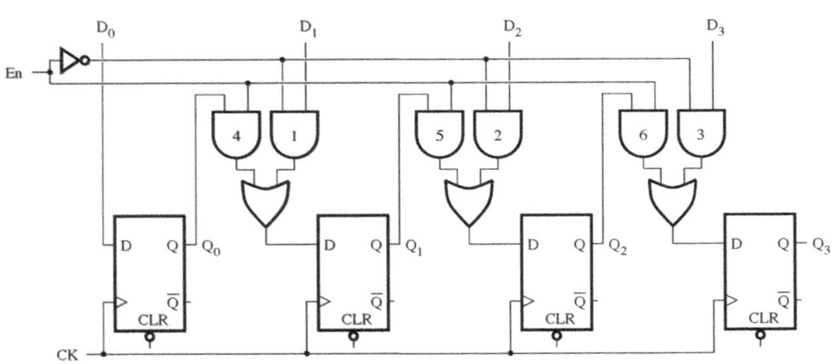

Figure 3.3. *Parallel-in serial-out shift register*

Table 3.1 depicts the functional truth table of the parallel-in serial-out shift register. A timing diagram is given in Figure 3.4 assuming that the flip-flops are initially set to 0.

Inputs		NS				
CK	En	Q_3^+	Q_2^+	Q_1^+	Q_0^+	
↑	0	D_3	D_2	D_1	D_0	Load
↑	1	Q_2	Q_1	Q_0	D_0	Right shift
0	x	Q_3	Q_2	Q_1	Q_0	Hold
1	x	Q_3	Q_2	Q_1	Q_0	Hold

Table 3.1. *Functional truth table (NS, next state)*

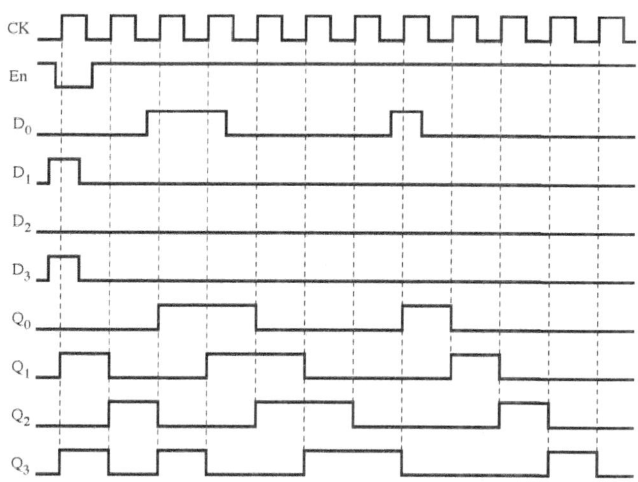

Figure 3.4. *Timing diagram*

Some applications require shift registers that can be enabled or disabled. Figure 3.5 depicts the logic circuit of a parallel-in serial-out shift register with an enable signal. To allow the realization of additional functions, the shift register is implemented by combining 2 : 1 multiplexers with D flip-flops.

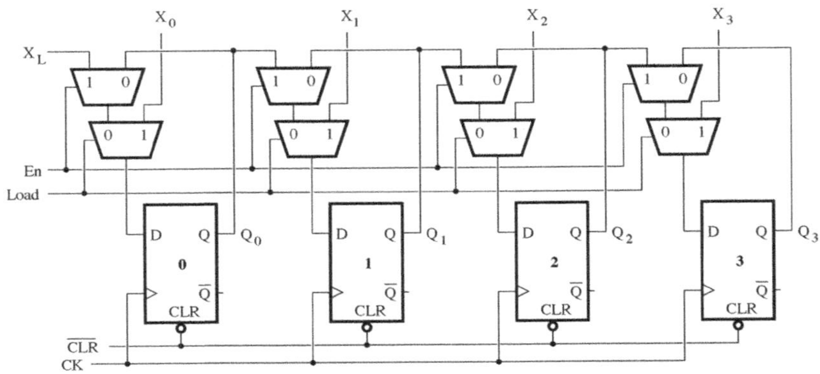

Figure 3.5. *Parallel-in serial-out shift register with enable signal*

The functional truth table is given in Table 3.2. The transfer of the input data takes place when the clock signal transitions from low to high. It must be noted that the input X_L can be considered as a SI and the output Q_3 as a SO.

Inputs				NS				
CK	$\overline{CLR}$	Load	En	Q_3^+	Q_2^+	Q_1^+	Q_0^+	
x	0	x	x	0	0	0	0	Reset
↑	1	0	0	Q_3	Q_2	Q_1	Q_0	Hold
↑	1	0	1	Q_2	Q_1	Q_0	X_L	Left shift
↑	1	1	x	X_3	X_2	X_1	X_0	Load

Table 3.2. *Functional truth table of the parallel-in serial-out shift register with an enable signal (NS, next state)*

3.4. Bidirectional shift register

A bidirectional shift register can be used to shift data bits toward the left or the right based on the logic state of a control signal.

A four-bit bidirectional shift register is illustrated in Figure 3.6. Its operation can be described as follows:

– when the control input Sh is at the high logic state, the gates 1, 2, 3 and 4 are activated, enabling the application of the input signal to the first flip-flop and the connection of the output Q of each flip-flop to the D input of the next flip-flop. Data bits are shifted by one position to the right at the rising edge of the clock signal;

– when the control input Sh is at the low logic state, the gates 5, 6, 7 and 8 are activated, enabling the connection of the output Q of each flip-flop to the D input of the preceding flip-flop. Data bits are shifted by one position to the left at the rising edge of the clock signal.

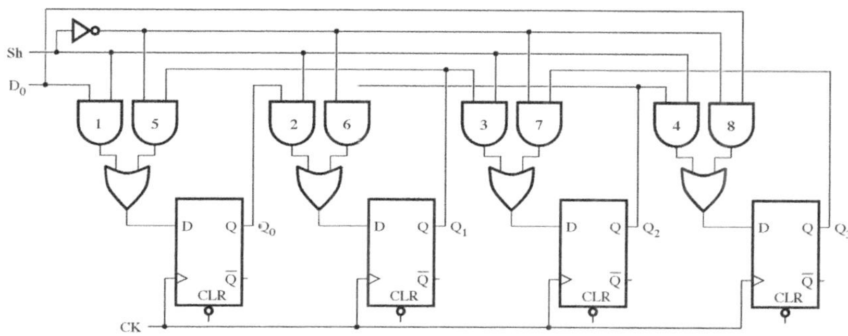

Figure 3.6. *Bidirectional shift register*

Table 3.1 depicts the functional truth table of the bidirectional shift register. Assuming that the flip-flops are initially set to 0, the timing diagram can be represented as shown in Figure 3.7.

Inputs		NS				
CK	Sh	Q_3^+	Q_2^+	Q_1^+	Q_0^+	
↑	0	D_0	Q_3	Q_2	Q_1	Left shift
↑	1	Q_2	Q_1	Q_0	D_0	Right shift
0	x	Q_3	Q_2	Q_1	Q_0	Hold
1	x	Q_3	Q_2	Q_1	Q_0	Hold

Table 3.3. *Functional truth table (NS, next state)*

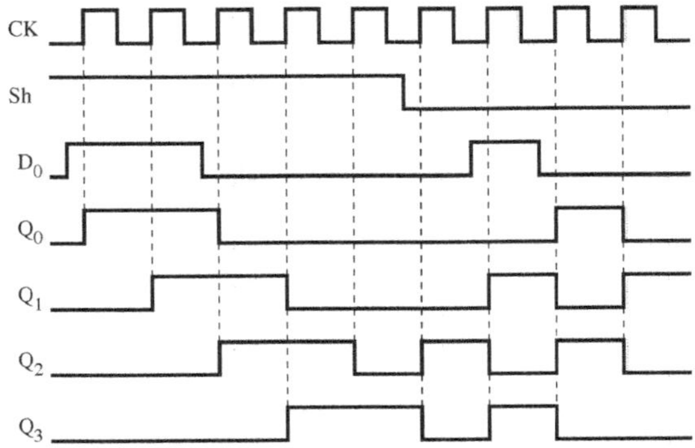

Figure 3.7. *Timing diagram*

3.5. Register file

A register file is composed of flip-flops organized into a two-dimensional array. It has the advantage of having a regular structure and can be used by the central unit of a processor to temporarily store data.

In general, a register file has 2^n lines of m flip-flops and each of these lines can be considered as a register that can contain a binary word.

The structure of a register file with a write input and two read outputs is shown in Figure 3.8. It uses flip-flops and 2 : 4 decoders with active-low enable signal (WE, REA and REB). The input data are words with four bits, D_3, D_2, D_1 and D_0, just like the output data, A_3, A_2, A_1 and A_0 for port A, or B_3, B_2, B_1 and B_0 for port B. Write address bits, WA_1 and WA_0, are decoded in order to identify which flip-flops to select for storage, while the decoding of the read address bits, RAA_1 and RAA_0 for port A, or RAB_1 and RAB_0 for port B, is useful for the selection of the flip-flops that supply the data to be read. The addressing table for data words can be represented as shown in Table 3.4.

	Write operation		Reading operation for port A		Read operation for port B	
	WA_1	WA_0	RAA_1	RAA_0	RAB_1	RAB_0
Word 0	0	0	0	0	0	0
Word 1	0	1	0	1	0	1
Word 2	1	0	1	0	1	0
Word 3	1	1	1	1	1	1

Table 3.4. *Data word addressing table*

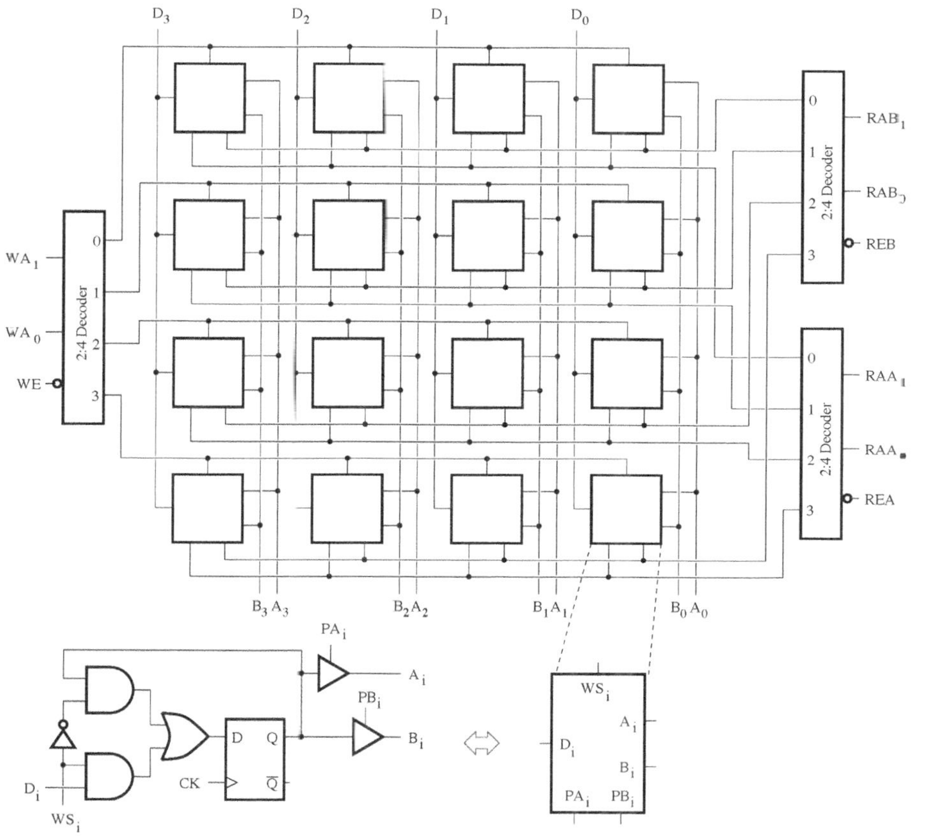

Figure 3.8. *Register file*

The input WE must be set to 0 to allow the write operation of a word whose address is specified. When this condition is not satisfied, the state of the flip-flops remains unchanged. At least one of the inputs, REA and REB, must be set to 0 in conjunction with a valid address to allow a word to be read. When one of the inputs, REA or REB, is set to 1, the corresponding output takes a high impedance state.

3.6. Shift register based counter

A counter can be implemented by connecting the input and output of a shift register to generate a given bit sequence.

3.6.1. *Ring counter*

A ring counter, which is depicted in Figure 3.9, uses a flip-flop for each state in its count sequence. The output of the right-most flip-flop is connected to the input of the left-most flip-flop, thus forming a ring or loop.

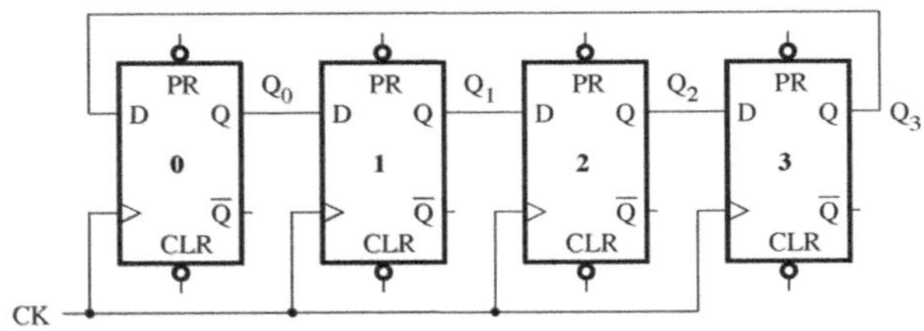

Figure 3.9. *Four-bit ring counter*

Assuming $Q_3Q_2Q_1Q_0 = 0001$ is the initial state of the counter, we can draw the timing diagram shown in Figure 3.10. The logic state 1 is circularly shifted from one flip-flop to another at the rising edge of each clock pulse. The four outputs of the flip-flops indicate the clock pulse number. Thus, $Q_0 = 1$ corresponds to 0, $Q_1 = 1$ to 1, $Q_2 = 1$ to 2 and $Q_3 = 1$ to 3.

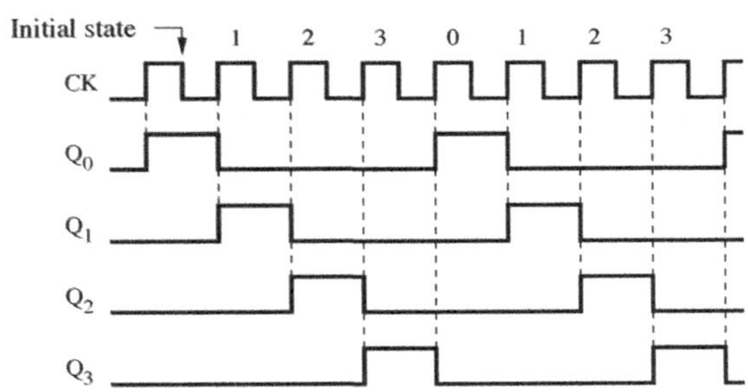

Figure 3.10. *Timing diagram illustrating two cycles of the counter*

Starting with the initial state, $Q_3Q_2Q_1Q_0 = 0011$ or $Q_3Q_2Q_1Q_0 = 0111$, we also obtain a count sequence with four states as shown in the timing diagram in Figure 3.11.

The maximum modulo is equal to the number of different states of the ring counter. For example, from the initial state $Q_3Q_2Q_1Q_0 = 0001$, four different states, 0001, 0010, 0100 and 1000, can be generated, while the initial state $Q_3Q_2Q_1Q_0 = 0101$ is associated with only two different states: 0101 and 1010.

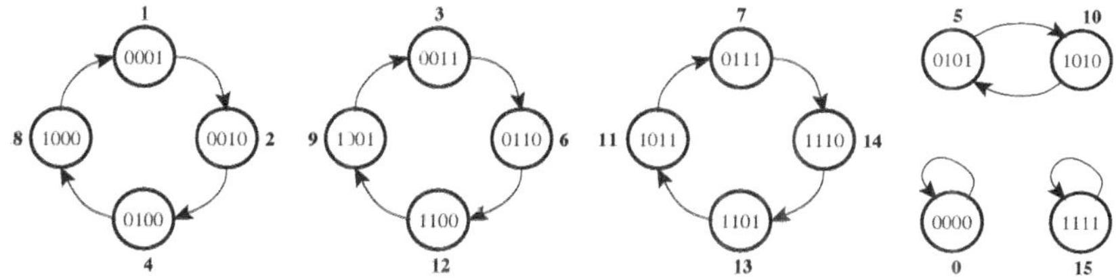

Figure 3.11. *State diagram of the four-bit ring counter*

However, the ring counter remains indefinitely in the state $Q_3Q_2Q_1Q_0 = 0000$ or $Q_3Q_2Q_1Q_0 = 1111$.

In general, a ring counter made up of n flip-flops has a maximum modulo equal to n.

3.6.2. *Johnson counter*

To implement a Johnson counter, the logical complement of the output of the last flip-flop is applied to the input of the first flip-flop. This feedback connection helps produce a characteristic count sequence.

A four-bit Johnson counter is represented in Figure 3.12. Assuming that all the flip-flops are initially reset, the count sequence is composed of eight distinct states, as shown in the timing diagram in Figure 3.13. Unlike the ring counter, a decoder must be connected to the Johnson counter in order to identify the numbers represented by the different logic states of the flip-flops. Table 3.5 represents the count sequence and the decoding equations for a four-bit Johnson counter (from the zero initial condition). For each number, there are always two flip-flops that form a unique combination in the sequence of states of the Johnson counter. For example, the combination $Q_3Q_0 = 00$ is obtained only when the count result is 0.

Based on the state diagram shown in Figure 3.14, the four-bit Johnson counter can operate following two possible count sequences depending on its initial state.

In general, a Johnson counter has a modulo of $2n$, where n is the number of flip-flops in the counter. The decoder has $2n$ two-input AND gates for a Johnson counter that uses n flip-flops. Additionally, the number of inputs for each gate of the decoder remains the same whatever the modulo of the counter.

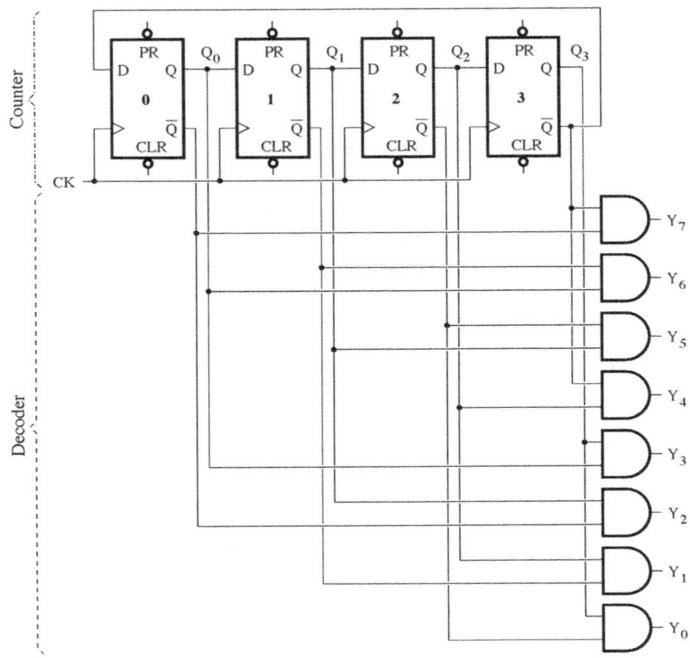

Figure 3.12. *Four-bit Johnson counter with a decoder*

Pulse	Q_3	Q_2	Q_1	Q_0	Output Y_i
	0	0	0	*0*	$Y_7 = \overline{Q_3} \cdot \overline{Q_0}$
1	0	0	*0*	*1*	$Y_6 = \overline{Q_1} \cdot Q_0$
2	0	*0*	*1*	1	$Y_5 = \overline{Q_2} \cdot Q_1$
3	*0*	*1*	1	1	$Y_4 = \overline{Q_3} \cdot Q_2$
4	*1*	1	1	*1*	$Y_3 = Q_3 \cdot Q_0$
5	1	1	*1*	*0*	$Y_2 = Q_1 \cdot \overline{Q_0}$
6	1	*1*	*0*	0	$Y_1 = Q_2 \cdot \overline{Q_1}$
7	*1*	*0*	0	0	$Y_0 = Q_3 \cdot \overline{Q_2}$

Table 3.5. *Count sequence and decoding equations for the four-bit Johnson counter*

3.6.3. *Linear feedback counter*

A linear feedback shift register (LFSR) counter, also known as a pseudo-random sequence generator, is generally used to generate signals for digital circuit testing and cryptography. The input for this counter is a linear function (for example the XOR function) of certain previous states.

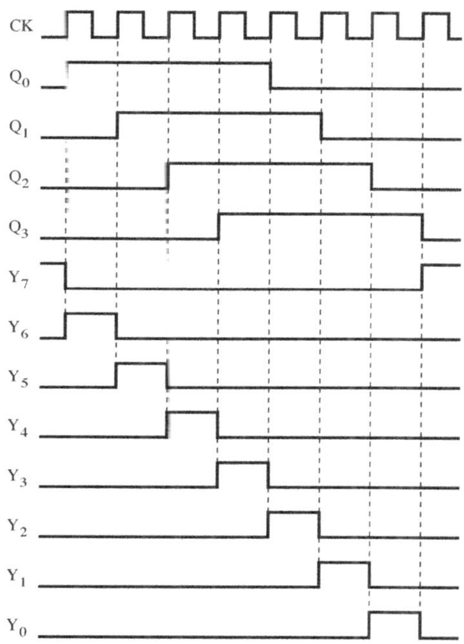

Figure 3.13. *Timing diagram of the four-bit Johnson counter with a decoder*

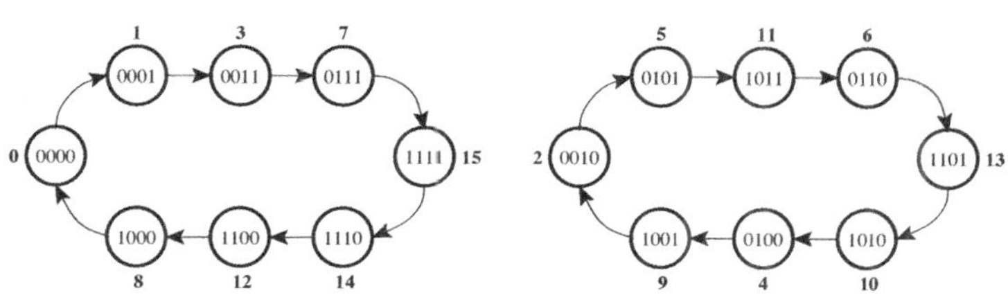

Figure 3.14. *State diagram for the four-bit Johnson counter*

3.6.3.1. *Four-bit counters*

Let us consider the four-bit LFSR counter shown in Figure 3.15. Beginning from the state 1000, the counter cycles through 14 states before returning to the initial state. Based on the state diagram shown in Figure 3.16, this counter has $2^4 - 1$ states, each state being specified by a combination $Q_3Q_2Q_1Q_0$.

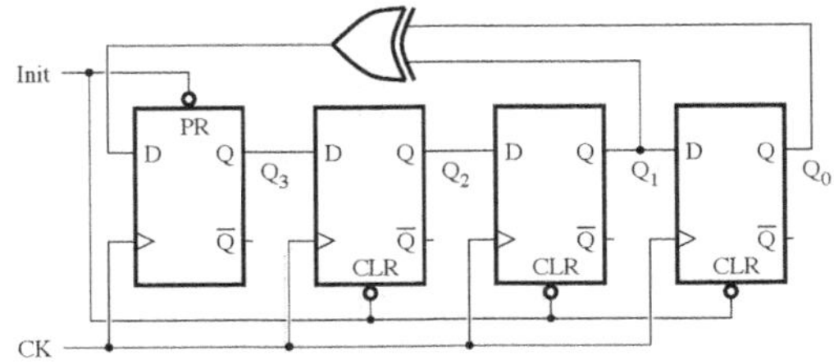

Figure 3.15. *Four-bit LFSR counter (external XOR) (initially, $\overline{PR} = 0$ for flip-flop 3 and $\overline{CLR} = 0$ for the other flip-flops)*

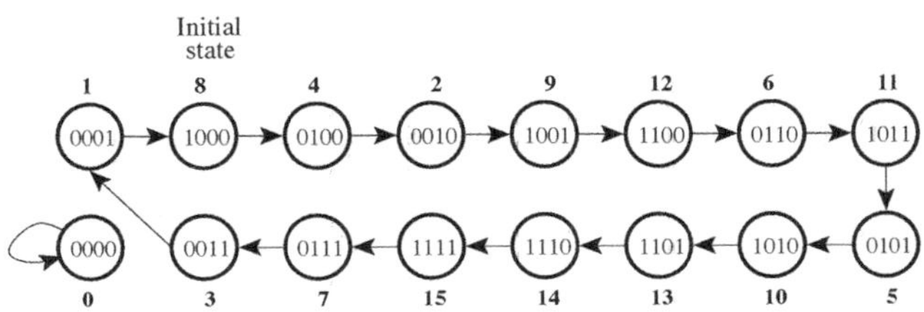

Figure 3.16. *State diagram of the four-bit LFSR counter (external XOR)*

Another version of the four-bit LFSR counter is represented in Figure 3.17. The operation of this counter is described by the state diagram shown in Figure 3.18, where each combination $Q_3Q_2Q_1Q_0$ designates a state.

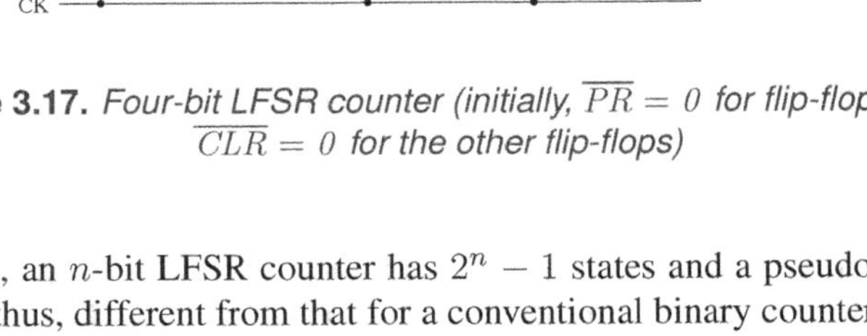

Figure 3.17. *Four-bit LFSR counter (initially, $\overline{PR} = 0$ for flip-flop 0, and $\overline{CLR} = 0$ for the other flip-flops)*

In general, an n-bit LFSR counter has $2^n - 1$ states and a pseudo-random count order that is, thus, different from that for a conventional binary counter.

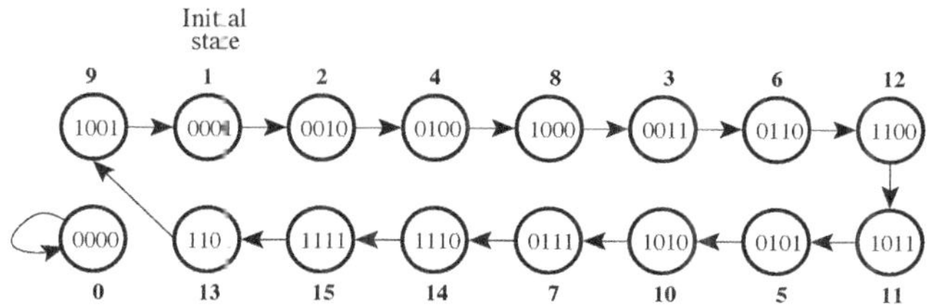

Figure 3.18. *State diagram of the four-bit LFSR counter (internal XOR)*

3.6.3.2. *Application: built-in self-test circuit*

With the increase in the number of transistors on a chip, using built-in self-test (BIST) circuits has become necessary in order to ensure quality and reliability and to reduce production costs.

In general, the signal delivered by the circuit being tested is assumed to carry some information corresponding to an n-bit sequence that can be represented as a polynomial: $I(x)$. It is applied at the input of a signature register, which is implemented using k flip-flops and which can be represented by the characteristic polynomial $G(x)$, with x^k as the monomial of the highest degree. After n shifts, the output bit sequence forms the quotient $Q(x)$, with a length of $n - k$, and the output bits from the flip-flops constitute the remainder, $R(x)$, with a length of k, which is then called the signature. Hence:

$$I(x) = Q(x)G(x) + R(x) \qquad [3.1]$$

where the degree of $R(x)$ is inferior to that of $G(x)$. When the circuit under test exhibits an error, a corresponding polynomial error, $E(x)$, will be added to $I(x)$. In this case, we have:

$$I(x) + E(x) = Q^*(x)G(x) + R^*(x) \qquad [3.2]$$

This error can only be detected if the remainder $R^*(x)$ differs from $R(x)$. It must be noted that the polynomials $I(x)$ and $I(x) + E(x)$ have the same remainder if the error $E(x)$ is a multiple of $G(x)$. The higher the degree of the polynomial $G(x)$, the less probable it is that the error $E(x)$ is a multiple of $G(x)$.

The characteristic polynomial, $G(x)$, associated with an LFSR counter or register, can be used to describe the behavior of the output signal from the last flip-flop. The coefficients of the monomials x^k and x^0 always take the value 1, while each of the other non-zero coefficients is linked to the presence of an XOR gate on the corresponding feedback path.

NOTE.– There are four different ways of representing an LFSR counter or register with a given characteristic polynomial because the variables and coefficients can be assigned from left to right or from right to left.

In principle, the BIST consists of applying a sequence of bits to the input of the *circuit under test* in order to analyze the output response. The sequence of input bits is generated by an *LFSR counter*. A *signature register* followed by a *signature analyzer* can be used to detect the presence of certain anomalies in the sequence of output bits.

Figure 3.19 depicts the implementation of a BIST for a combinational circuit consisting of four logic gates. The initialization signal is represented by *Init*, the $T/\overline{T}$ signal is used to select the circuit's operating mode (BIST or normal), and the *OK* signal is set to 1 when no error is detected. The state diagram of the signal generator (three-bit LFSR counter) is illustrated in Figure 3.20.

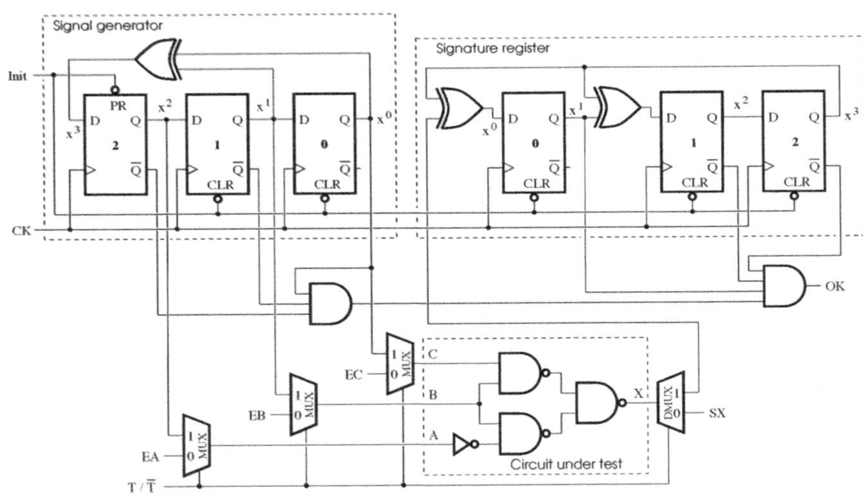

Figure 3.19. *Example of a BIST for a combinational circuit*

APPLICATION.– The characteristic polynomial for the signature register is written as:

$$G(x) = x^3 + x + 1$$

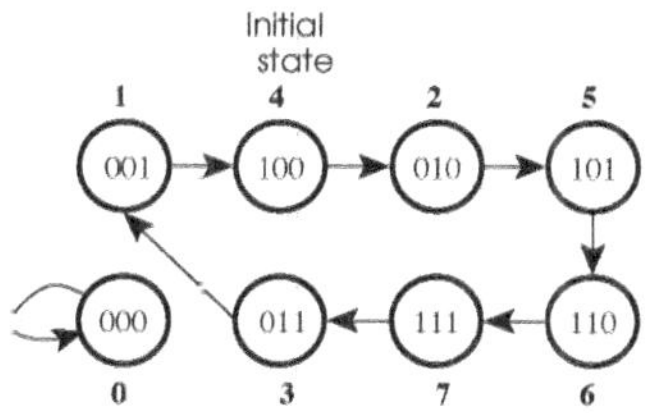

Figure 3.20. *State diagram for the three-bit LFSR counter*

The bit sequence successively applied to the input X is 0110010 and corresponds to the polynomial:

$$\begin{aligned} I(x) &= 0 \cdot x^0 + x + x^2 + 0 \cdot x^3 + 0 \cdot x^4 + x^5 + 0 \cdot x^6 \\ &= x^5 + x^2 + x \end{aligned}$$

Dividing $I(x)$ by $G(x)$, we can obtain the quotient, $Q(x)$, and the remainder $R(x)$.

$$\begin{array}{lll} x^5 + \quad x^2 + x & \big| \; x^3 + x + 1 & \\ \underline{x^5 + x^3 + x^2} & x^2 + 1 & \leftarrow \text{Quotient} \\ \quad x^3 + \quad x & & \\ \quad \underline{x^3 + \quad x + 1} & & \\ \qquad\qquad 1 & \leftarrow \text{Remainder} & \end{array}$$

Thus, $Q(x) = x^2 + 1$ and $R(x) = 1$. The remainder corresponds to the signature in polynomial form.

If, on the other hand, a short-circuit always sets the output of the inverter to 1 regardless of the input bit, A, the bit sequence at the input X becomes 0111010. The corresponding polynomial then takes the following form:

$$\begin{aligned} I(x) + E(x) &= 0 \cdot x^0 - x + x^2 + x^3 + 0 \cdot x^4 + x^5 + 0 \cdot x^6 \\ &= x^5 + x^3 + x^2 + x \end{aligned}$$

Dividing $I(x) + E(x)$ by $G(x)$ is carried out in the following manner:

$$\begin{array}{lll} x^5 + x^3 + x^2 + x & \big| \; x^3 + x + 1 & \\ \underline{x^5 + x^3 + x^2} & x^2 & \leftarrow \text{Quotient} \\ \qquad\qquad x & \leftarrow \text{Remainder} & \end{array}$$

The quotient can be put into the form, $Q^*(x) = x^2$, and the resulting remainder, $R^*(x) = x$, is different from $R(x)$.

It can be observed that the division is carried out by using the addition modulo 2 for the coefficients of the polynomials $(0 + 0 = 0,\ 1 + 0 = 1,\ 1 + 1 = 0)$.

Starting from the initial state $Q_2Q_1Q_0 = 100$ for the generator and $Q_2Q_1Q_0 = 000$ for the signature register, the bit sequence after each clock signal pulse is entered in Table 3.6. After six pulses, the signature register has the sequence 001, which is the signature for a circuit without error. And when the circuit under test is affected by a short circuit that always sets the output of the inverter to 1, whatever the state of the input A, another sequence is obtained, that is: 010.

				Without error				With error			
	Generator			Register				Register			
Pulse	Q_2	Q_1	Q_0	X	Q_2	Q_1	Q_0	X	Q_2	Q_1	Q_0
	1	0	0	0	0	0	0	0	0	0	0
1	0	1	0	1	0	0	1	1	0	0	1
2	1	0	1	0	0	1	0	0	0	1	0
3	1	1	0	0	1	0	0	1	1	0	1
4	1	1	1	1	0	1	0	1	0	0	0
5	0	1	1	1	1	0	1	1	0	0	1
6	0	0	1	0	*0*	*0*	*1*	0	0	1	0

Table 3.6. *Bit sequence for the generator and the signature register*

Thus, a fault that causes a modification in the input bit sequence of the signature register can, at the last clock signal pulse, result in a response with the flip-flop output bits differing from those of the signature.

3.6.3.3. *Built-in logic block observer register*

The built-in logic block observer (BILBO) register is often used to reduce the hardware cost of BIST circuits. It can be configured to operate in test mode or normal mode.

The logic circuit for a three-bit BILBO register is represented in Figure 3.21. It has an SI, an SO, parallel inputs, D_i $(i = 0, 1, 2)$, parallel outputs, Q_i $(i = 0, 1, 2)$, and control inputs, C_1 and C_0. The function table for the BILBO register is given in Table 3.7.

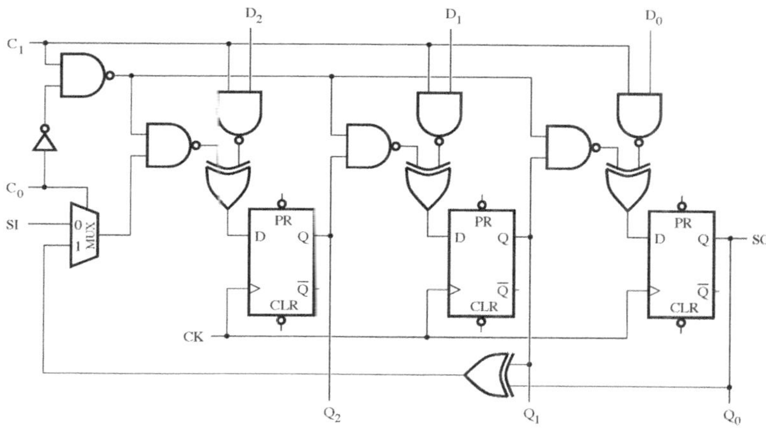

Figure 3.21. *Logic circuit of a BILBO register*

C_1	C_C	Operating mode
0	0	Shift register, serial input, serial output
0	1	LFSR generator
1	0	D flip-flops
1	1	Signature register

Table 3.7. *Function table of the BILBO register*

In a circuit, the same BILBO register can generate signals that are stored in the memory and can then be configured as a signature register. And if no error is detected during the BIST phase, the control circuit will configure the BILBO register to operate in normal mode.

3.7. Exercises

EXERCISE 3.1.– Implement a five-bit serial-in serial out shift register.

Represent the timing diagram illustrating the operation of this register, assuming that the flip-flops are initially set to 0.

EXERCISE 3.2.– Complete the timing diagram for the modulo 6 Johnson counter shown in Figure 3.22(a).

Construct and compare the state diagrams for the counters shown in Figure 3.22(a) and 3.23.

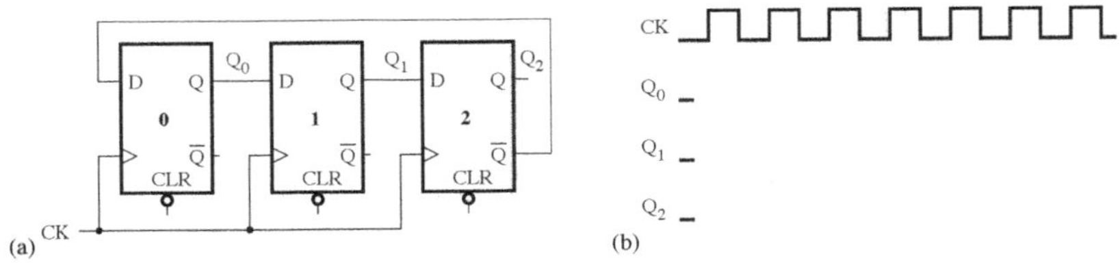

Figure 3.22. *a) Modulo 6 Johnson counter (circuit 1); b) timing diagram*

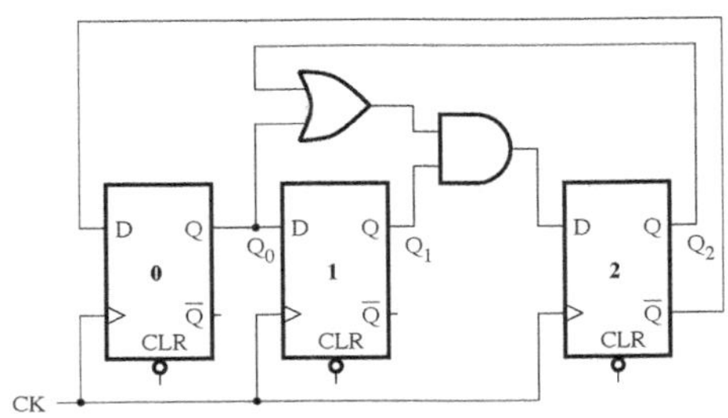

Figure 3.23. *Modulo 6 Johnson counter (circuit 2)*

EXERCISE 3.3.– Using a shift register, a counter and logic gates, implement a logic circuit of a detector for the sequence 1101, assuming that the bits can or cannot overlap between consecutive four-bit words.

EXERCISE 3.4.– Consider the shift register shown in Figure 3.24, which is implemented using D flip-flops and 2 : 1 multiplexers.

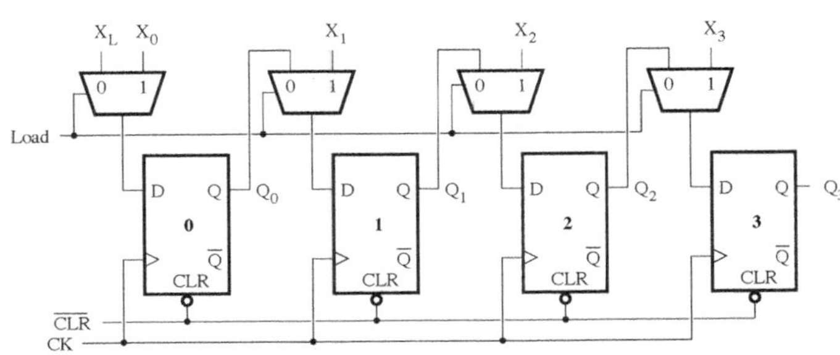

Figure 3.24. *Four-bit shift register*

Complete the functional truth table shown in Table 3.8.

Inputs			NS			
CK	$\overline{CLR}$	Load	Q_3^+	Q_2^+	Q_1^+	Q_0^+
x	0	x				
↑	1	0				
↑	1	1				

Table 3.8. *Functional truth table (NS, next state)*

Complete (Q_0, Q_1, Q_2 and Q_3) the timing diagram in Figure 3.25 assuming that $X_3X_2X_1X_0 = 0101$.

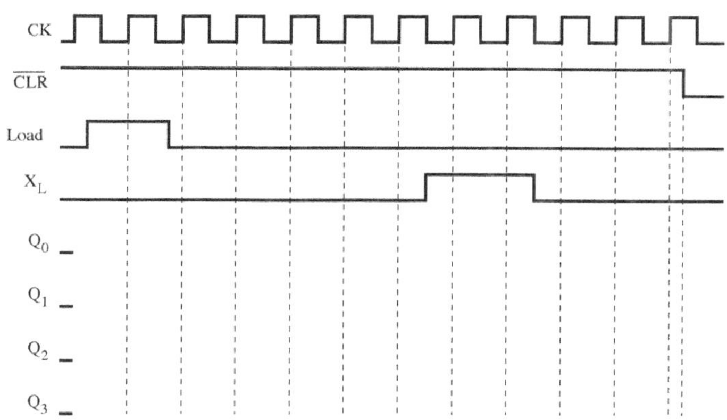

Figure 3.25. *Timing diagram ($X_3X_2X_1X_0 = 0101$)*

Figure 3.26 depicts the logic circuit for another type of shift register that is also based on D flip-flops and 2 : 1 multiplexers.

Complete the functional truth table represented in Table 3.9.

EXERCISE 3.5.– The universal shift register shown in Figure 3.27 is composed of D flip-flops and 4 : 1 multiplexers. During a normal operation, the synchronous *reset* input is set to 1 and does not affect the outputs.

Complete the functional truth table given in Table 3.10.

Complete the timing diagram shown in Figure 3.28.

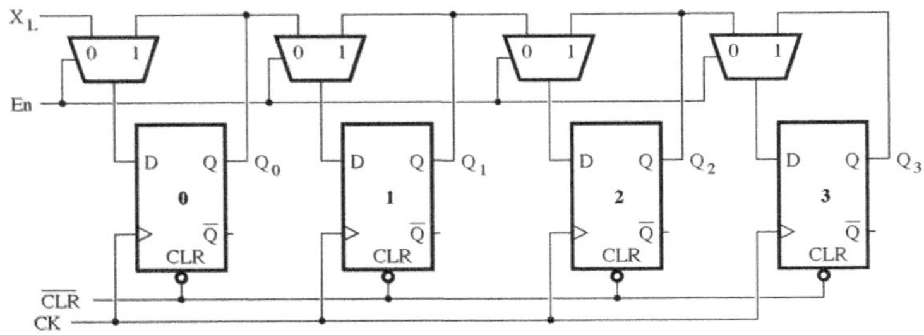

Figure 3.26. *Shift register with enable signal*

Inputs			NS			
CK	$\overline{CLR}$	En	Q_3^+	Q_2^+	Q_1^+	Q_0^+
x	0	x				
↑	1	0				
↑	1	1				

Table 3.9. *Functional truth table (NS, next state)*

Inputs				NS			
CK	$\overline{Reset}$	C_1	C_0	Q_3^+	Q_2^+	Q_1^+	Q_0^+
x	0	x	x				
x	1	0	0				
↑	1	0	1				
↑	1	1	0				
↑	1	1	1				

Table 3.10. *Functional truth table*

EXERCISE 3.6.– Determine the count sequence for the logic circuit shown in Figure 3.29 assuming that the initial state is $Q_2Q_1Q_0 = 111$, and the count sequence of the circuit shown in Figure 3.30 for a case where the initial state is $Q_2Q_1Q_0 = 000$.

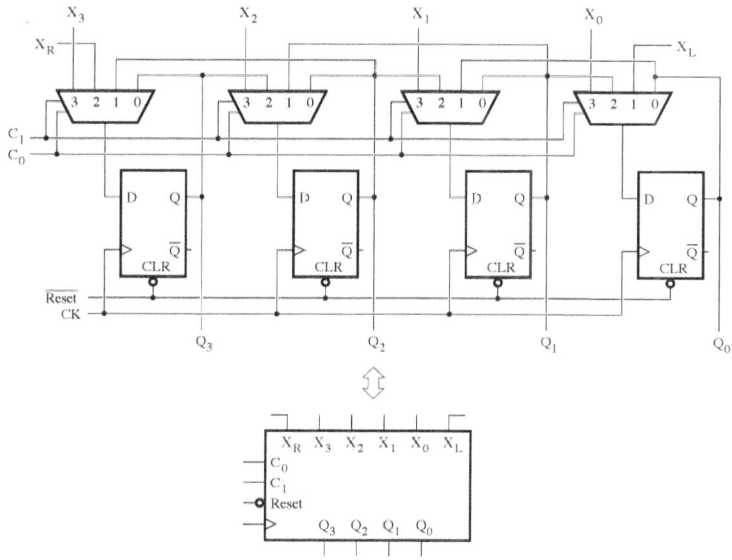

Figure 3.27. *Logic circuit and symbol for the four-bit universal shift register*

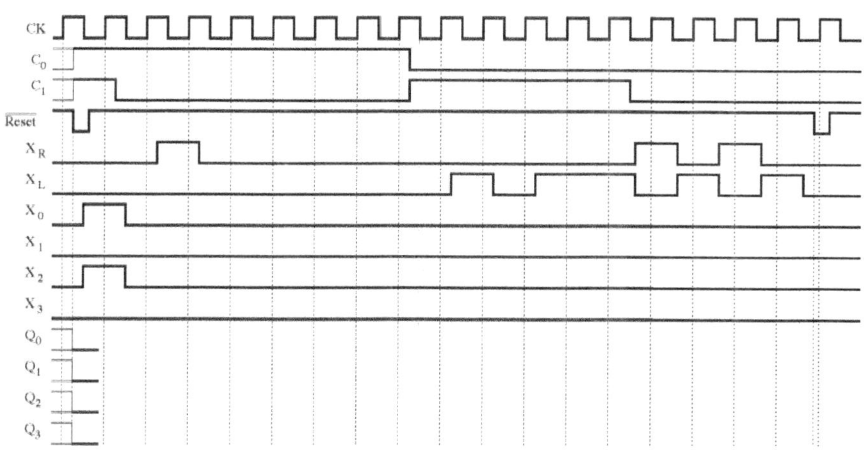

Figure 3.28. *Timing diagram for a universal register*

EXERCISE 3.7.– Modulo m or $m - 1$ counter.

Consider the counter using two flip-flops shown in Figure 3.31(a).

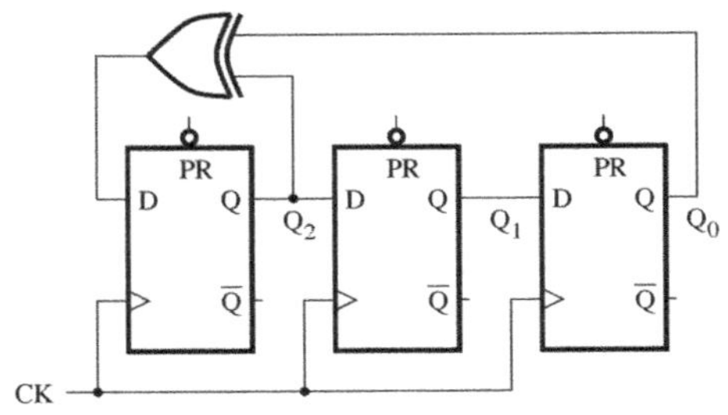

Figure 3.29. *Three-bit LFSR counter with flip-flops initially set to 1*

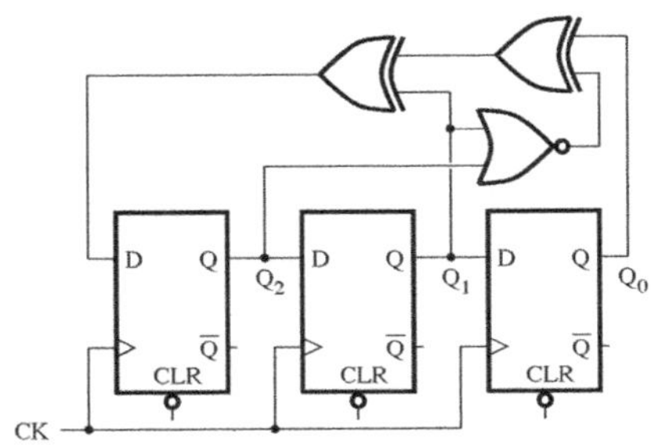

Figure 3.30. *Three-bit LFSR counter with flip-flops initially set to 0*

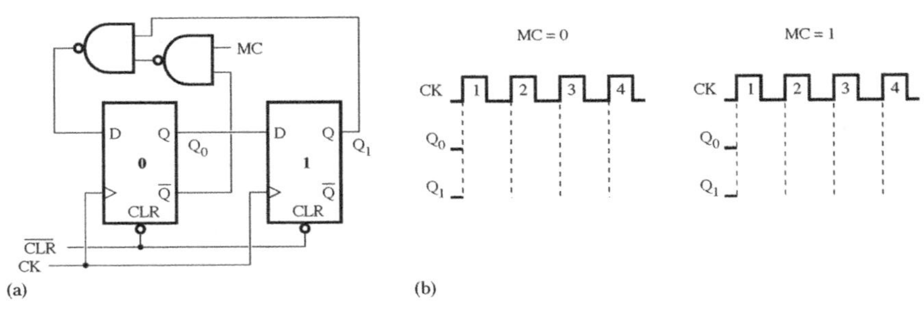

Figure 3.31. *Modulo m or $m - 1$ counter with two flip-flops: a) logic circuit; b) timing diagram*

Complete the timing diagrams of Figure 3.31(b) when MC $= 0$ and MC $= 1$. We will assume that the initial state is 0.

Determine the value of the modulo when MC $= 0$ and MC $= 1$.

Answer the same questions for the counter using three flip-flops and the timing diagram shown in Figure 3.32.

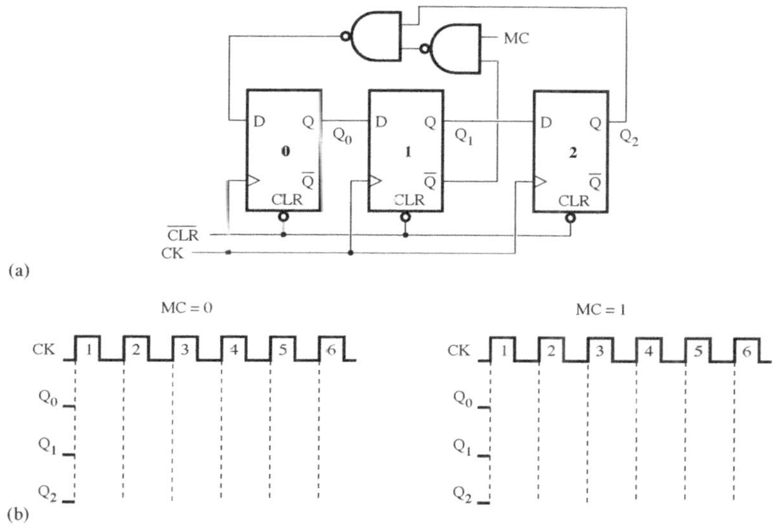

Figure 3.32. *Modulo m or m − 1 counter with three flip-flops: a) logic circuit; b) timing diagram*

3.8. Solutions

SOLUTION 3.1.– Five-bit shift register.

Figure 3.33 depicts the logic circuit of a five-bit shift register. The timing diagram shown in Figure 3.34 is constructed assuming that the initial state is 0.

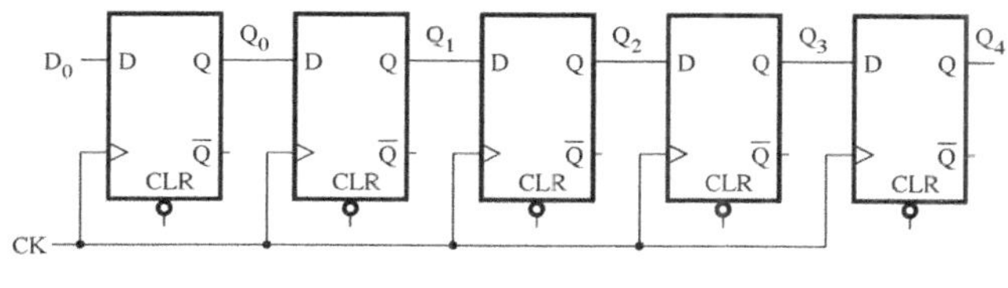

Figure 3.33. *Five-bit shift register*

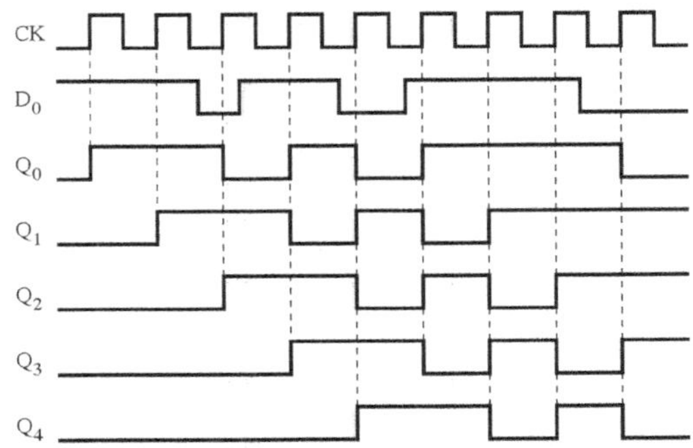

Figure 3.34. *Timing diagram*

SOLUTION 3.2.– Modulo 6 Johnson counter.

The analysis of the counter shown in Figure 3.35(a) yields the following equations:

$$D_0 = \overline{Q_2}, \quad D_1 = Q_0 \quad \text{and} \quad D_2 = Q_1$$

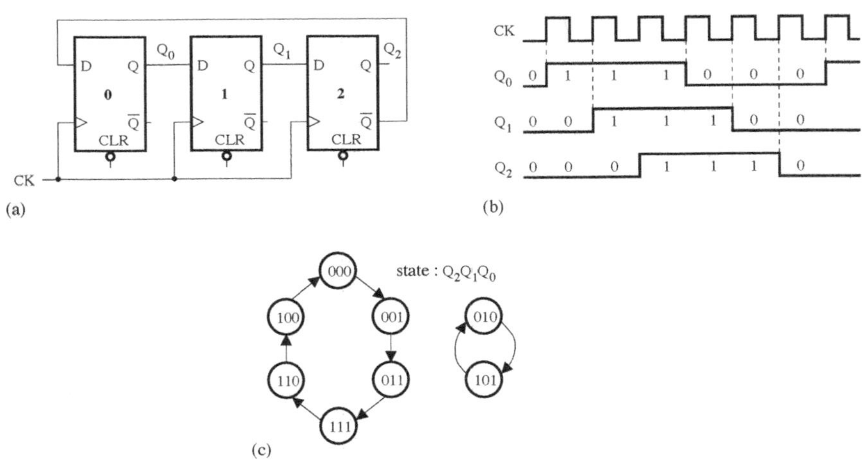

Figure 3.35. *Modulo 6 Johnson counter: a) circuit 1; b) timing diagram; c) state diagram*

Assuming that the counter is initially set to 0, the timing diagram can be represented as shown in Figure 3.35(b). The count cycle is made up of six different states. However, starting from the state 010, the counter goes to the state 101, and vice versa. The state diagram of the counter is given in Figure 3.35(c).

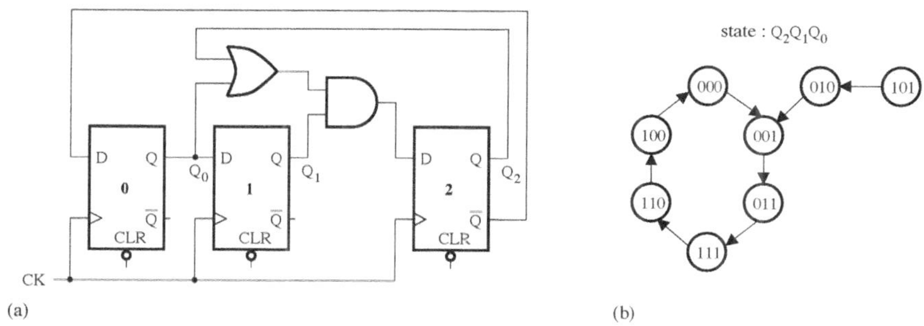

Figure 3.36. *Modulo 6 Johnson counter: a) circuit 2; b) state diagram*

For the counter shown in Figure 3.36(a), we have:

$$D_0 = \overline{Q_2}, \quad D_1 = Q_0 \quad \text{and} \quad D_2 = (Q_2 + Q_0) \cdot Q_1$$

Starting from the initial state $Q_2Q_1Q_0 = 000$, the counter successively takes the states 001, 011, 111, 110, 100, and returns to 000. When the counter is set to the state 101, it successively goes to 010 and 001. Figure 3.36(b) depicts the state diagram of the counter. In this case, the counter returns to the same count cycle regardless of the initial state.

SOLUTION 3.3.– Detector of the sequence 1101.

The logic circuit for the 1101 sequence detector with overlap is represented in Figure 3.37. The input data bits, X, being applied to the register with X_3 first, the sequence 1101 is detected after three clock pulses. Assuming $X = X_3X_2X_1X_0$, the output logic equation can take the form:

$$Y = Q_2 \cdot Q_1 \cdot \overline{Q_0} \cdot X_0$$

where $Q_2 = X_3$, $Q_1 = X_2$ and $Q_0 = X_1$.

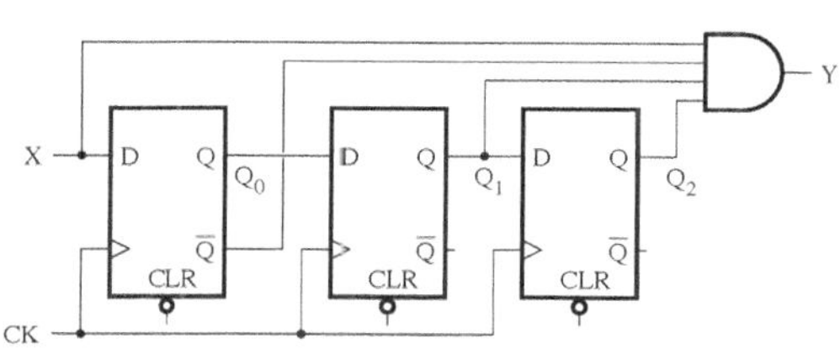

Figure 3.37. *Logic circuit for the 1101 sequence detector with overlap*

The logic circuit for the 1101 sequence detector without overlap is illustrated in Figure 3.38. The counter is initially reset to 0 and the output Z can take the logic state 1 only if $Q_B = Q_A = 1$. Each four-bit word is processed independently.

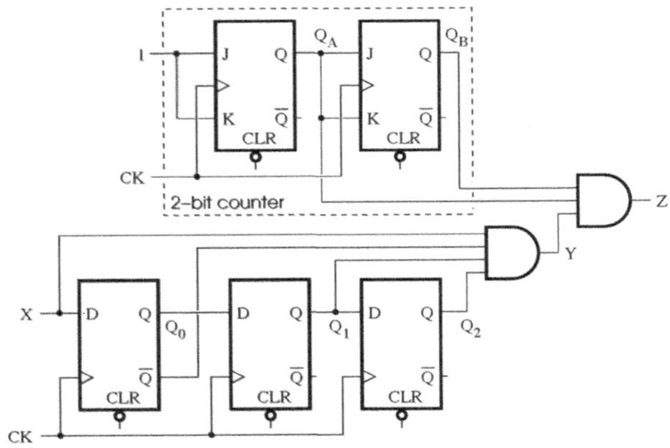

Figure 3.38. *Logic circuit of the 1101 sequence detector without overlap*

SOLUTION 3.4.– Shift register.

The analysis of the shift register logic circuit provides the following logic equations:

$$Q_0^+ = D_0 = X_L \cdot \overline{\text{Load}} + X_0 \cdot \text{Load} \tag{3.3}$$

and:

$$Q_i^+ = D_i = Q_{i-1} \cdot \overline{\text{Load}} + X_i \cdot \text{Load}, \qquad i = 1, 2, 3 \tag{3.4}$$

Table 3.11 gives the functional truth table for the shift register.

The timing diagram of the shift register is represented in Figure 3.39, for a case where $X_3X_2X_1X_0 = 0101$.

In this case, the logic equations for the shift register are written as follows:

$$Q_0^+ = D_0 = X_L \cdot \overline{\text{En}} + Q_0 \cdot \text{En} \tag{3.5}$$

and:

$$Q_i^+ = D_i = Q_{i-1} \cdot \overline{\text{En}} + Q_i \cdot \text{En}, \qquad i = 1, 2, 3 \tag{3.6}$$

Inputs			NS				
CK	$\overline{CLR}$	Load	Q_3^+	Q_2^+	Q_1^+	Q_0^+	
x	0	x	0	0	0	0	Reset
↑	1	0	Q_2	Q_1	Q_0	X_L	Left shift
↑	1	1	X_3	X_2	X_1	X_0	Load

Table 3.11. *Functional truth table (NS: next state)*

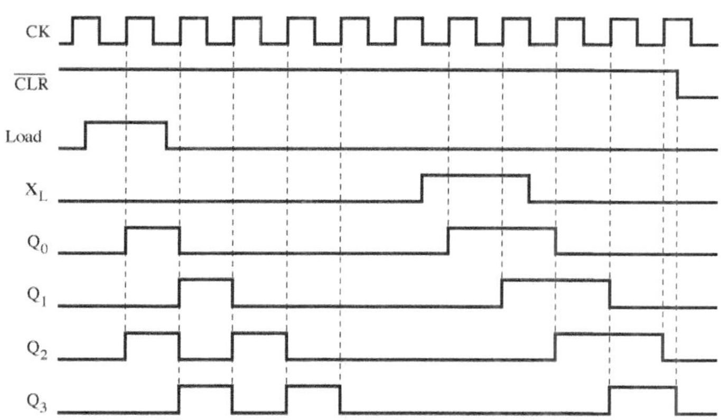

Figure 3.39. *Timing diagram of the shift register*

The functional truth table can, thus, be represented as shown in Table 3.12.

SOLUTION 3.5.– Four-bit universal shift register.

By analyzing the logic circuit of the shift register, we can obtain the following logic equations:

$$Q_0^+ = D_0 = C_1 \cdot C_0 \cdot X_0 + C_1 \cdot \overline{C_0} \cdot Q_1 + \overline{C_1} \cdot C_0 \cdot X_L + \overline{C_1} \cdot \overline{C_0} \cdot Q_0 \qquad [3.7]$$

$$Q_i^+ = D_i = C_1 \cdot C_0 \cdot X_i + C_1 \cdot \overline{C_0} \cdot Q_{i+1} + \overline{C_1} \cdot C_0 \cdot Q_{i-1} + \overline{C_1} \cdot \overline{C_0} \cdot Q_i \quad [3.8]$$

$$Q_3^+ = D_3 = C_1 \cdot C_0 \cdot X_3 + C_1 \cdot \overline{C_0} \cdot X_R + \overline{C_1} \cdot C_0 \cdot Q_2 + \overline{C_1} \cdot \overline{C_0} \cdot Q_3 \quad [3.9]$$

where $i = 1, 2$.

Inputs			NS				
CK	$\overline{CLR}$	En	Q_3^+	Q_2^+	Q_1^+	Q_0^+	
x	0	x	0	0	0	0	Reset
↑	1	0	Q_2	Q_1	Q_0	X_L	Left shift
↑	1	1	Q_3	Q_2	Q_1	Q_0	Hold

Table 3.12. *Functional truth table (NS, next state)*

Table 3.13 gives the functional truth table of the universal register.

Inputs				NS				
CK	$\overline{Reset}$	C_1	C_0	Q_3^+	Q_2^+	Q_1^+	Q_0^+	
x	0	x	x	0	0	0	0	Reset
x	1	0	0	Q_3	Q_2	Q_1	Q_0	Hold
↑	1	0	1	Q_2	Q_1	Q_0	X_L	Left shift
↑	1	1	0	X_R	Q_3	Q_2	Q_1	Right shift
↑	1	1	1	X_3	X_2	X_1	X_0	Load

Table 3.13. *Functional truth table (NS, next state)*

The functional truth table helps complete the timing diagram of the universal register, as shown in Figure 3.40.

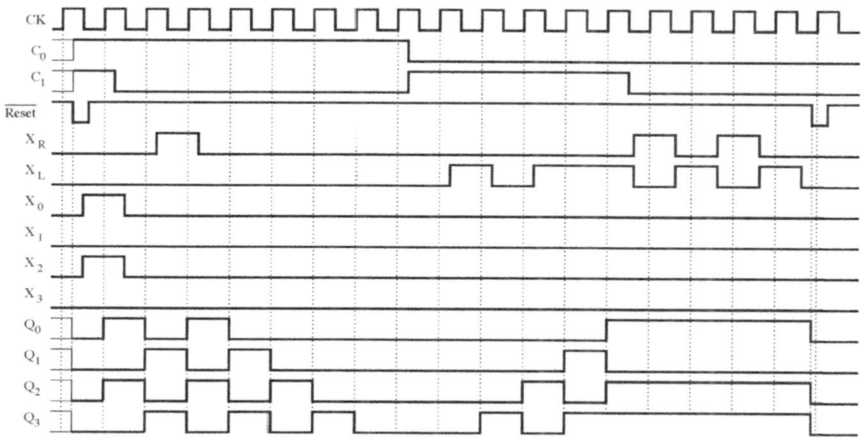

Figure 3.40. *Timing diagram of the universal register*

SOLUTION 3.6.– Three-bit linear-feedback shift register.

The flip-flops of the first counter operate according to the following characteristic equations:

$$Q_0^+ = D_0 = Q_1 \quad [3.10]$$

$$Q_1^+ = D_1 = Q_2 \quad [3.11]$$

$$Q_2^+ = D_2 = Q_2 \oplus Q_0 \quad [3.12]$$

Table 3.14 gives the transition table of the counter. When the counter is reset (or $Q_2Q_1Q_0 = 000$), it remains in the same state. Hence, the state $Q_2Q_1Q_0 = 000$ is not part of the following count cycle: *111* $\rightarrow$ 011 $\rightarrow$ 101 $\rightarrow$ 010 $\rightarrow$ 001 $\rightarrow$ 100 $\rightarrow$ 110 $\rightarrow$ *111*.

The count cycle is composed of seven different states and the counter modulo is, thus, equal to 7.

The characteristic equations for the flip-flops of the second counter are given as follows:

$$Q_0^+ = D_0 = Q_1 \quad [3.13]$$

$$Q_1^+ = D_1 = Q_2 \quad [3.14]$$

$$Q_2^+ = D_2 = (\overline{Q_2 + Q_1}) \oplus Q_1 \oplus Q_0 = (\overline{Q_2} + Q_1) \oplus Q_0 \quad [3.15]$$

EA			Inputs			ES		
Q_2	Q_1	Q_0	D_2	D_1	D_0	Q_2^+	Q_1^+	Q_0^+
1	1	1	0	1	1	0	1	1
0	1	1	1	0	1	1	0	1
1	0	1	0	1	0	0	1	0
0	1	0	0	0	1	0	0	1
0	0	1	1	0	0	1	0	0
1	0	0	1	1	0	1	1	0
1	1	0	1	1	1	1	1	1
0	0	0	0	0	0	0	0	0

Table 3.14. *Transition table (PS, present state, NS, next state)*

PS			Inputs			NS		
Q_2	Q_1	Q_0	D_2	D_1	D_0	Q_2^+	Q_1^+	Q_0^+
0	0	0	1	0	0	1	0	0
1	0	0	0	1	0	0	1	0
0	1	0	1	0	1	1	0	1
1	0	1	1	1	0	1	1	0
1	1	0	1	1	1	1	1	1
1	1	1	0	1	1	0	1	1
0	1	1	0	0	1	0	0	1
0	0	1	0	0	0	0	0	0

Table 3.15. *Transition table (PS, present state; NS, next state)*

The transition table of the counter is given in Table 3.15.

Beginning with the initial state $Q_2Q_1Q_0 = 000$, the count sequence is as follows: *000* $\rightarrow$ 100 $\rightarrow$ 010 $\rightarrow$ 101 $\rightarrow$ 110 $\rightarrow$ 111 $\rightarrow$ 011 $\rightarrow$ 001 $\rightarrow$ *000*.

The count cycle is made up of eight different states and the counter modulo is, thus, equal to 8.

SOLUTION 3.7.– Modulo m or $m - 1$ counter.

Based on the logic circuit shown in Figure 3.41(a), we can obtain the following logic equations:

$$D_0 = \overline{Q_1 \cdot \overline{\overline{Q_0}} \cdot \mathrm{MC}} = \overline{Q_1} + \overline{Q_0} \cdot \mathrm{MC} \quad [3.16]$$

$$D_1 = Q_0 \quad [3.17]$$

Figure 3.41(b) shows the timing diagrams of the counter. The counter modulo is equal to either 4, when MC = 0, or 3, when MC = 1.

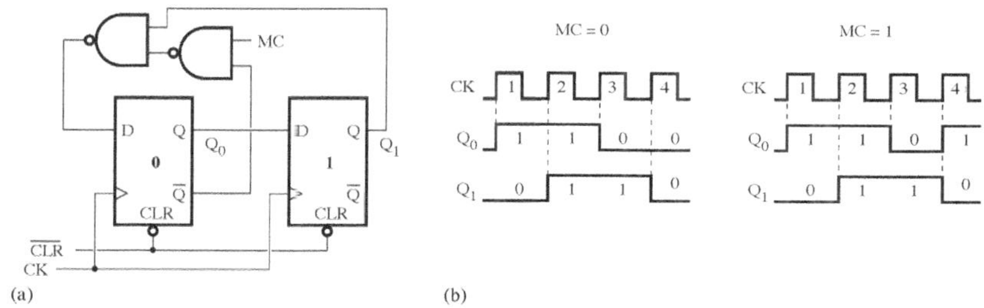

Figure 3.41. *Modulo m or $m-1$ counter using two flip-flops: a) logic circuit; b) timing diagram*

For the logic circuit shown in Figure 3.42(a), the inputs of the flip-flops are characterized by the following logic equations:

$$D_0 = \overline{Q_2 \cdot \overline{\overline{Q_1} \cdot \text{MC}}} = \overline{Q_2} + \overline{Q_1} \cdot \text{MC} \qquad [3.18]$$

$$D_1 = Q_0 \qquad [3.19]$$

$$D_2 = Q_1 \qquad [3.20]$$

Figure 3.42(b) presents the timing diagram for the counter. The counter modulo is equal to either 6, when MC = 0, or 5, when MC = 1.

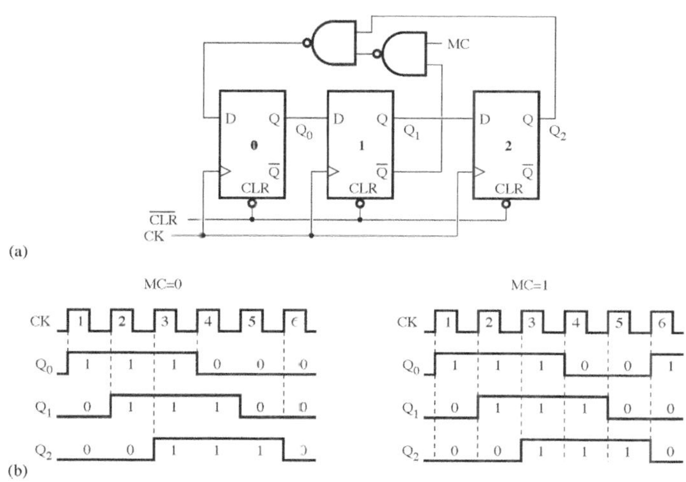

Figure 3.42. *Modulo m or $m-1$ counter using three flip-flops: a) logic circuit; b) timing diagram*

4

Arithmetic and Logic Circuits

4.1. Introduction

Arithmetic circuits are essential in the implementation of microprocessors and circuits for digital signal processing.

As the complexity of the direct approach (construct a truth table and then derive and simplify the output logic equations) increases with an increase in the size of data, the modular approach is often chosen to implement arithmetic circuits. A given digital circuit is thus implemented by assembling modules that are designed for numbers with small word lengths.

Despite the increasing number of arithmetic operations that are becoming routine, most microprocessors only contain circuits for basic operations, such as an adder (or summer), comparator, multiplier and divider. In addition to arithmetic circuits, there is also a need for circuits that can perform bitwise logic and shift operations. One of the main components of a microprocessor, therefore, is the arithmetic and logic unit (ALU), which encases all the circuits required for carrying out operations on digital data.

4.2. Adder

Adders are used to perform a large number of digital operations. In spite of the apparent simplicity of an addition operation, there are several approaches to designing adders.

4.2.1. *Half adder*

A half adder (HA) is a circuit that generates the sum, S, and the carry, C, resulting from the addition of two 1-bit numbers, A and B.

An example of a simple addition, $A + B$, is given in Figure 4.1. The truth table shown in Table 4.1 is constructed by considering bits C and S to be the MSB and LSB in the 2-bit representation of the addition result. The logic equations for the two outputs are given by:

$$S = \overline{A} \cdot B + A \cdot \overline{B} = A \oplus B \quad [4.1]$$

$$C = A \cdot B \quad [4.2]$$

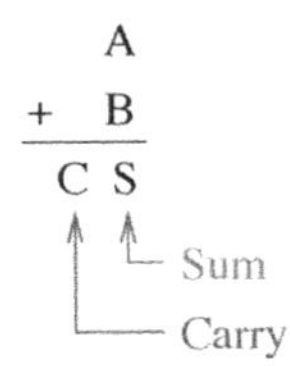

Figure 4.1. *Example of a simple addition. For a color version of the figure, see www.iste.co.uk/ndjountche/electronics2.zip*

A	B	S	C
0	0	0	0
0	1	1	0
1	0	1	0
1	1	0	1

Table 4.1. *Truth table for a half adder*

The circuit and the symbol of an HA are given in Figures 4.2(a) and 4.2(b), respectively.

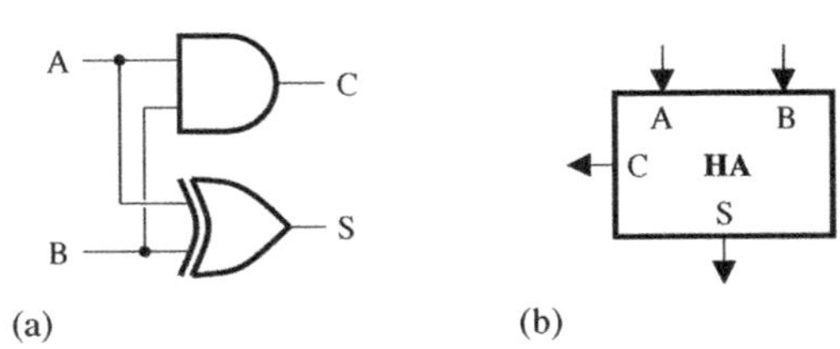

Figure 4.2. *Half adder: a) logic circuit; b) symbol*

4.2.2. *Full adder*

A full adder (FA) is a circuit that generates the sum, S, and the carry-out C_C resulting from the addition of two 1-bit numbers, A and B, and a 1-bit carry-in C_i.

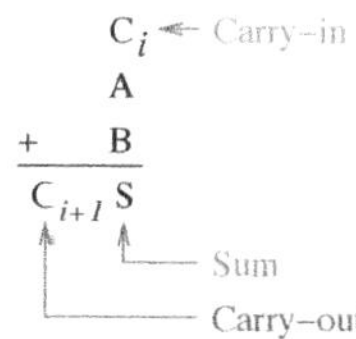

Figure 4.3. *Example of an addition. For a color version of the figure, see www.iste.co.uk/ndjountche/electronics2.zip*

The working of an FA is based on the arithmetic operation given in Figure 4.3. The truth table is constructed as shown in Table 4.2. The output logic equations cannot be simplified and are therefore written as follows:

$$S = \overline{A} \cdot \overline{B} \cdot C_i + \overline{A} \cdot B \cdot \overline{C_i} + A \cdot \overline{B} \cdot \overline{C_i} + A \cdot B \cdot C_i \quad [4.3]$$

$$= (\overline{A} \cdot \overline{B} + A \cdot B) \cdot C_i + (A \cdot \overline{B} + \overline{A} \cdot B) \cdot \overline{C_i} \quad [4.4]$$

$$= (A \oplus B) \oplus C_i$$

$$C_{i+1} = A \cdot B \cdot \overline{C_i} + A \cdot \overline{B} \cdot C_i + \overline{A} \cdot B \cdot C_i + A \cdot B \cdot C_i \quad [4.5]$$

$$= A \cdot B + (\overline{A} \cdot B + A \cdot \overline{B}) \cdot C_i$$

$$= A \cdot B + (A \ominus B) \cdot C_i \quad [4.6]$$

A	B	C_i	S	C_{i+1}
0	0	0	0	0
0	0	1	1	0
0	1	0	1	0
0	1	1	0	1
1	0	0	1	0
1	0	1	0	1
1	1	0	0	1
1	1	1	1	1

Table 4.2. *Truth table of a full adder*

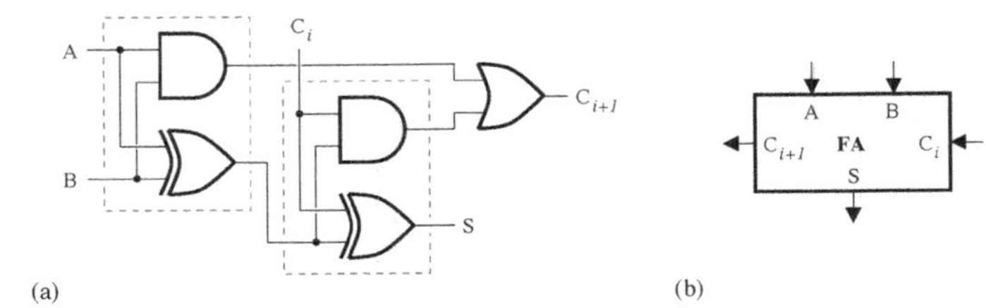

Figure 4.4. *Full adder: a) logic circuit; b) symbol*

An FA can be implemented using two HAs and an OR gate, as shown in Figure 4.4(a). The symbol of an FA is represented in Figure 4.4(b).

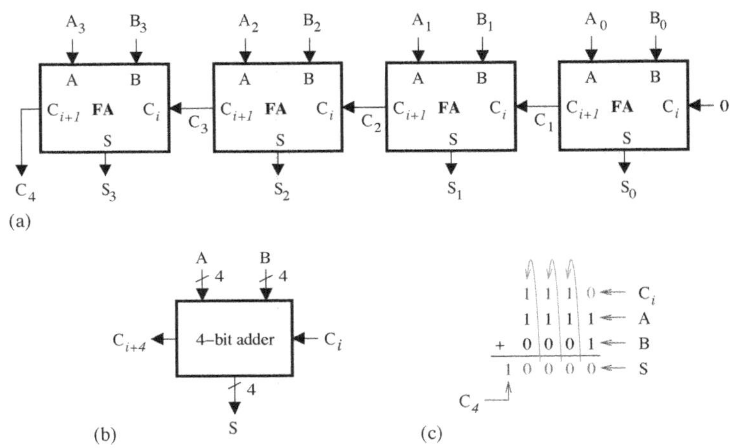

Figure 4.5. *Four-bit ripple-carry adder: a) logic circuit; b) symbol; c) example of addition with ripple-carry. For a color version of the figure, see www.iste.co.uk/ndjountche/electronics2.zip*

4.2.3. *Ripple-carry adder*

A ripple-carry adder is implemented by using an FA stage for each bit and connecting the carry-out of a given stage to the carry-in of the next stage, as shown in Figure 4.5(a) for 4-bit numbers. The FA that provides the LSB of the sum and whose carry-in is set to 0 may be replaced by an HA. Figure 4.5(b) shows the symbol of a ripple-carry adder. An example of addition with ripple-carry is illustrated in Figure 4.5(c).

Subtraction can be performed by expressing the subtrahend (or number to be subtracted) with a negative sign in the two's complement representation and then performing an addition operation without taking into account the last carry-out. The logic circuit and the symbol of an adder configured for subtraction are given in Figures 4.6(a) and 4.6(b). An example of a subtraction operation is illustrated in Figure 4.6(c).

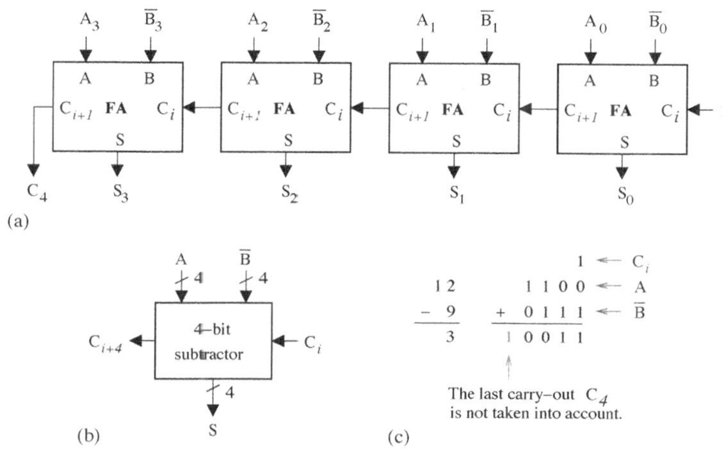

Figure 4.6. *Four-bit ripple-carry adder configured for subtraction: a) logic circuit; b) symbol; c) example*

Figure 4.7 depicts a 4-bit ripple-carry adder/subtractor. Each XOR gate operates as a programmable inverter whose output can take either the state or the logic complement of the input variable.

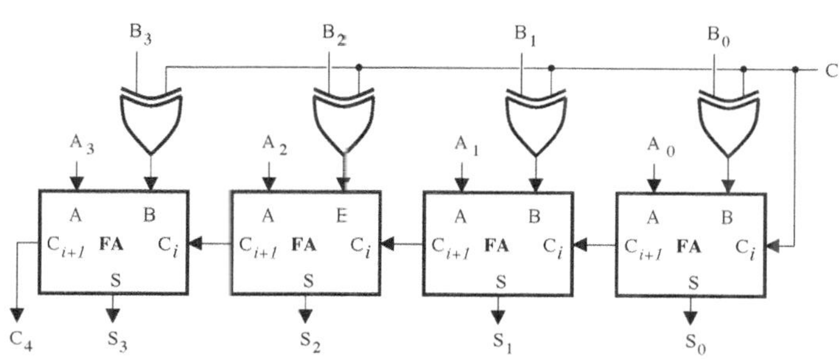

Figure 4.7. *Four-bit ripple-carry adder/subtractor ($C = 0$: adder; $C = 1$: subtractor)*

A 4-bit ripple-carry adder is made up of four FAs that can singly exhibit the timing parameters defined as follows:

– τ_i, propagation delay $A, B \rightarrow C_{i+1}$;

– τ_c, propagation delay $C_i \rightarrow C_{i+1}$;

– τ_s, propagation delay $C_i \rightarrow S$.

The critical path for a ripple-carry adder corresponds to the propagation path of the carry. The first carry is generated with a propagation delay τ_i. The carry is then propagated from one FA to another with a propagation delay τ_c, and the last FA can, depending on the input data, either produce a carry-out or not. The propagation delay is thus given by:

$$\tau = \tau_i + 2\tau_c + \max(\tau_c, \tau_s) \qquad [4.7]$$

Assuming that the delay introduced by the XOR gate, τ_{XOR}, is superior to the propagation delay of the AND gates, τ_{AND}, and OR gates, τ_{OR}, we have [1]:

$$\tau = \tau_{XOR} + 3(\tau_{AND} + \tau_{OR}) + \max(\tau_{XOR}, \tau_{AND} + \tau_{OR}) \qquad [4.8]$$

The propagation delay for a ripple-carry adder increases with an increase in the number of input data bits and can put a limitation on the operating speed. One solution that can be used to reduce the propagation delay is to use a carry-lookahead adder.

4.2.4. *Carry-lookahead adder*

In a carry-lookahead adder, the carry-out of each stage is computed independently.

An analysis of the 4-bit carry-lookahead adder shown in Figure 4.8 provides the following logic equations:

$$g_i = A_i \cdot B_i \qquad [4.9]$$

$$p_i = A_i \oplus B_i \qquad [4.10]$$

and:

$$C_{i+1} = g_i + p_i \cdot C_i \qquad [4.11]$$

1 For an n-bit ripple-carry adder the propagation delay is of the form:

$$\tau = \tau XOR + (n-1)(\tau AND + \tau OR) + \max(\tau XOR, \tau AND + \tau OR)$$

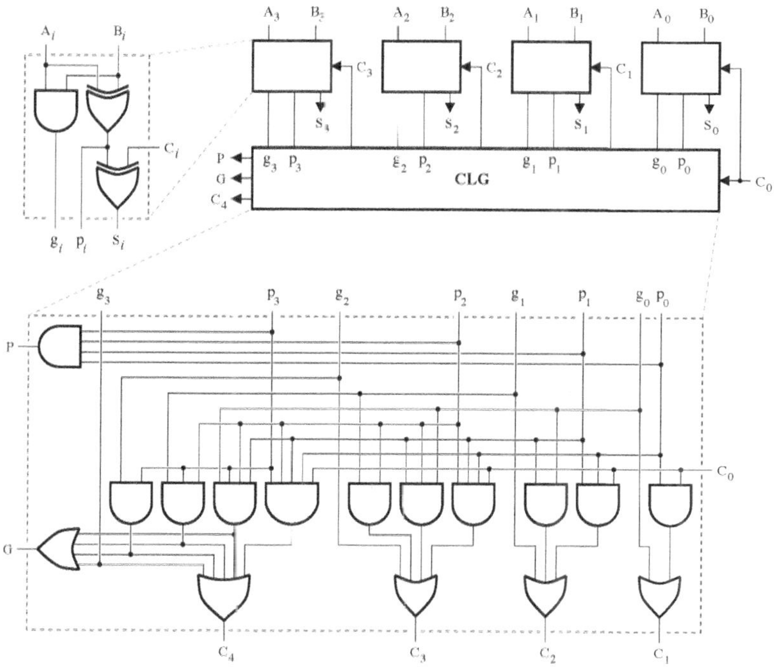

Figure 4.8. *Four-bit carry-lookahead adder*

Upon successive substitutions, we can obtain:

$$C_1 = g_0 + p_0 \cdot C_0 \quad [4.12]$$

$$\begin{aligned} C_2 &= g_1 + p_1 \cdot C_1 \\ &= g_1 + g_0 \cdot p_1 + p_0 \cdot p_1 \cdot C_0 \end{aligned} \quad [4.13]$$

$$\begin{aligned} C_3 &= g_2 + p_2 \cdot C_2 \\ &= g_2 + g_1 \cdot p_2 + g_0 \cdot p_1 \cdot p_2 + p_0 \cdot p_1 \cdot p_2 \cdot C_0 \end{aligned} \quad [4.14]$$

$$\begin{aligned} C_4 &= g_3 + p_3 \cdot C_3 \\ &= g_3 + g_2 \cdot p_3 + g_1 \cdot p_2 \cdot p_3 + g_0 \cdot p_1 \cdot p_2 \cdot p_3 + p_0 \cdot p_1 \cdot p_2 \cdot p_3 \cdot C_0 \end{aligned} \quad [4.15]$$

The outputs G and P, produced by the carry-lookahead generator, are used to determine whether a given bit pair will generate a carry or propagate a carry:

$$P = p_0 \cdot p_1 \cdot p_2 \cdot p_3 \quad [4.16]$$

$$G = g_3 + g_2 \cdot p_3 + g_1 \cdot p_3 \cdot p_2 + g_0 p_3 \cdot p_2 \cdot p_1 \quad [4.17]$$

The propagation delay (path $A_0 - S_3$) in a 4-bit carry-lookahead adder is:

$$\tau = 2\tau_{XOR} + \tau_{AND} + \tau_{OR} \tag{4.18}$$

To facilitate the design of an n-bit adder, the carry-lookahead adder has two outputs P and G:

– P indicates the propagation of the value of the carry-in by a module (C_i);

– G indicates the generation of a carry-out ($C_{i+1} = 1$) by a module, regardless of the value of the carry-in.

Hence:

$$C_{i+1} = G + P \cdot C_i \tag{4.19}$$

$$P = \begin{cases} 1 & \text{if} \quad X + Y = 2^n - 1 \\ 0 & \text{otherwise} \end{cases} \tag{4.20}$$

$$G = \begin{cases} 1 & \text{if} \quad X + Y \geq 2^n \\ 0 & \text{otherwise} \end{cases} \tag{4.21}$$

$$P = \prod_{i=0}^{n-1} p_i \tag{4.22}$$

$$G = g_{n-1} + g_{n-2}p_{n-1} + g_{n-3}p_{n-1}p_{n-2} + \cdots + g_0 p_{n-1}p_{n-2} \cdots p_1 \tag{4.23}$$

As the number of input data bits increases, the structure of the carry-lookahead adder becomes more complex. In this case, a better trade-off between the hardware cost and the operating speed can be achieved by using other approaches such as the carry-select adder or the carry-skip adder.

4.2.5. *Carry-select adder*

The working principle of a carry-select adder is to perform the same calculation in parallel: one with the carry-in set to 0 and the other with the carry-in set to 1. The correct result is then chosen when the exact value of the carry-in becomes available. A carry select adder is thus composed of several sections of identical or different sizes and each section, with the exception of the one related to the least significant bits, carries two additions in parallel.

Figure 4.9 depicts the logic circuit of a 4-bit carry-select adder. Two circuits are used to perform the addition involving the two most significant bits and the selection of the result depends on the logic state of the carry, which is generated by the adder of the two least significant bits and is applied to the multiplexers. The circuit that determines the carry, C_4, which is composed of an OR gate and an AND gate, can be replaced by a 2 : 1 multiplexer whose selection signal is the carry C_2.

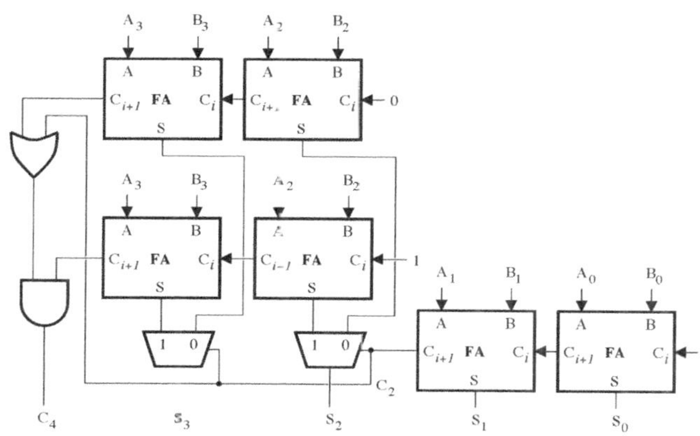

Figure 4.9. *Four-bit carry-select adder*

4.2.6. *Carry-skip adder*

Another approach that can be used to accelerate the propagation of the carry consists of using a carry-skip adder. It can be implemented using consecutive adder groups of identical or variable size. The addition of a propagation path enables the carry to skip a group of $k - l$ FAs whose propagation signal, $P_{k:l}$, is at the logic state 1. An FA with a propagation signal, P, is represented in Figure 4.10. Figure 4.11 depicts a carry-skip adder cell. The expressions for the propagation signals for cells having between two to four bits are given in Table 4.3, where $P_i = X_i \oplus Y_i$ $(i = 0, 1, 2, 3)$, and the input data are represented by X_i and Y_i.

The logic circuit for an 8-bit carry-skip adder is represented in Figure 4.12. It is composed of four identical groups, each of which performs a two-bit addition. In general, the optimal size of each group is determined based on the number of input data bits.

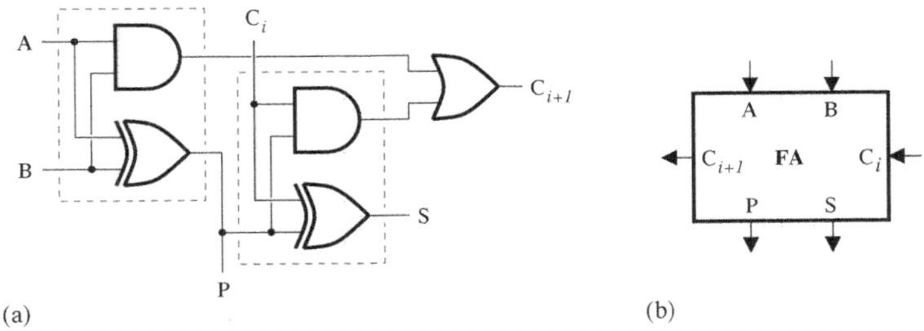

Figure 4.10. *Full adder*

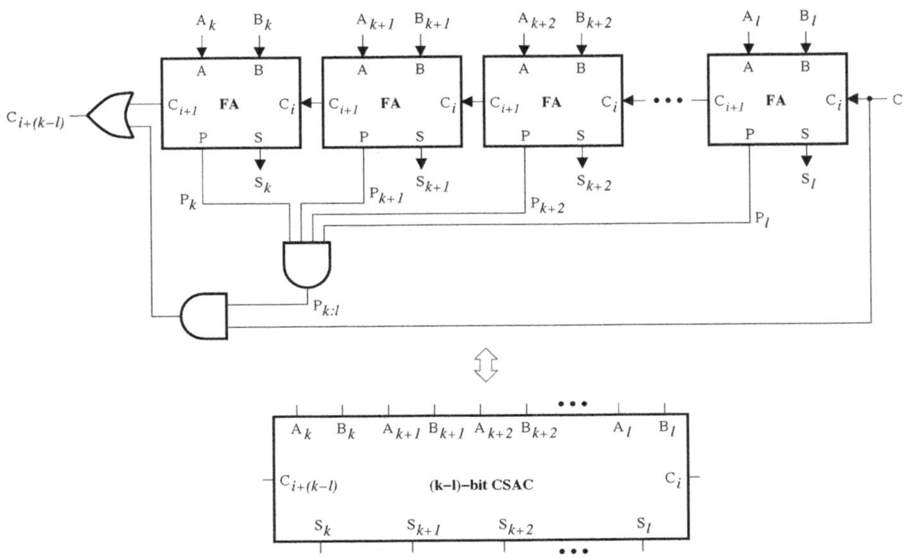

Figure 4.11. *Carry-skip adder cell*

2-bit cell	$P_{1:0} = P_1 \cdot P_0$
3 -bit cell	$P_{2:0} = P_2 \cdot P_1 \cdot P_0$
4-bit cell	$P_{3:0} = P_3 \cdot P_2 \cdot P_1 \cdot P_0$

Table 4.3. *Expressions of the propagation signal for cells having from two to four bits*

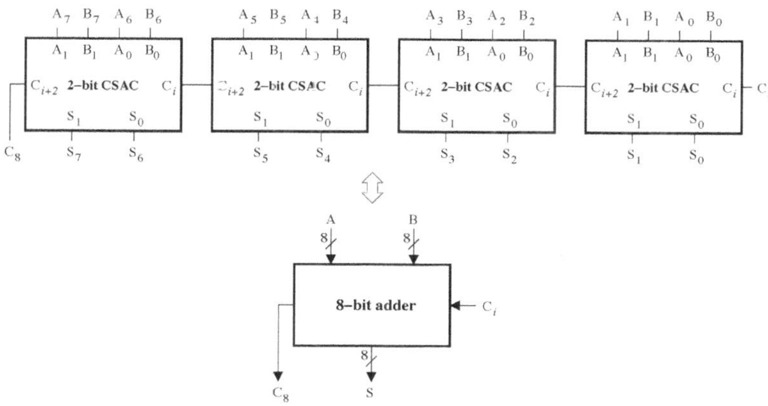

Figure 4.12. *Eight-bit carry-skip adder*

4.3. Comparator

A digital comparator is a logic circuit that compares two binary numbers. The comparator can be used to determine whether a given number is smaller than, equal to or larger than another number.

A 1-bit comparator has two inputs (A, B) and three outputs ($O_{A<B}$, $O_{A=B}$, $O_{A>B}$). Table 4.4 gives the truth table describing the comparator operation. The output logic equations can be written as:

$$O_{A>B} = A \cdot \overline{B} \qquad [4.24]$$

$$O_{A=B} = A \cdot B + \overline{A} \cdot \overline{B} = \overline{(A \oplus B)} = A \odot B \qquad [4.25]$$

$$O_{A<B} = \overline{A} \cdot B \qquad [4.26]$$

A	B	$O_{A>B}$	$O_{A=B}$	$O_{A<B}$
0	0	0	1	0
0	1	0	0	1
1	0	1	0	0
1	1	0	1	0

Table 4.4. *Truth table for a 1-bit comparator*

The logic circuit and symbol of a 1-bit comparator are represented in Figure 4.13.

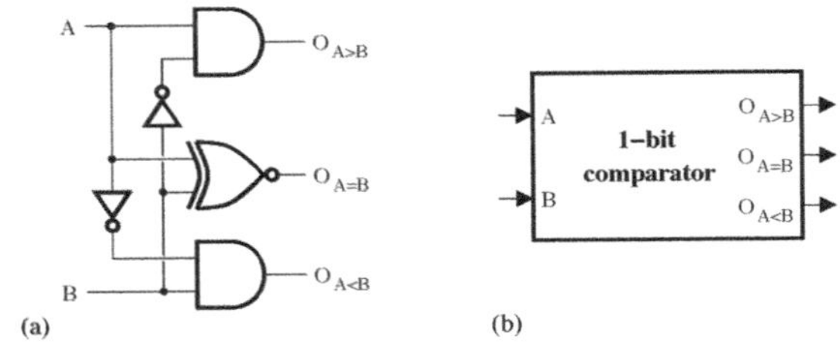

Figure 4.13. *One-bit comparator: a) logic circuit; b) symbol*

More inputs ($I_{A<B}$, $I_{A=B}$, $I_{A>B}$) can be added to a 1-bit comparator to enable the cascade connection of several circuits of the same type. The truth table of the *cascadable* 1-bit comparator is given in Table 4.14. The Karnaugh maps associated with each output are represented in Figure 4.15. The output logic expressions are factorized instead of being simplified, so that the XOR logic gate performing the $\overline{A \oplus B}$ function can be shared by the outputs. Thus:

$$O_{A>B} = (A + \overline{B}) \cdot I_{A>B} + A \cdot \overline{B} \quad [4.27]$$

$$= \overline{(A \oplus B)} \cdot I_{A>B} + A \cdot \overline{B} \quad [4.28]$$

$$O_{A=B} = \overline{(A \oplus B)} \cdot I_{A=B} \quad [4.29]$$

$$O_{A<B} = (\overline{A} + B) \cdot I_{A<B} + \overline{A} \cdot B \quad [4.30]$$

$$= \overline{(A \oplus B)} \cdot I_{A<B} + \overline{A} \cdot B \quad [4.31]$$

A	B	$O_{A>B}$	$O_{A=B}$	$O_{A<B}$
0	0	$I_{A>B}$	$I_{A=B}$	$I_{A<B}$
0	1	0	0	1
1	0	1	0	0
1	1	$I_{A>B}$	$I_{A=B}$	$I_{A<B}$

Figure 4.14. *Truth table of a cascadable 1-bit comparator*

The n-bit comparator shown in Figure 4.17 is implemented by cascading 1-bit comparators. The comparison is carried out in an iterative manner beginning with the most significant bits of the input data. The logic state 1 for the outputs, $O_{A>B}$ or $O_{A<B}$, generated by a given stage is propagated to the following stages while if the output $O_{A=B}$ of a given stage is set to 1, the same comparison operation is repeated in the next stage.

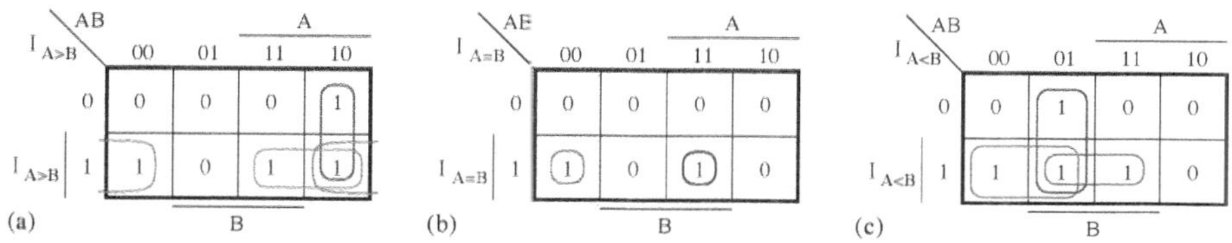

Figure 4.15. *Karnaugh maps: a) $O_{A>B}$; b) $O_{A=B}$; c) $O_{A<B}$. For a color version of the figure, see www.iste.co.uk/ndjountche/electronics2.zip*

The logic circuit and symbol for the *cascadable* 1-bit comparator are given in Figures 4.16(a) and 4.16(b), respectively.

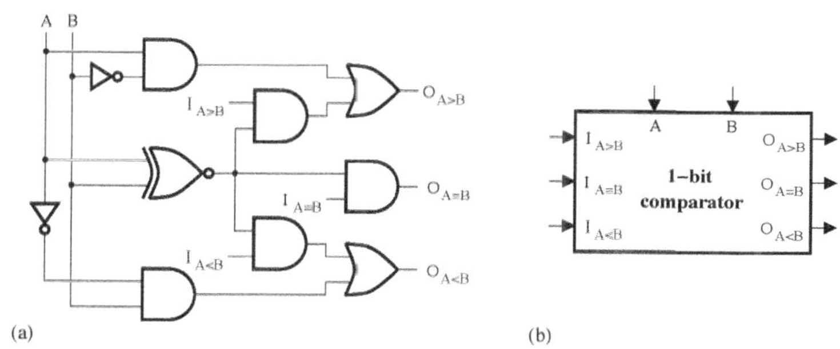

Figure 4.16. *One-bit cascadable comparator: a) logic circuit; b) symbol*

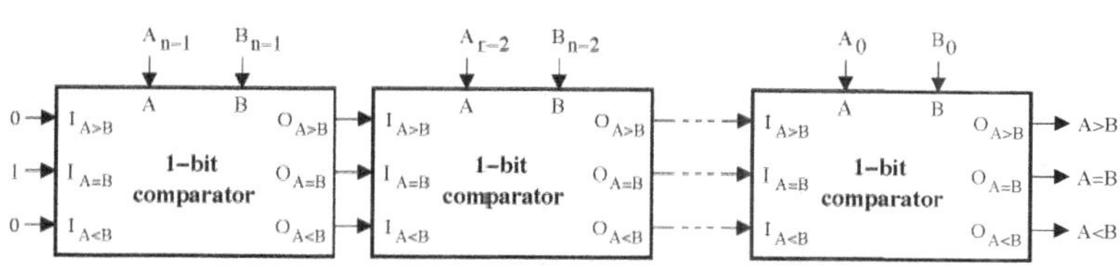

Figure 4.17. *Structure of the n-bit comparator*

4.4. Arithmetic and logic unit

The ALU is the microprocessor component that carries out the arithmetic, logic and comparison operations. In addition to the data inputs and an output for the result, it has inputs that are used to select the desired operation.

The ALU can be designed in a hierarchical manner, as illustrated in Figure 4.18, where $\overline{A} = \overline{A_3}\ \overline{A_2}\ \overline{A_1}\ \overline{A_0}$, $\overline{B} = \overline{B_3}\ \overline{B_2}\ \overline{B_1}\ \overline{B_0}$ and $\overline{F} = \overline{F_3}\ \overline{F_2}\ \overline{F_1}\ \overline{F_0}$. The ALU is implemented by connecting the appropriate logic circuits (AND and OR function generators) to the inputs of an adder whose operation is governed by a control signal. Input data are thus coded depending on the operation to be executed before being applied to the adder. AND and OR function generators can be implemented as shown in Figure 4.19. Their outputs, for input bits with the same index, A_k and B_k $(k = 0, 1, 2, 3)$, are characterized by the following equations:

$$\begin{aligned} F_{AND} &= \overline{\overline{A_k} + B_k \cdot S_1 + \overline{B_k} \cdot S_0} \\ &= A_k \cdot B_k \cdot \overline{S_1} + A_k \cdot \overline{B_k} \cdot \overline{S_0} + A_k \cdot \overline{S_1} \cdot \overline{S_0} \end{aligned} \qquad [4.32]$$

and

$$\begin{aligned} F_{OR} &= \overline{\overline{A_k} \cdot \overline{B_k} \cdot S_3 + \overline{A_k} \cdot B_k \cdot S_2} \\ &= A_k + B_k \cdot \overline{S_2} + \overline{B_k} \cdot \overline{S_3} + \overline{S_2} \cdot \overline{S_3} \end{aligned} \qquad [4.33]$$

where the selection inputs are represented by S_0, S_1, S_2 and S_3. Tables 4.5 and 4.6 give the function tables of the AND and OR function generators, respectively.

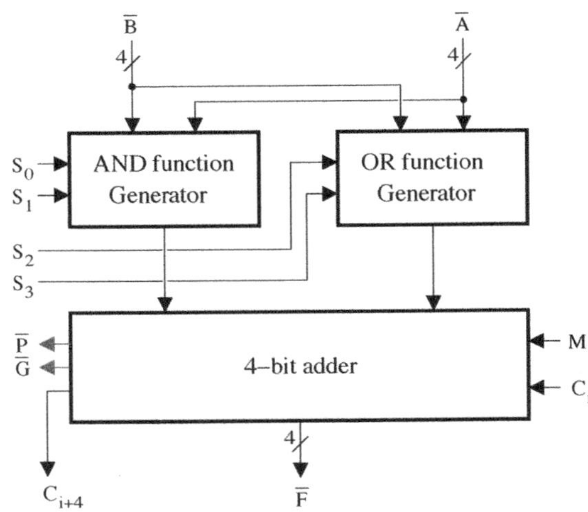

Figure 4.18. *Diagram illustrating the operation principle of a 4-bit arithmetic and logic unit*

The ALU (74HC181 integrated circuit) shown in Figure 4.20 is designed for 4-bit data. It has select inputs, S_0, S_1, S_2 and S_3, and a control input M. It can carry out all sixteen logic operations possible with two variables or sixteen different arithmetic operations on high-active or low-active input data, as shown in Figures 4.7 and 4.8.

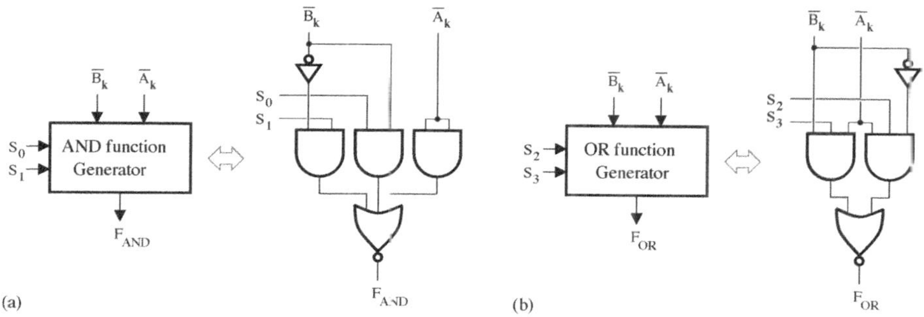

Figure 4.19. *a) OR function generator; b) AND function generator*

S_1	S_0	F_{AND}
0	0	A_k
0	1	$A_k \cdot B_k$
1	0	$A_k \cdot \overline{B_k}$
1	1	0

Table 4.5. *Function table of the F_{AND} generator*

S_3	S_2	F_{OR}
0	0	1
0	1	$A_k + \overline{B_k}$
1	0	$A_k + B_k$
1	1	A_k

Table 4.6. *Function table of the F_{OR} generator*

Functions can be generated using either one of the logic operators NOT, AND, NAND, OR, NOR, XOR and XNOR or using an operation like addition, subtraction, incrementing, decrementing, multiplication by two, data transfer and comparison of two numbers. To carry out the logic operations, all the internal carries must be deactivated by setting the control input, M, to 1, while for arithmetic operations, the carries are activated by resetting the control input, M, to 0.

As the ALU is based on a carry-lookahead adder, it can supply a carry-out C_{i+4} or signals for the propagation ($\overline{P}$) and the generation ($\overline{G}$) of the carry, which are not affected by the carry-in, C_i. To process data larger than four bits it is necessary to cascade several ALUs by connecting the carry-out and carry-in or by using $\overline{P}$ and $\overline{G}$ signals in conjunction with a carry-lookahead generator when high-speed operation is desired.

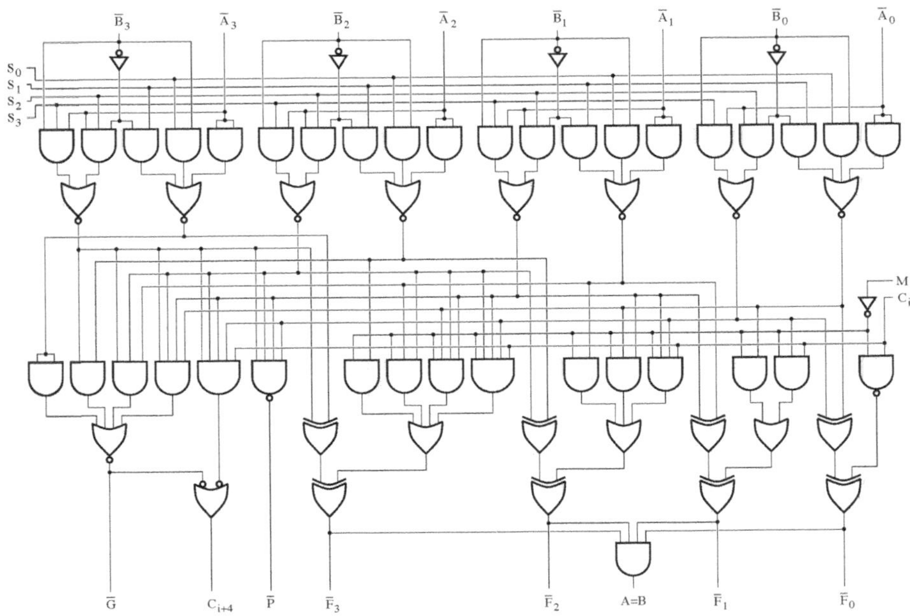

Figure 4.20. *Four-bit arithmetic and logic units*

Two's complement representation is used to perform arithmetic operations. The function table lists the arithmetic operations that are executed with or without the carry-in C_i. The carry-in is taken into account by adding 1 to a result. In this manner, for the select code $S_3S_2S_1S_0 = 0110$, the operation carried out is either $A - B$, if there is a carry-in, otherwise it is $A - B - 1$. Subtraction is executed like addition with the 1's complement of the number to be subtracted, to which 1 is added, and the carry-out, if any, must be ignored.

The output $A = B$ takes the logic state 1 when all outputs from F_3 to F_0 are at the state 1 and is used to indicate the logic equivalence of the four output bits when the ALU operates as a subtractor with $C_i = 1$. It is supplied by an open drain gate whose output can be cabled with other outputs from the same type of gates in order to allow a comparison of data larger than four bits. It requires an external polarization resistance to take the high logic level.

Select inputs				Active-low data		
				$M = 1$	$M = 0$: Arithmetic operations	
S_3	S_2	S_1	S_0	Logic functions	$C_i = 0$ (without carry)	$C_i = 1$ (with carry)
0	0	0	0	$F = \overline{A}$	$F = A - 1$	$F = A$
0	0	0	1	$F = \overline{A \cdot B}$	$F = A \cdot B - 1$	$F = A \cdot B$
0	0	1	0	$F = \overline{A} + B$	$F = A \cdot \overline{B} - 1$	$F = A \cdot \overline{B}$
0	0	1	1	$F = 1$	$F = -1$	$F = 0$
0	1	0	0	$F = \overline{A + B}$	$F = A$ plus $(A + \overline{B})$	$F = A$ plus $(A + \overline{B})$ plus 1
0	1	0	1	$F = \overline{B}$	$F = A \cdot B$ plus $(A + \overline{B})$	$F = A \cdot B$ plus $(A + \overline{B})$ plus 1
0	1	1	0	$F = \overline{A \oplus B}$	$F = A - B - 1$	$F = A - B$
0	1	1	1	$F = A + \overline{B}$	$F = A + \overline{B}$	$F = (A + \overline{B})$ plus 1
1	0	0	0	$F = \overline{A} \cdot B$	$F = A$ plus $(A + B)$	$F = A$ plus $(A + B)$ plus 1
1	0	0	1	$F = A \oplus B$	$F = A$ plus B	$F = A$ plus B plus 1
1	0	1	0	$F = B$	$F = A \cdot \overline{B}$ plus $(A + B)$	$F = A \cdot \overline{B}$ plus $(A + B)$ plus 1
1	0	1	1	$F = A + B$	$F = A + B$	$F = (A + B)$ plus 1
1	1	0	0	$F = 0$	$F = A$ plus A	$F = A$ plus A plus 1
1	1	0	1	$F = A \cdot \overline{B}$	$F = A \cdot B$ plus A	$F = A \cdot B$ plus A plus 1
1	1	1	0	$F = A \cdot B$	$F = A \cdot \overline{B}$ plus A	$F = A \cdot \overline{B}$ plus A plus 1
1	1	1	1	$F = A$	$F = A$	$F = A$ plus 1

Table 4.7. *Function table for the 4-bit arithmetic and logic unit*

When the ALU operates as a subtractor (or with the select code $S_3S_2S_1S_0 = 0110$), the logic level of the carry-out, C_{i+4}, can also be used to establish a comparison between the input data. The possible results are given in Table 4.9.

Status flag bits, which are associated with an ALU, supply information on the nature of the results related to the manipulation of data with a limited word-length.

Figure 4.21 depicts a 4-bit ALUs with status flag bits. The inputs and outputs can be defined as follows:

– inputs:

A and B are two 4-bit numbers ($A = A_3A_2A_1A_0$ and $B = B_3B_2B_1B_0$),
S is a 4-bit select code ($S = S_3S_2S_1S_0$),
M is a control bit,

$C_i \in \{0, 1\}$;

– outputs:

F is a 4-bit number ($F = F_3F_2F_1F_0$),

$$\text{Bit } N = \begin{cases} 1 & if \quad F < 0 \\ 0 & otherwise \end{cases}$$

$$\text{Bit } Z = \begin{cases} 1 & if \quad F = 0 \\ 0 & otherwise \end{cases}$$

$$\text{Bit } V = \begin{cases} 1 & \text{if there is overflow } (-8 \leq R \leq 7) \\ 0 & otherwise. \end{cases}$$

Select inputs				Active-high data		
				$M = 1$	$M = 0$: Arithmetic operation	
S_3	S_2	S_1	S_0	Logic functions	$C_i = 1$ (without carry)	$C_i = 0$ (with carry)
0	0	0	0	$F = \overline{A}$	$F = A$	$F = A$ plus 1
0	0	0	1	$F = \overline{A + B}$	$F = A + B$	$F = (A + B)$ plus 1
0	0	1	0	$F = \overline{A} \cdot B$	$F = A + \overline{B}$	$F = (A + \overline{B})$ plus 1
0	0	1	1	$F = 0$	$F = -1$	$F = 0$
0	1	0	0	$F = \overline{A \cdot B}$	$F = A$ plus $(A \cdot \overline{B})$	$F = A$ plus $(A \cdot \overline{B})$ plus 1
0	1	0	1	$F = \overline{B}$	$F = (A + B)$ plus $A \cdot \overline{B}$	$F = (A + B)$ plus $A \cdot \overline{B}$ plus 1
0	1	1	0	$F = A \oplus B$	$F = A - B - 1$	$F = A - B$
0	1	1	1	$F = A \cdot \overline{B}$	$F = A \cdot \overline{B} - 1$	$F = A \cdot \overline{B}$
1	0	0	0	$F = \overline{A} + B$	$F = A$ plus $(A \cdot B)$	$F = A$ plus $A \cdot B$ plus 1
1	0	0	1	$F = \overline{A \oplus B}$	$F = A$ plus B	$F = A$ plus B plus 1
1	0	1	0	$F = B$	$F = (A + \overline{B})$ plus $A \cdot B$	$F = (A + \overline{B})$ plus $A \cdot B$ plus 1
1	0	1	1	$F = A \cdot B$	$F = A \cdot B - 1$	$F = A \cdot B$
1	1	0	0	$F = 1$	$F = A$ plus A	$F = A$ plus A plus 1
1	1	0	1	$F = A + \overline{B}$	$F = (A + B)$ plus A	$F = (A + B)$ plus A plus 1
1	1	1	0	$F = A + B$	$F = (A + \overline{B})$ plus A	$F = (A + \overline{B})$ plus A plus 1
1	1	1	1	$F = A$	$F = A - 1$	$F = A$

Table 4.8. *Function table of the 4-bit arithmetic and logic unit*

The ALU uses two's representation:

– the sign bit is determined based on the most significant bit in the result F:

$$F \geq 0 \text{ if } F_3 = 0$$
$$F < 0 \text{ if } F_3 = 1$$

– the overflow bit V is obtained using the relationship:

$$V = C_4 \oplus C_3$$

C_i	C_{i+4}	Active-low data	Active-high data
0	0	$A \leq B$	$A \geq B$
0	1	$A > B$	$A < B$
1	0	$A < B$	$A > B$
1	1	$A \geq B$	$A \leq B$

Table 4.9. *Comparison results*

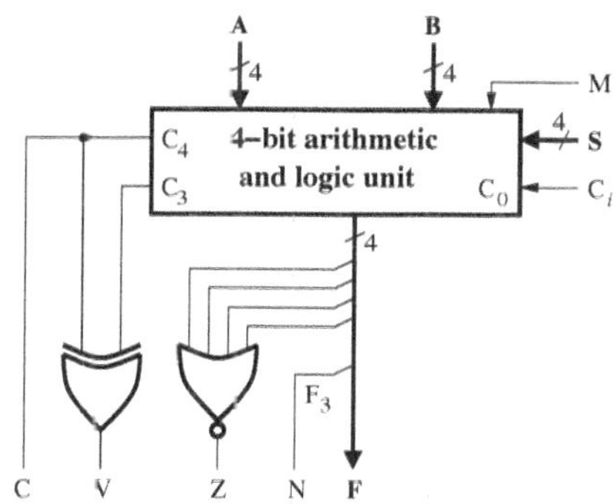

Figure 4.21. *Aritnmetic unit: status flag bits*

In practice, we can also compare two unsigned numbers, A and B, by carrying out a subtraction followed by a comparison of the difference $A - B$ with respect to 0. We thus have:

– for $A - B = 0$, $C = 1$ and $Z = 1$, it follows that $A = B$;

– for $A - B > 0$, $C = 1$ and $Z = 0$, it follows that $A > B$;

– for $A - B < 0$, $C = 0$ and $Z = 0$, it follows that $A < B$.

An ALU can be designed to carry out any operation. The more complex the operation, the greater the unit's required size and power consumption. In practice, a tradeoff is made to implement an ALU whose size remains moderate so as not to critically affect the power consumption and operating speed. As a consequence, more complex operations may be carried out in an iterative manner or using a specialized unit.

4.5. Multiplier

A cellular digital multiplier is a combinational circuit that calculates and sums the partial products that make up the product of two numbers, X and Y. The product of two n-bit numbers, X and Y, is a number of $2n$ bits, $P = X \times Y$.

4.5.1. *Multiplier of 2-bit unsigned numbers*

Considering $X = X_1X_0$ and $Y = Y_1Y_0$ as two 2-bit unsigned binary numbers, the multiplication is carried out as shown in Figure 4.22. The logic circuit of a 2 × 2-bit multiplier is given in Figure 4.23, where the partial products are generated by the four AND logic gates and the product bits, with the exception of the least significant bit, are supplied by two HAs whose role is to combine the partial products.

		X_1	X_0
	$\times$	Y_1	Y_0
		Y_0X_1	Y_0X_0
$+$	Y_1X_1	Y_1X_0	
P_3	P_2	P_1	P_0

Figure 4.22. *Multiplication of two 2-bit numbers*

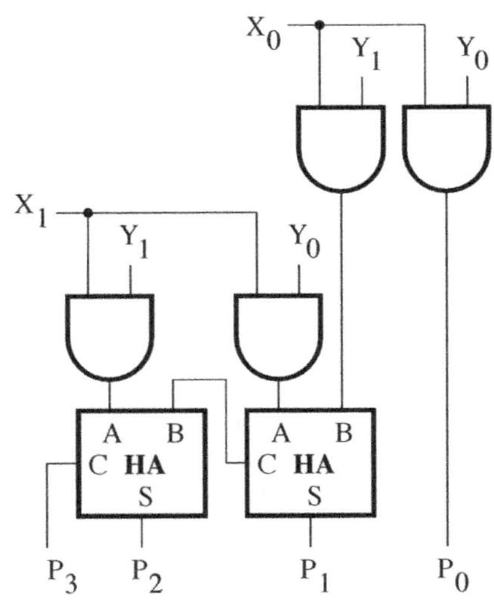

Figure 4.23. *Logic circuit of a 2×2-bit multiplier*

4.5.2. *Multiplier of 4-bit unsigned numbers*

A simple cellular multiplier is based on shift and add algorithm. The multiplication of two 4-bit unsigned binary numbers, $X = X_3X_2X_1X_0$ and $Y = Y_3Y_2Y_1Y_0$, is carried out as illustrated in Figure 4.24.

					X_3	X_2	X_1	X_0
				$\times$	Y_3	Y_2	Y_1	Y_0
					Y_0X_3	Y_0X_2	Y_0X_1	Y_0X_0
				Y_1X_3	Y_1X_2	Y_1X_1	Y_1X_0	
			Y_2X_3	Y_2X_2	Y_2X_1	Y_2X_0		
$+$		Y_3X_3	Y_3X_2	Y_3X_1	Y_3X_0			
	P_7	P_6	P_5	P_4	P_3	P_2	P_1	P_0

Figure 4.24. *Multiplication of two 4-bit numbers*

The logic circuit for a 4×4-bit cellular multiplier is given in Figure 4.25. The AND logic gates are used to compute the partial products Y_iX_j, where $i, j \in \{0, 1, 2, 3\}$.

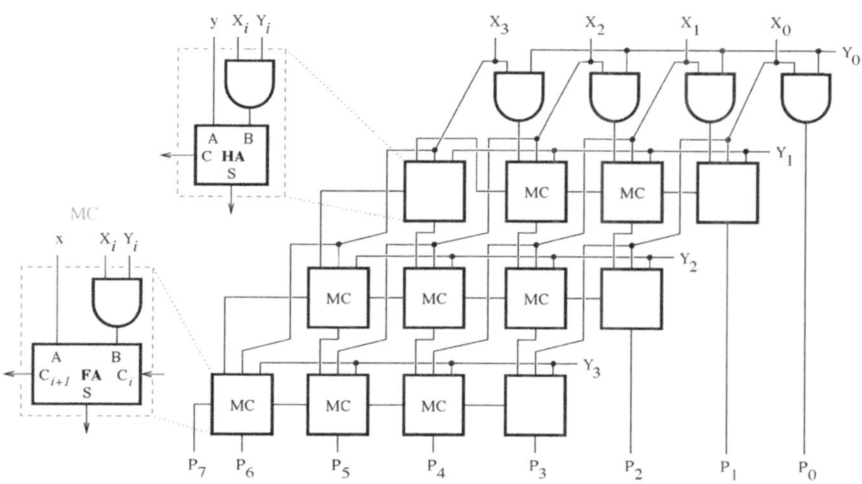

Figure 4.25. *4×4-bit cellular multiplier*

In general, an $n \times n$ cellular multiplier requires n^2 AND gates, n HAs and $n^2 - 2n$ FAs. The propagation delay of the multiplier is caused by the propagation of carries between the adders. There are several critical pathways of identical length that pass through an AND gate and $3n - 4$ adders; this is because of the matrix structure of the multiplier. Assuming that $\tau_{sum} \geq \tau_{carry}$, where τ_{carry} and τ_{sum} represent the propagation delays between the adder inputs and each output C_{i+1} and S, respectively, and that τ_{AND} designates the AND gate propagation delay, the propagation delay of the multiplier can be expressed as follows:

$$\tau = (2n - 3)\tau_{carry} + (n - 1)\tau_{sum} + \tau_{AND} \quad [4.34]$$

Using other adder architecture that is faster will, thus, reduce the propagation delay and increase the operating speed of the multiplier.

Another approach used to implement a multiplier is given in Figure 4.26, where all multiplier cells are identical. It is based on the use of a systolic network that offers the advantage of modularity and is suitable for integrated-circuit implementation.

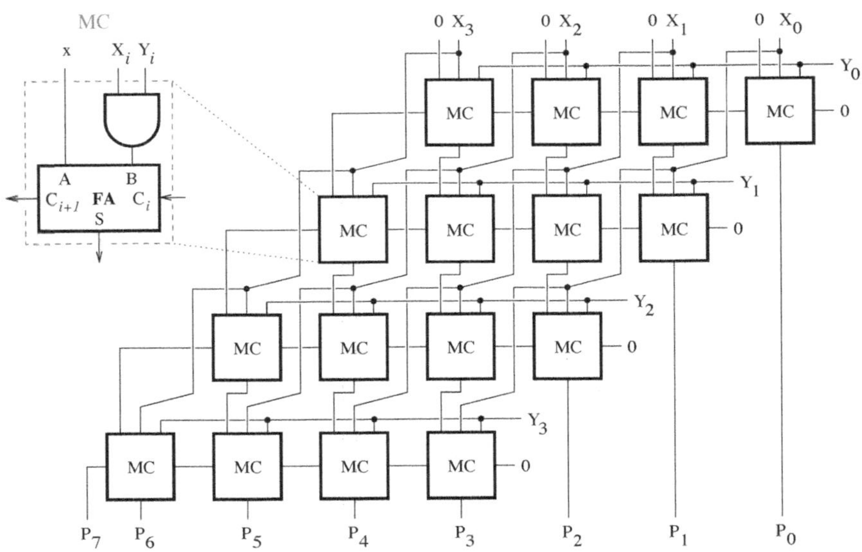

Figure 4.26. *4 × 4-bit cellular multiplier based on a systolic network*

4.5.3. *Multiplier for signed numbers*

The multiplication of signed numbers is quite simple in a sign-magnitude representation. It is carried out as shown earlier for bits representing the magnitudes, while the sign bits are applied to the inputs of a XOR logic gate to determine the product sign. However, in two's complement representation, multiplication is carried out using other algorithms.

Consider the signed numbers X and Y, which can be expressed in n-bit two's complement representation, as follows:

$$X = -X_{n-1}2^{n-1} + \sum_{i=0}^{n-2} X_i 2^i \qquad [4.35]$$

and:

$$Y = -Y_{n-1}2^{n-1} + \sum_{j=0}^{n-2} Y_j 2^j \qquad [4.36]$$

where the sign-bits are represented by X_{n-1} and Y_{n-1}. The product P is given by the following equations:

$$P = X \cdot Y \qquad [4.37]$$

$$= \left(-X_{n-1}2^{n-1} + \sum_{i=0}^{n-2} X_i 2^i\right)\left(-Y_{n-1}2^{n-1} + \sum_{j=0}^{n-2} Y_j 2^j\right) \qquad [4.38]$$

$$= X_{n-1}Y_{n-1}2^{2n-2} + \sum_{i=0}^{n-2}\sum_{j=0}^{n-2} X_i Y_j 2^{i+j} - Y_{n-1}2^{n-1}\sum_{i=0}^{n-2} X_i 2^i - X_{n-1}2^{n-1}\sum_{j=0}^{n-2} Y_j 2^j \qquad [4.39]$$

The multiplication of two 4-bit signed numbers can be executed as illustrated in Figure 4.27. It can be seen that it is necessary to add a sign extension to each line of partial products. In addition, to obtain the right value of the product, the addition of the first lines of partial products must be followed by the subtraction of the last line of partial products. In this case, the direct approach to implementing the multiplier results in an increase of the hardware cost.

The sum of a number, A, and it's one's complement, $\overline{A}$, is a number whose n bits are all set at 1, or $2^n - 1$. In other words, we have:

$$A + \overline{A} = 2^n - 1 \qquad [4.40]$$

					X_3	X_2	X_1	X_0
				$\times$	Y_3	Y_2	Y_1	Y_0
	X_3Y_0	X_3Y_0	X_3Y_0	X_3Y_0	X_3Y_0	X_2Y_0	X_1Y_0	X_0Y_0
+	X_3Y_1	X_3Y_1	X_3Y_1	X_3Y_1	X_2Y_1	X_1Y_1	X_0Y_1	
+	X_3Y_2	X_3Y_2	X_3Y_2	X_2Y_2	X_1Y_2	X_0Y_2		
−	X_3Y_3	X_3Y_3	X_2Y_3	X_1Y_3	X_0Y_3			
	P_7	P_6	P_5	P_4	P_3	P_2	P_1	P_0

Figure 4.27. *Multiplication of two 4-bit signed numbers*

This leads to:

$$-Y_{n-1}2^{n-1}\sum_{i=0}^{n-2} X_i 2^i = 2^{n-1}\left(-2^{n-1} + \sum_{i=0}^{n-2} \overline{X_i Y_{n-1}} 2^i + 1\right) \quad [4.41]$$

$$-X_{n-1}2^{n-1}\sum_{j=0}^{n-2} Y_j 2^j = 2^{n-1}\left(-2^{n-1} + \sum_{j=0}^{n-2} \overline{X_{n-1} Y_j} 2^j + 1\right) \quad [4.42]$$

The product, P, then takes the following form:

$$P = X_{n-1}Y_{n-1}2^{2n-2} + \sum_{i=0}^{n-2}\sum_{j=0}^{n-2} X_i Y_j 2^{i+j}$$
$$+ 2^{n-1}\sum_{i=0}^{n-2} \overline{X_i Y_{n-1}} 2^i + 2^{n-1}\sum_{j=0}^{n-2} \overline{X_{n-1} Y_j} 2^j - 2^{2n-1} + 2^n \quad [4.43]$$

The multiplication of two 4-bit signed numbers is illustrated in Figure 4.28. The partial products are transformed and reorganized so that the multiplier implementation can now require only one arithmetic operation (addition).

					X_3	X_2	X_1	X_0
				$\times$	Y_3	Y_2	Y_1	Y_0
				1	$\overline{Y_0X_3}$	Y_0X_2	Y_0X_1	Y_0X_0
				$\overline{Y_1X_3}$	Y_1X_2	Y_1X_1	Y_1X_0	
			$\overline{Y_2X_3}$	Y_2X_2	Y_2X_1	Y_2X_0		
+	1	Y_3X_3	$\overline{Y_3X_2}$	$\overline{Y_3X_1}$	$\overline{Y_3X_0}$			
	P_7	P_6	P_5	P_4	P_3	P_2	P_1	P_0

Figure 4.28. *Multiplication of two 4-bit signed numbers*

The logic circuit for the corresponding multiplier (Baugh–Wooley) is represented in Figure 4.29, where the MC and MC^* cells are based on an AND gate and a NAND gate, respectively.

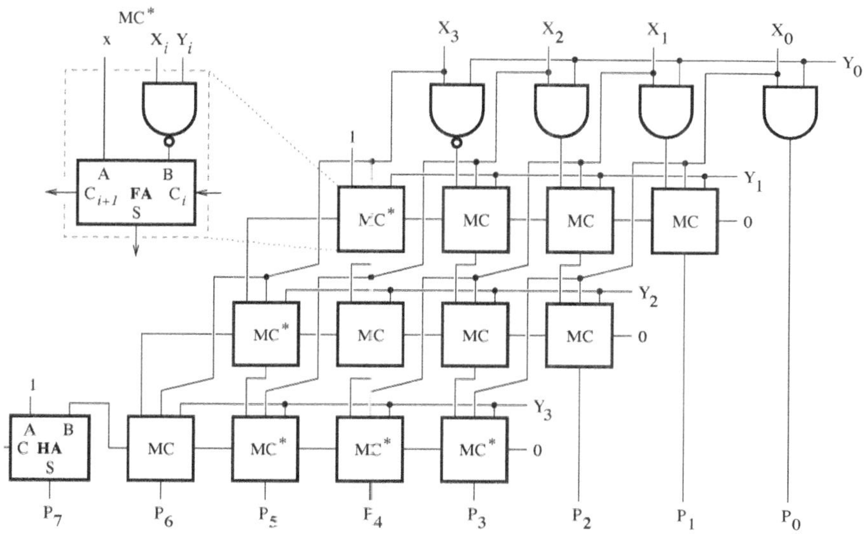

Figure 4.29. *4×4-bit (Baugh–Wooley) multiplier for signed numbers*

In a multiplication, the product is obtained as the sum of the partial products, which increase in number as the word-lengths of the multiplicand and the multiplier become large. It is possible to reduce the number of partial products using Booth's algorithm to encode the multiplicand bits. This encoding has the advantages of resulting in a multiplier implementation with a reduced hardware cost and operating at a high speed.

In the two's complement representation, the multiplicand can be expressed as follows:

$$Y = -Y_{n-1}2^{n-1} + \sum_{j=0}^{n-2} Y_j 2^j \qquad [4.44]$$

By separating terms of odd and even ranks we have:

$$Y = -Y_{n-1}2^{n-1} + \sum_{j=0}^{n/2-1} Y_{2j-1}2^{2j-1} + \sum_{j=0}^{n/2-1} Y_{2j}2^{2j} \qquad [4.45]$$

With the application of the following identity:

$$Y_k 2^k = \left(2Y_k - \frac{1}{2}2Y_k\right)2^k = Y_k 2^{k+1} - 2Y_k 2^{k-1} \quad [4.46]$$

to an odd term, we arrive at:

$$Y = -Y_{n-1}2^{n-1} + \sum_{j=0}^{n/2-1} Y_{2j-1}2^{2j} + \sum_{j=0}^{n/2-1} Y_{2j}2^{2j} - 2\sum_{j=0}^{n/2-1} Y_{2j-1}2^{2j-2} \quad [4.47]$$

The transformation $j \mapsto k+1$ can be used to combine the first term and the last term in the above-mentioned expression for Y as follows:

$$-Y_{n-1}2^{n-1} - 2\sum_{j=0}^{n/2-1} Y_{2j-1}2^{2j-2} = -Y_{n-1}2^{n-1} - 2\sum_{k=-1}^{n/2-2} Y_{2k+1}2^{2k} \quad [4.48]$$

$$= -2\sum_{k=-1}^{n/2-1} Y_{2k+1}2^{2k} \quad [4.49]$$

Assuming that $Y_{-1} = 0$, we finally obtain:

$$Y = \sum_{j=0}^{n/2-1} (Y_{2j-1} + Y_{2j} - 2Y_{2j+1})2^{2j} \quad [4.50]$$

The product of the numbers X and Y can thus take the following form:

$$P = X \cdot Y = \sum_{j=0}^{n/2-1} X(Y_{2j-1} + Y_{2j} - 2Y_{2j+1})2^{2j} \quad [4.51]$$

Hence, three bits are simultaneously involved in the encoding used to obtain each partial product, PP_j, which may be added or deducted from the result. The value of the expression in parentheses is either 0, or ± 1, or ± 2. The partial products generated are multiples of the multiplicand, that is $-2X$, $-X$, 0, X and $2X$, and their number, when compared with an ordinary multiplication, is practically halved. The encoding of the multiplicand bits based on Booth's algorithm is presented in Table 4.10. It must be noted that a left shift operation can be used to obtain a partial product of the form,

$2X$, or to implement the multiplication by a factor of 2, and the negative form of a partial product is obtained by first forming the one's complement and then adding 1 to the result.

Multiplier Y_{2j+1} Y_{2j} Y_{2j-1}	Selection PP_j
0 0 0	0
0 0 1	$+X$
0 1 0	$+X$
0 1 1	$+2X$
1 0 0	$-2X$
1 0 1	$-X$
1 1 0	$-X$
1 1 1	0

Table 4.10. *Encoding of the multiplicand bits based on Booth's algorithm*

The application of Booth's encoding helps reduce the number of the partial product lines, as illustrated in the examples of signed-number multiplication presented in Figure 4.30. A Y_{-1} bit, which is always equal to 0, is added to the binary sequence of the multiplicand. In addition, a partial product line is completed by the sign extension each time the value of the left-most bit is equal to 1.

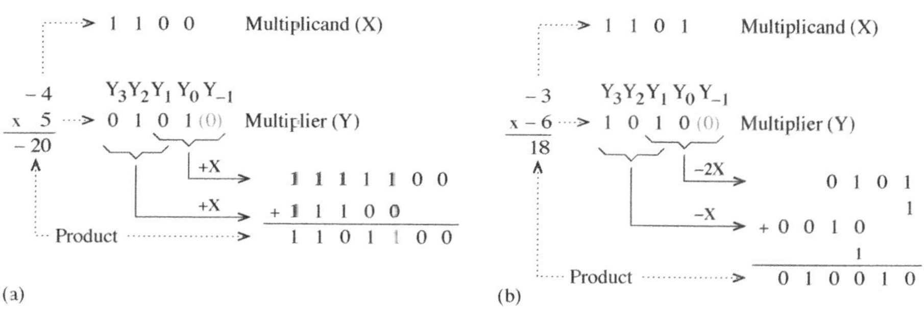

Figure 4.30. *Examples of multiplication based on Booth's algorithm. For a color version of the figure, see www.iste.co.uk/ndjountche/electronics2.zip*

4.6. Divider

The division of an m-bit number or *dividend*, A, by an n-bit ($n \leq m$) number, or the *divisor* D, generally yields a quotient Q and a remainder R. Thus, $A = Q \times D + R$, where $D \neq 0$, $R < D$, and R has the same sign as A.

With a dividend A of $2n$ bits and a divisor D of n bits, the division with restoration may be executed according to the algorithm 4.1.

Algorithm 4.1. *Division with restoration*

Result: quotient $Q = Q_{n-1} \cdots Q_2 Q_1 Q_0$ and remainder $R(0)$

[1] Initially, assign the value of the dividend to the partial remainder, $R(0) = A$, and set the most significant bit of the quotient to 1 ($Q_{n-1} = 1$);

[2] For $i = 0, 1, 2, \cdots, n-1$:

calculate the partial remainder:

$R(i+1) = 2R(i) - Q_{n-(i+1)} \times D$

where $0 \leq R(i) < D$ and D is the divisor;

if $R(i+1) \geq 0$, then

$Q_{n-(i+1)} = 1$;

else

$Q_{n-(i+1)} = 0$, and the previous value of the remainder must be restored partial;

After the last iteration, the quotient $Q = Q_{n-1} \cdots Q_2 Q_1 Q_0$ and $R(0)$ represent the final remainder.

The logic circuit of a cellular divider based on the algorithm for division with restoration is illustrated in Figure 4.31, in the case where $n = 4$. It consists of controlled-subtract (CS) cells, comprising a full subtractor and a $2:1$ multiplexer, OR logic gates and inverters.

Initially, the partial remainder is assumed to be identical to the dividend. The multiplication by 2 is carried out by shifting the most significant bit of the partial remainder one position to the left with respect to that of the divisor. After each of the four subtractions, the partial remainder takes the value of the difference obtained if the output borrow of a row of S cells is set to 0, otherwise the previous value of the partial remainder is restored. The corresponding bit in the quotient is obtained as the OR logic function of the output borrow complement and the most significant bit of the partial remainder.

In general, performing a division with a dividend of $2n$ bits and a divisor of n bits requires n^2 CS cells and n OR gates and inverters.

Figure 4.32 gives two examples of the division operation executions. Each time that the result of the subtraction is negative, the partial remainder is restored to its previous value.

Another approach that can be adopted to carry out cellular division is based on the algorithm for division without restoration. Compared to the division with restoration method, this offers the advantage of reducing the number of operation (addition/subtraction) required, especially in cases where the quotient contains several zeros. However, it has the disadvantage of requiring an additional stage for the correction of the final remainder.

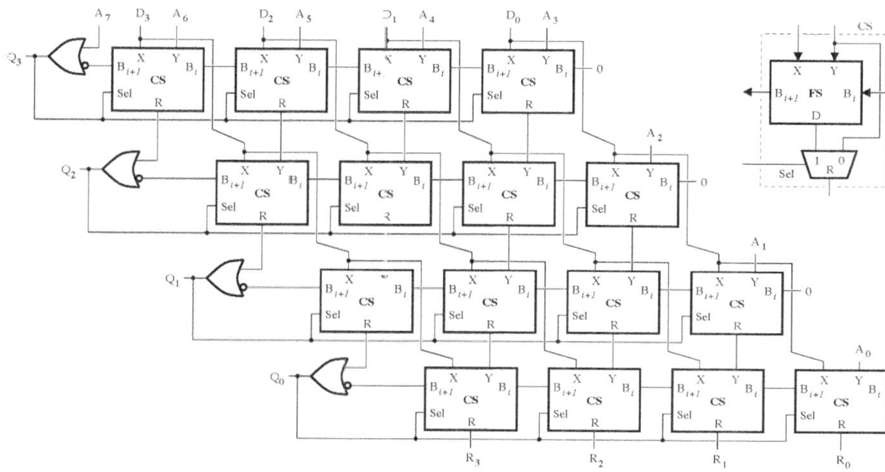

Figure 4.31. *Cellular divider based on the algorithm for division with restoration of the dividend*

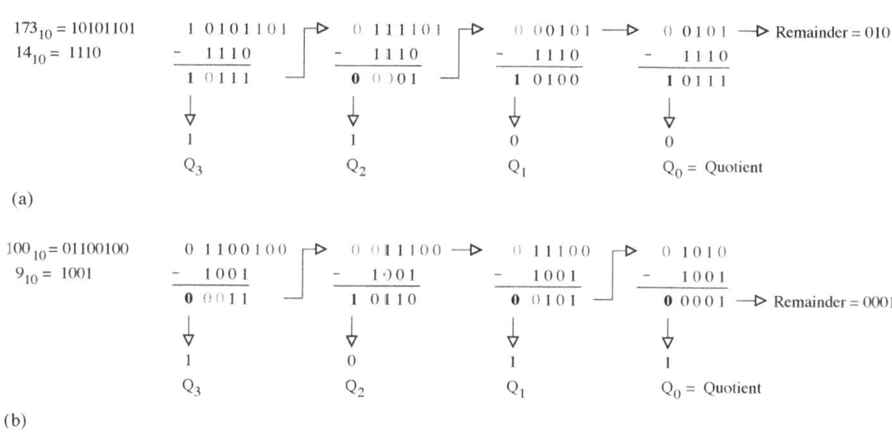

Figure 4.32. *Examples of division with restoration: a)* $173 \div 14$ *(*$Q = 12$*,* $R = 5$*); b)* $100 \div 9$ *(*$Q = 11$*,* $R = 1$*). For a color version of the figure, see www.iste.co.uk/ndjountche/electronics2.zip*

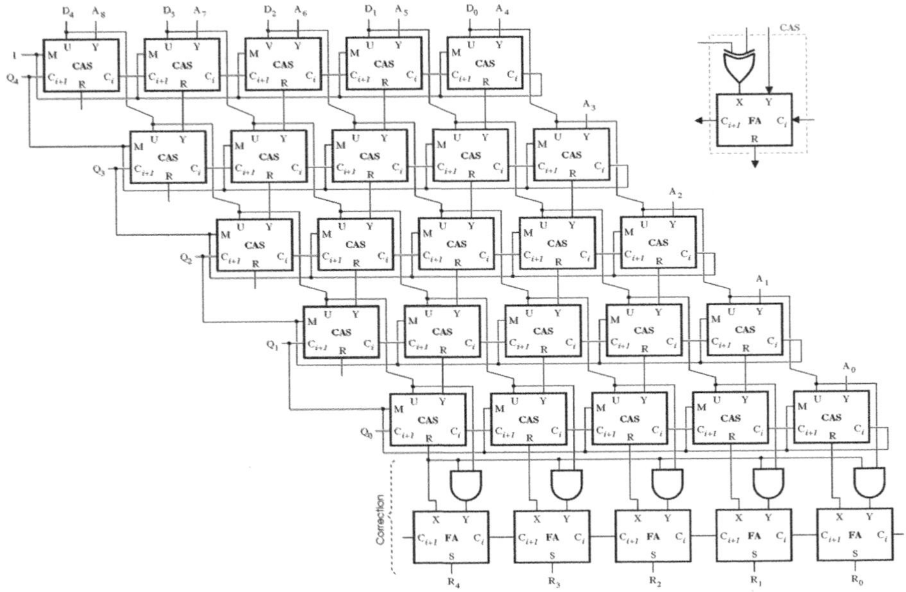

Figure 4.33. *Cellular division based on the algorithm for division without restoration of the dividend*

Initially, the partial remainder is supposed to take the value of the dividend. Assuming that $D > 0$ and $|R(i+1)| < D$, the partial remainders $R(i+1)$ $(i = 0, 1, 2, \cdots, n-1)$ can be positive or negative.

The operation to be carried out at each iteration can be a subtraction or an addition, depending on the value of the partial remainder whose expression is given by:

$$R(i+1) = \begin{cases} 2R(i) - D & \text{if} \quad 2R(i) > 0 \\ 2R(i) + D & \text{if} \quad 2R(i) < 0 \end{cases} \qquad [4.52]$$

The corresponding quotient bits are determined as follows:

$$Q_{n-(i+1)} = \begin{cases} 1 & \text{if} \quad 0 < 2R(i) < D \\ -1 & \text{if} \quad -D < 2R(i) < 0 \end{cases} \qquad [4.53]$$

The quotient is represented by a sequence of signed digits[2], whose value may be either 1 or -1 but not 0, and must then be converted to two's complement.

For division without restoration the partial remainder is negative when $2R(i) < D$, but, instead of being restored to its previous value (that is, $2R(i)$, in order to then be transformed according to the relationship $2[2R(i)] - D = 4R(i) - D$, as in the case of division with restoration) it is kept unchanged and is only corrected in the next iteration. It thus becomes $2[2R(i) - D] + D = 4R(i) - D$. The result of both types of division are, therefore, identical. However, a correction of the final remainder value is required in the case of division without restoration when the last partial remainder is negative. This corresponds to the addition of the divisor to the remainder, whose correct value can then be obtained as follows:

$$R = R_n + D = 2R_{n-1} \qquad [4.54]$$

The algorithm for division without restoration can be transformed into a version that can directly generate a quotient and a remainder of $n + 1$ bits, both represented in two's complement, from a dividend of $2n$ bits and a divisor of n bits. Its different steps are defined by algorithm 4.2.

Algorithm 4.2. *Division without restoration*

[1] When the dividend and divisor are positive, the first step consists of subtracting the divisor from the dividend with both most significant bits aligned. If the obtained partial remainder is thus positive, one bit of the quotient is set to 1 and the next operation is a subtraction; if, on the other hand, the obtained partial remainder is negative, one bit of the quotient is set to 0 and the next operation is an addition.

[2] For each of the following steps, the execution of the previously-determined operation is carried out after the most significant bit of the partial remainder is shifted to the left with respect to that of the divisor. If the obtained partial remainder is positive, the corresponding bit in the quotient is set to 1 and the next operation is a subtraction; otherwise, the corresponding bit of the quotient is set to 1 and the next operation is an addition.

The previously mentioned version of the algorithm for division without restoration is most suitable for cellular divider implementation. The logic circuit for a 8 × 4

2 The following steps can be used to convert a binary sequence of n signed digits (1 or -1) to two's complement:

– transform each -1 to 0 to form a code $P = p_{n-1}p_{n-2}\cdots p_2p_1p_0$, where the value of each bit, p_k $(k = 0, 1, 2, \cdots, n-1)$, is either 0 or 1;

– obtain the two's complement representation by shifting the bits one position to the left and by inserting 1 in the position of the least significant bit, that is $Q = p_{n-2}\cdots p_2p_1p_0 1$.

cellular divider is represented in Figure 4.33. The regular section, comprising CAS (controlled adder/subtract) cells, is used to generate the partial remainders and the stage composed of FAs and AND logic gates is used for the correction of the final remainder.

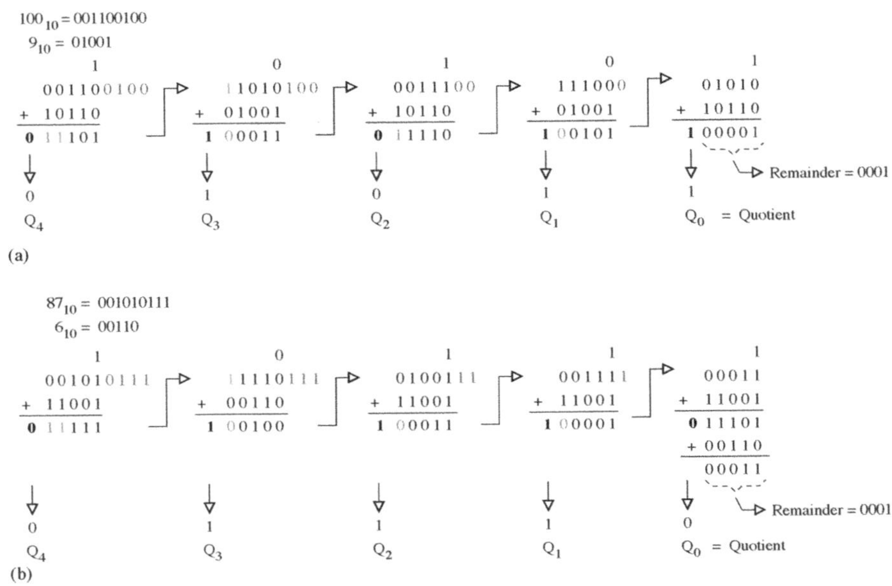

Figure 4.34. *Examples of division without restoration: a)* $100 \div 9$ *(*$Q = 11, R = 1$*); b)* $87 \div 6$ *(*$Q = 14$*,* $R = 3$*). For a color version of the figure, see www.iste.co.uk/ndjountche/electronics2.zip*

Each line of CAS cells, each of which comprises an FA and an XOR logic gate, carries out either a subtraction or an addition based on the state, 1 or 0, applied to one of the two inputs of the XOR logic gates. Subtraction is performed in two's complement representation by adding each partial remainder to the one's complement of the divisor and by setting the input carry of the right-most CAS cell to 1. Each bit of the quotient corresponds to the carry-out generated by the left-most CAS cell, and which represents the complement of the sign-bit of the corresponding partial remainder.

When the last bit of the quotient is 0, the last partial remainder is negative and a correction based on a conditional addition is necessary to obtain the correct value of the final remainder.

To divide a 2-bit number, A, by an n-bit number, D, and to obtain a quotient Q and a remainder R of n-bits, $(n+1)^2$ CAS cells and $n+1$ FAs and AND logic gates are required. This is due to the fact that subtractions are carried out on $n+1$ bits in a cellular divider based on the algorithm for division without restoration of the dividend.

Two examples for the execution of division operations are illustrated in Figure 4.34. The division is carried out as a succession of left-shift operations by one bit followed by either a subtraction or an addition.

4.7. Exercises

EXERCISE 4.1.– Design a half subtractor and a full subtractor.

EXERCISE 4.2.– Two's complement generator.

A two's complement generator, that can provide the two's complement representation of a four-bit binary number after having changed its sign, is to be designed. This circuit has the inputs B_3, B_2, B_1 and B_0, and the outputs T_3, T_2, T_1 and T_0. Its operation is described by the truth table given in Table 4.11.

Inputs				Outputs			
B_3	B_2	B_1	B_0	T_3	T_2	T_1	T_0
0	0	0	0	0	0	0	0
0	0	0	1	1	1	1	1
0	0	1	0	1	1	1	0
0	0	1	1	1	1	0	1
0	1	0	0	1	1	0	0
0	1	0	1	1	0	1	1
0	1	1	0	1	0	1	0
0	1	1	1	1	0	0	1
1	0	0	0	1	0	0	0
1	0	0	1	0	1	1	1
1	0	1	0	0	1	1	0
1	0	1	1	0	1	0	1
1	1	0	0	0	1	0	0
1	1	0	1	0	0	1	1
1	1	1	0	0	0	1	0
1	1	1	1	0	0	0	1

Table 4.11. *Truth table*

a) simplify the functions T_0, T_1, T_2 and T_3 using the Karnaugh map;

b) for each of the functions T_1, T_2 and T_3, determine X_{i-1} so that the next logic expression is verified:

$$T_i = X_{i-1} \oplus B_i, \qquad i = 1, 2, 3.$$

c) implement this circuit using only 2-input OR and XOR gates.

EXERCISE 4.3.– Let P and Q be two 2-bit binary numbers; X, Y and Z are three 1-bit binary numbers. Propose a logic circuit using a comparator and multiplexer to implement the following loop:

If $P = Q$
  then $Z = Y$;
  else $Z = X$;
End

EXERCISE 4.4.– Express the logic functions E and F as functions of X, Y and C_i.

Show that the logic circuit depicted in Figure 4.35 is an FA.

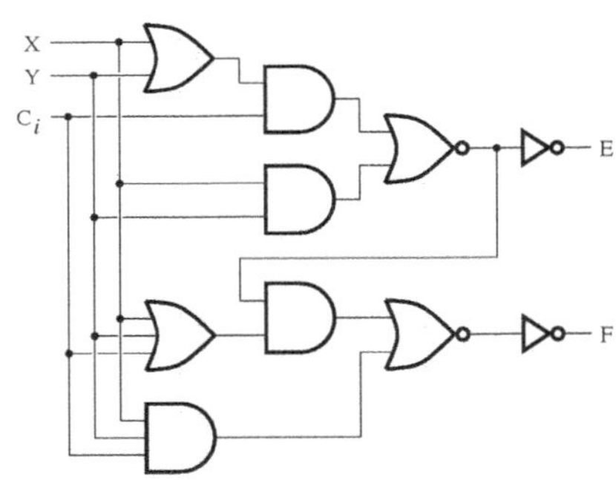

Figure 4.35. *Logic circuit*

EXERCISE 4.5.– Implement an FA using $4:1$ multiplexers.

EXERCISE 4.6.– Propose a circuit to add five 1-bit numbers, D, E, F, G and H, using an HA and two FAs.

EXERCISE 4.7.– Logic unit.

Consider a logic unit implemented from a $4 : 1$ multiplexer, as shown in Figure 4.36. The select inputs are represented by S_0, S_1, S_2 and S_3, and A and B designate the input variables.

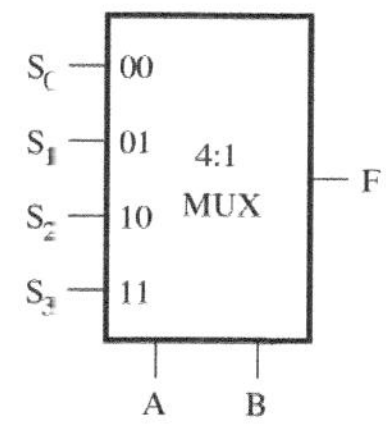

Figure 4.36. *Multiplexer logic circuit*

How many logic functions of a maximum of two variables can be implemented with this logic unit?

EXERCISE 4.8.– Arithmetic unit.

Using an FA (input carry bit C_i), AND and OR logic gates and inverter, and two select inputs, S_0 and S_1, design a logic circuit that can perform the operations given in Table 4.12.

S_1S_0	$C_i = 0$	$C_i = 1$
00	$F = A + B$ (addition)	$F = A + B + 1$
01	$F = A$ (transfer)	$F = A + 1$
10	$F = \overline{B}$ (inversion)	$F = \overline{B} + 1$ (negation)
11	$F = A + \overline{B}$	$F = A + \overline{B} + 1$ (subtraction)

Table 4.12. *Arithmetic operations*

EXERCISE 4.9.– BCD adder.

Consider the BCD adder shown in Figure 4.37, some of which can be connected in cascade to carry out an n-bit operation. BCD digits vary from 0000 to 1001 (or 0 to 9). Two BCD digits are required for the representation of numbers larger than 1001 (or 9). The output, F, of the correction circuit permits the detection of a sum, $C_4S_3S_2S_1S_0$, larger than 01001 (or 9) and smaller than 10011 (or 19), and the second adder adds 0110 (or 6) to this sum depending on F's state.

Propose an implementation of the correction circuit using logic gates.

Verify the operation of the BCD adder using the following numbers:

– $X = 9$ and $Y = 7$;

– $X = 34$ and $Y = 19$.

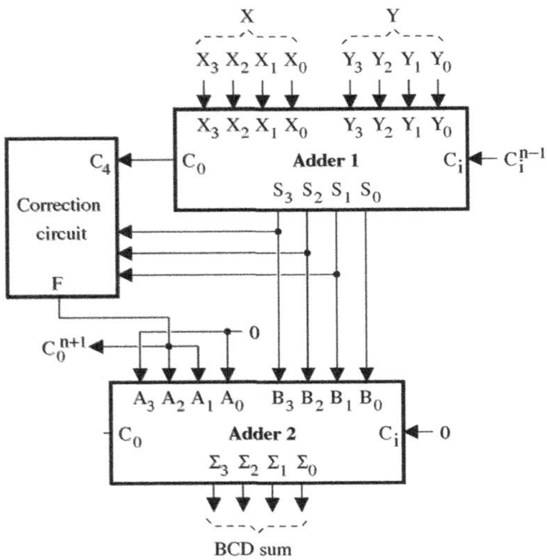

Figure 4.37. *BCD adder*

EXERCISE 4.10.– Analysis of a logic circuit.

The logic circuit shown in Figure 4.38 is implemented using 2 : 1 multiplexers and logic gates (inverter, NOR and XOR). Two unsigned 2-bit numbers, $X = X_1X_0$ and $Y = Y_1Y_0$, are applied to the inputs, and A, B and C designate the outputs.

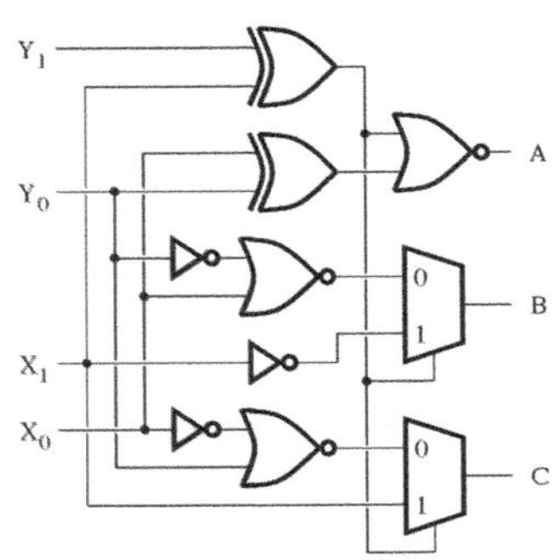

Figure 4.38. *Logic circuit*

Determine the logic equation for each of the outputs.

Deduce the function implemented by this circuit.

EXERCISE 4.11.– *Cascadable* comparator cell.

Implement a 1-bit *cascadable* comparator using only NAND gates.

EXERCISE 4.12.– *Cascadable* comparator.

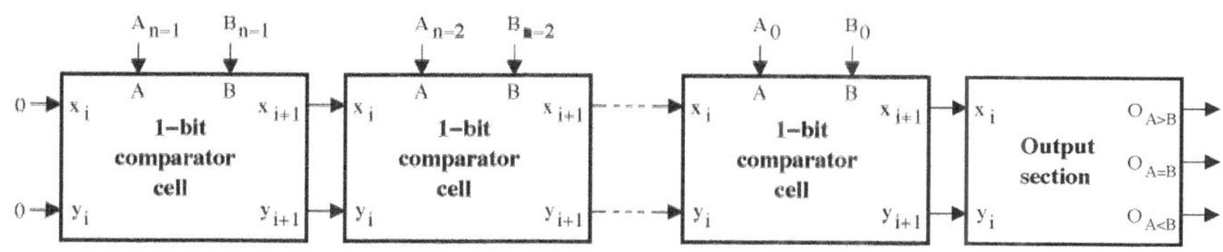

Figure 4.39. *n-bit cascadable comparator*

x_i	y_i	$O_{A>B}$	$O_{A=B}$	$O_{A<B}$
0	0	0	1	0
0	1	1	0	0
1	0	0	0	1
1	1	0	0	0

Table 4.13. *Function table of the comparator*

Another approach used to implement a *cascadable* comparator for two numbers (A_{n-1} and B_{n-1} are the most significant bits):

$$A = A_{n-1}A_{n-2}\cdots A_0 \quad \text{and} \quad B = B_{n-1}B_{n-2}\cdots B_0$$

consists of encoding the three possible outputs using two variables. In this case, as shown in Figure 4.39, each comparator cell only supplies two output signals and one output section is used to generate the comparison result in terms of equality, larger, or smaller. Table 4.13 gives the function table where the combination of variables $x_i y_i$ take the values 00, 01 and 10, respectively, for the outputs $O_{A=B}$, $O_{A>B}$ and $O_{A<B}$. The code $x_i y_i = 11$ is not associated with any comparison result and is, thus, considered as a don't-care code for the outputs x_{i+1} and y_{i+1}.

Draw up the truth table of the comparator, considering x_i, y_i, A and B, as inputs and x_{i+1}, y_{i+1}, $O_{A=B}$, $O_{A>B}$ and $O_{A<B}$, as outputs.

Determine the output logic equations.

Represent the logic circuits for a comparator cell and the output section.

EXERCISE 4.13.– Consider the logic circuit shown in Figure 4.40, that is implemented using two FAs and logic gates and where $A = A_1A_0$ and $B = B_1B_0$ are two numbers.

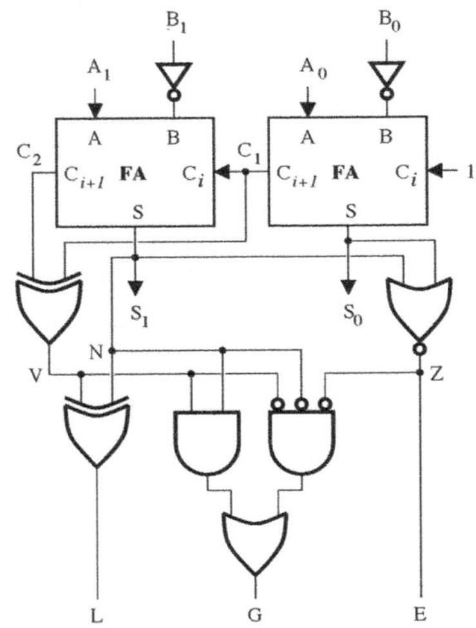

Figure 4.40. *Logic circuit*

Determine the logic expressions for G, E and L as functions of N, V and Z.

Construct a truth table where A_1, A_0, B_1 and B_0 are considered as inputs and N, V and Z, as outputs.

Establish a logic relationship between E, G and L.

What is the role of this circuit?

Modify the proposed logic circuit to implement a comparator for 4-bit signed numbers.

EXERCISE 4.14.– Arithmetic and logic unit.

A 4-bit ALU can be implemented by cascading four 1-bit stages, as shown in Figure 4.41. It can be used to carry out addition and subtraction operations and NOT, AND, OR and XOR logic functions.

Determine the logic equations for the outputs F and C_{i+1}.

Complete the function table given in Table 4.14.

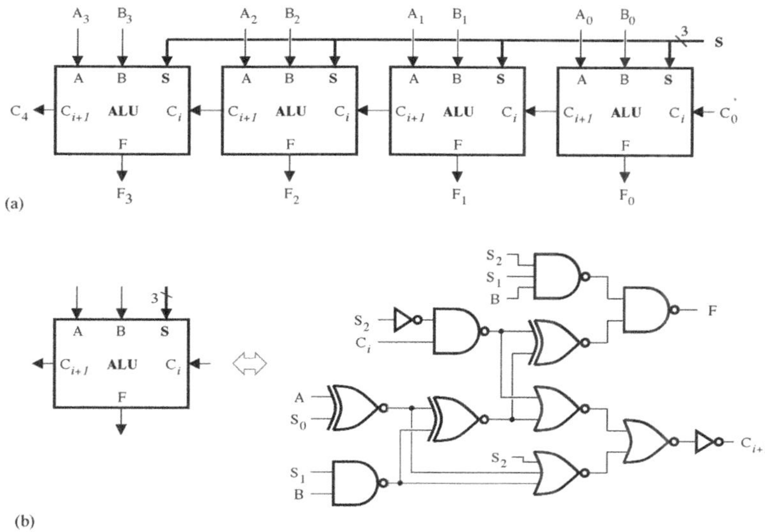

Figure 4.41. *a) Four-bit arithmetic and logic unit; b) 1-bit arithmetic and logic unit*

$S_2S_1S_0$	F	C_{i+1}	Operation
0 0 0			Transfer ($C_i = 0$) or Incrementing ($C_i = 1$) of A
0 0 1			1's complement ($C_i = 0$) or 2's complement ($C_i = 1$) of A
0 1 0	$A \oplus B \oplus C_i$		A plus B if $C_i = 0$ or A plus B plus 1 if $C_i = 1$
0 1 1		$C_i \cdot (\overline{A \oplus B}) + \overline{A} \cdot B$	$\overline{A}$ plus B if $C_i = 0$ or B minus A if $C_i = 1$
1 0 0			Transfer of A
1 0 1			Complement of A
1 1 0			A OR B
1 1 1			Complement of A OR B

Table 4.14. *Operations implemented by the arithmetic and logic unit*

EXERCISE 4.15.– Programmable logic circuit.

The programmable logic circuit shown in Figure 4.42 can be used to carry out different arithmetic operations. It comprises an adder, two multiplexers, AND, OR, XOR logic gates and inverters.

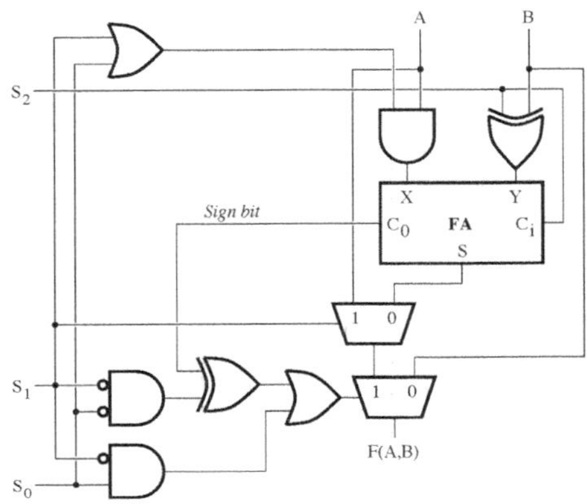

Figure 4.42. *Programmable logic circuit*

Complete the function table given in Table 4.15, by determining the arithmetic operations carried out depending on the select code $S_2S_1S_0$.

S_2	S_1	S_0	Operation $F(A,B)$
0	0	1	
1	0	0	
1	0	1	
1	1	0	
1	1	1	

Table 4.15. *Function table*

EXERCISE 4.16.– Bidirectional counter.

A bidirectional counter as shown in Figure 4.43 is made up of D flip-flops, 2 : 1 multiplexers and a half adder/subtractor (HAS). Figure 4.44 depicts the HAS circuit and an application for the counter.

During the normal operation, the asynchronous *Reset* input is set to 1 and does not affect the outputs; determine the logic equation of the next state, $Q^+ = D$, for each flip-flop and the logic equation for the RCO.

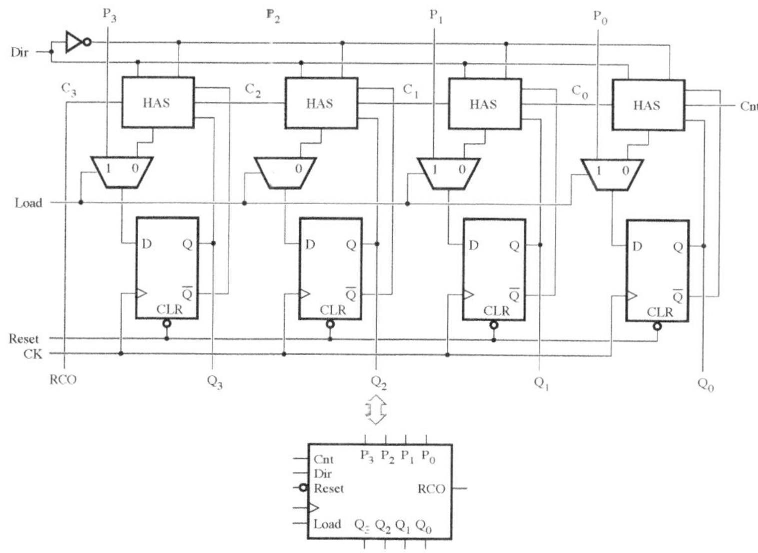

Figure 4.43. *Logic circuit of the bidirectional counter*

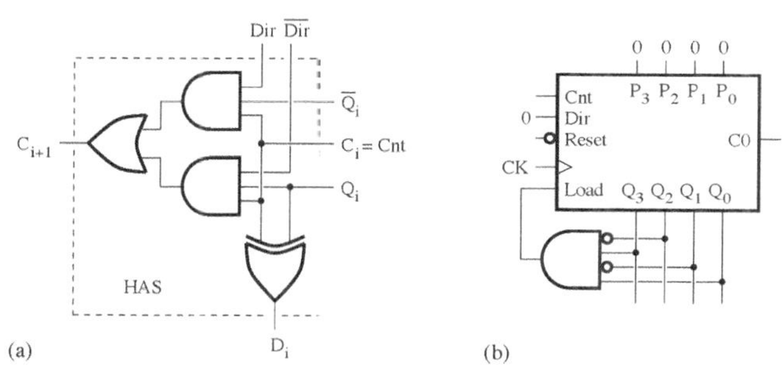

Figure 4.44. *a) Logic circuit of the half adder/subtractor (HAS); b) application*

Complete the function table given in Table 4.16.

Specify the type of counter implemented by the circuit of Figure 4.44(b).

Inputs					Operations
CK	$\overline{\text{Reset}}$	Load	Cnt	Dir	
x	0	x	x	x	
x	1	0	0	x	
↑	1	0	1	0	
↑	1	0	1	1	
↑	1	1	x	x	

Table 4.16. *Function table*

4.8. Solutions

SOLUTION 4.1.– Subtractor.

A full subtractor can be implemented using half subtractors and logic gates.

– Half subtractor:

Figure 4.45 illustrates the execution of a simple subtraction operation. Referring to the truth table given in Table 4.17, we can deduce the logic equations of the half subtractor as follows:

$$D = \overline{X} \cdot Y + X \cdot \overline{Y} = X \oplus Y \qquad [4.55]$$

$$B = \overline{X} \cdot Y \qquad [4.56]$$

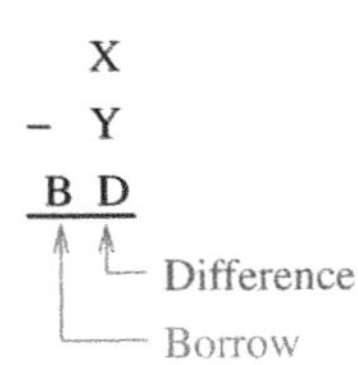

Figure 4.45. *Example of a subtraction operation. For a color version of the figure, see www.iste.co.uk/ndjountche/electronics2.zip*

Figure 4.46 depicts the logic circuit and symbol for a half subtractor.

X	Y	D	B
0	0	0	0
0	1	1	1
1	0	1	0
1	1	0	0

Table 4.17. *Truth table of a half subtractor*

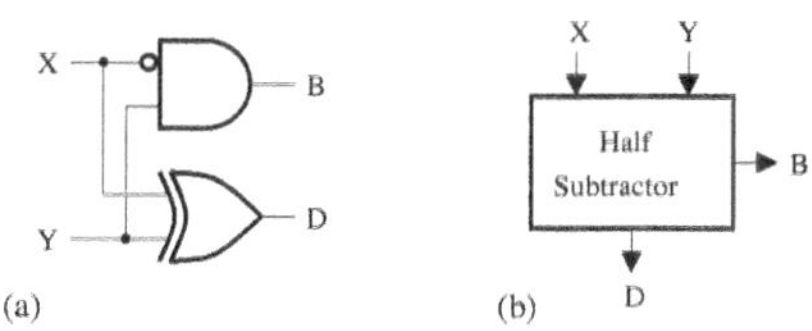

Figure 4.46. *Half subtractor (HS): a) logic circuit; b) symbol*

– Full subtractor:

A subtraction with a borrow-in and a borrow-out is represented in Figure 4.47. The truth table given in Table 4.18 can be used to derive the logic equations of the full subtractor, as follows:

$$\begin{aligned} D &= \overline{X}\cdot\overline{Y}\cdot B_i + \overline{X}\cdot Y\cdot\overline{B_i} + X\cdot\overline{Y}\cdot\overline{B_i} + X\cdot Y\cdot B_i \\ &= (\overline{X}\cdot\overline{Y} + X\cdot Y)\cdot B_i + (\overline{X}\cdot Y + X\cdot\overline{Y})\cdot\overline{B_i} \\ &= (X\oplus Y)\oplus B_i \end{aligned} \quad [4.57]$$

$$\begin{aligned} B_0 &= \overline{X}\cdot\overline{Y}\cdot B_i + \overline{X}\cdot Y\cdot\overline{B_i} + \overline{X}\cdot Y\cdot\overline{B_i} + X\cdot Y\cdot B_i \\ &= \overline{X}\cdot Y + (\overline{X}\cdot\overline{Y} + X\cdot Y)\cdot B_i \\ &= \overline{X}\cdot Y + (\overline{X\oplus Y})\cdot B_i \end{aligned} \quad [4.58]$$

Figure 4.48 depicts the logic circuit and symbol of a full subtractor.

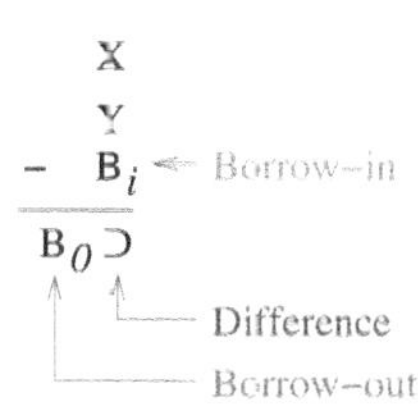

Figure 4.47. *Example of a subtraction operation. For a color version of the figure, see www.iste.co.uk/ndjountche/electronics2.zip*

X	Y	B_i	D	B_0
0	0	0	0	0
0	0	1	1	1
0	1	0	1	1
0	1	1	0	1
1	0	0	1	0
1	0	1	0	0
1	1	0	0	0
1	1	1	1	1

Table 4.18. *Truth table of a full subtractor*

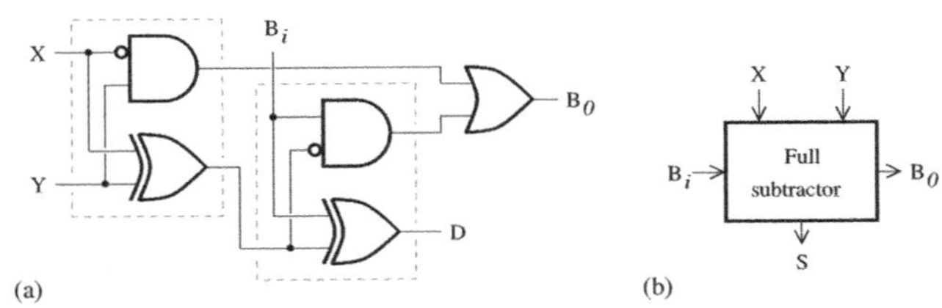

Figure 4.48. *Full subtractor (FS): a) logic circuit; b) symbol*

SOLUTION 4.2.– Two's complement generator.

Based on the truth table of the circuit that can generate two's complement, we construct the corresponding Karnaugh map for each output, as shown in Figure 4.49. The logic equations required for the circuit implementation can be written as follows:

– output T_0:

$$T_0 = B_0 \quad [4.59]$$

– output T_1:

$$\begin{aligned} T_1 &= B_0 \cdot \overline{B_1} + \overline{B_0} \cdot B_1 \\ &= B_0 \oplus B_1 \end{aligned} \quad [4.60]$$

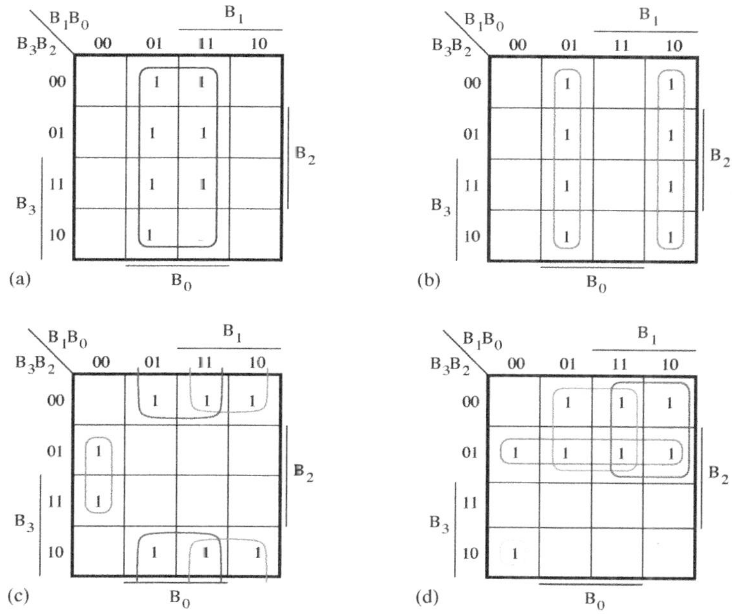

Figure 4.49. *Karnaugh maps: (a) T_0; (b) T_1; (c) T_2; (d) T_3. For a color version of the figure, see www.iste.co.uk/ndjountche/electronics2.zip*

– output T_2:

$$T_2 = B_0 \cdot \overline{B_2} + B_1 \cdot \overline{B_2} + \overline{B_0} \cdot \overline{B_1} \cdot B_2$$
$$= (B_0 + B_1) \cdot \overline{B_2} + \overline{(B_0 + B_1)} \cdot B_2$$
$$= (B_0 + B_1) \oplus B_2 \qquad [4.61]$$

– output T_3:

$$T_3 = B_0 \cdot \overline{B_3} + B_1 \cdot \overline{B_3} + B_2 \cdot \overline{B_3} + \overline{B_0} \cdot \overline{B_1} \cdot \overline{B_2} \cdot B_3$$
$$= (B_0 + B_1 + B_2) \cdot \overline{B_3} + \overline{(B_0 + B_1 + B_2)} \cdot B_3$$
$$= (B_0 + B_1 + B_2) \oplus B_3 \qquad [4.62]$$

The logic circuit of the two's complement generator is represented in Figure 4.50.

SOLUTION 4.3.– Implementation of an *If-then-else* loop.

An *If-then-else* loop can be implemented by using a multiplexer and logic gates (XNOR and AND).

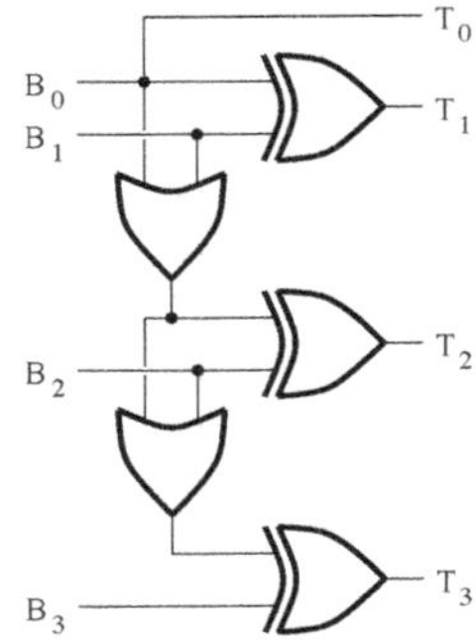

Figure 4.50. *Logic circuit*

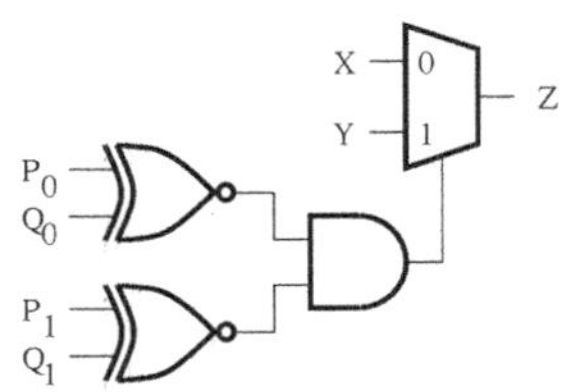

Figure 4.51. *Implementation of the If-then-else loop*

The corresponding logic circuit is represented in Figure 4.51, where the comparison section is composed of XNOR and AND logic gates.

SOLUTION 4.4.– Full adder.

Analyzing the proposed logic circuit we can obtain:

$$E = \overline{\overline{X \cdot Y + (X + Y) \cdot C_i}} = X \cdot Y + (X + Y) \cdot C_i \qquad (= C_0) \qquad [4.63]$$

$$\begin{aligned} F &= \overline{\overline{X \cdot Y \cdot C_i + (X + Y + C_i) \cdot \overline{E}}} = X \cdot Y \cdot C_i + (X + Y + C_i) \cdot \overline{E} \\ &= X \cdot Y \cdot C_i + (X + Y + C_i) \cdot [\overline{X \cdot Y + (X + Y) \cdot C_i}] \\ &= X \cdot Y \cdot C_i + (X + Y + C_i) \cdot (\overline{X} \cdot \overline{Y} + \overline{X} \cdot \overline{C_i} + \overline{Y} \cdot \overline{C_i}) \\ &= X \cdot Y \cdot C_i + X \cdot \overline{Y} \cdot \overline{C_i} + \overline{X} \cdot Y \cdot \overline{C_i} + \overline{X} \cdot \overline{Y} \cdot C_i \qquad (= S) \end{aligned} \qquad [4.64]$$

because $E = C_0$ and $F = S$, it is an FA.

SOLUTION 4.5.– Implementation of an FA using 4 : 1 multiplexers.

The logic equations of an FA are given by:

$$S = (\overline{A} \cdot \overline{B} + A \cdot B) \cdot C_i + (A \cdot \overline{B} + \overline{A} \cdot B) \cdot \overline{C_i}$$
$$C_{i+1} = A \cdot B + (\overline{A} \cdot B + A \cdot \overline{B}) \cdot C_i \quad [4.65]$$

They can be implemented using 4 : 1 multiplexers if they are decomposed as follows:

$$S = (\overline{A} \cdot \overline{B}) \cdot C_i + (\overline{A} \cdot B) \cdot \overline{C_i} + (A \cdot \overline{B}) \cdot \overline{C_i} + (A \cdot B) \cdot \overline{C_i}$$
$$C_0 = (\overline{A} \cdot \overline{B}) \cdot 0 + (\overline{A} \cdot B) \cdot C_i + (A \cdot \overline{B}) \cdot C_i + (A \cdot B) \cdot 1 \quad [4.66]$$

Figure 4.52 depicts the logic circuit of an FA based on 4 : 1 multiplexers.

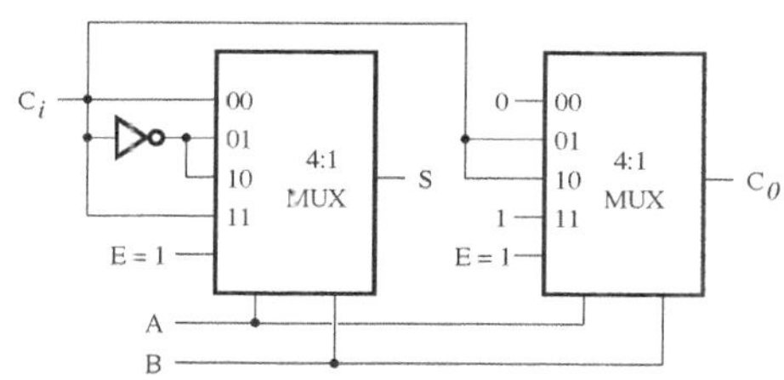

Figure 4.52. *Logic circuit of a full adder*

SOLUTION 4.6.– Adder for five 1-bit number.

To add five 1-bit numbers, as shown in Figure 4.53(a), we can use the logic circuit of Figure 4.53(b), which consists of two FAs and one HA.

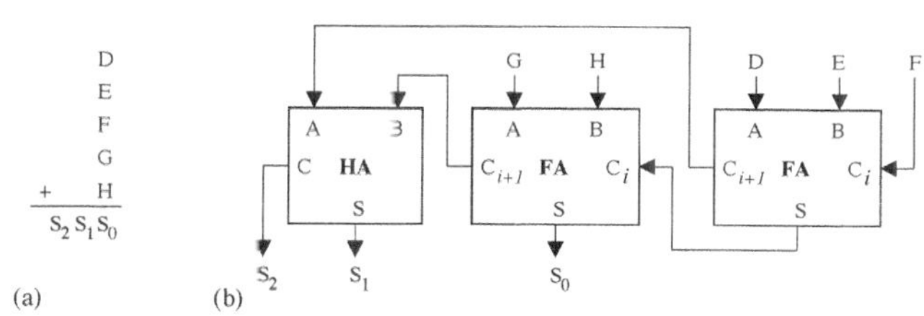

Figure 4.53. *a) Addition operation; b) logic circuit of the adder*

SOLUTION 4.7.– Logic unit.

The analysis of the logic unit circuit can yield the following equation:

$$F = S_0 \cdot \overline{A} \cdot \overline{B} + S_1 \cdot \overline{A} \cdot B + S_2 \cdot A \cdot \overline{B} + S_3 \cdot A \cdot B \qquad [4.67]$$

Table 4.19 shows the function table of the logic unit. By judiciously choosing the select signals S_0, S_1, S_2 and S_3, we can implement all two-variable logic functions. That is:

$$2^{2^2} = 16 \quad \text{functions}$$

S_3	S_2	S_1	S_0	$F(A,B)$
0	0	0	0	0
0	0	0	1	$\overline{A} \cdot \overline{B}$
0	0	1	0	$\overline{A} \cdot B$
0	0	1	1	$\overline{A}$
0	1	0	0	$A \cdot \overline{B}$
0	1	0	1	$\overline{B}$
0	1	1	0	$A \oplus B$
0	1	1	1	$\overline{A} + \overline{B}$
1	0	0	0	$A \cdot B$
1	0	0	1	$A \odot B$
1	0	1	0	B
1	0	1	1	$\overline{A} + B$
1	1	0	0	A
1	1	0	1	$A + \overline{B}$
1	1	1	0	$A + B$
1	1	1	1	1

Table 4.19. *Function table of the logic unit*

SOLUTION 4.8.– Arithmetic unit.

The arithmetic unit is implemented using an FA whose logic equations take the form:

$$\begin{aligned} S = F &= (\overline{X} \cdot \overline{Y} + X \cdot Y) \cdot C_i + (X \cdot \overline{Y} + \overline{X} \cdot Y) \cdot \overline{C_i} \\ C_0 &= X \cdot Y + (\overline{X} \cdot Y + X \cdot \overline{Y}) \cdot C_i \end{aligned} \qquad [4.68]$$

where:

$$X = A \cdot \overline{S_1} + A \cdot S_0$$

$$Y = B \cdot \overline{S_1} \cdot \overline{S_0} + \overline{B} \cdot S_1$$

The logic circuit of the arithmetic unit is depicted in Figure 4.54.

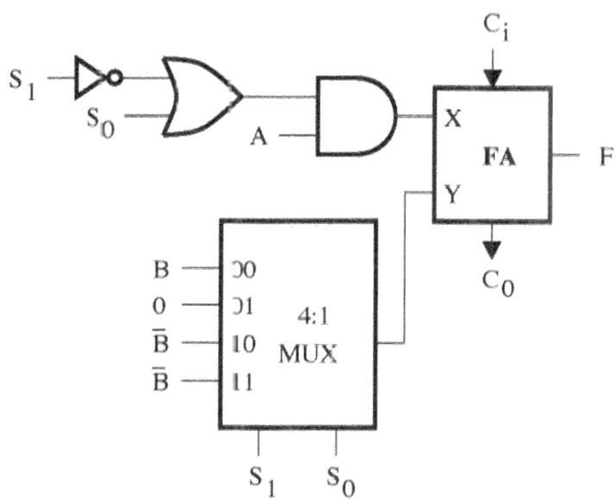

Figure 4.54. *Arithmetic circuit*

SOLUTION 4.9.– BCD adder.

The sum of two BCD numbers varies from $0_{10} = 00000_2$ $(0000+0000)$ to $19_{10} = 10011_2$ $(1001 + 1001 + 1)$, as shown in the function table of Table 4.20.

A BCD number only varies from $0_{10} = 0000_2$ to $9_{10} = 1001_2$, and a correction is necessary for numbers greater than 9.

The correction consists of adding $6_{10} = 0110_2$ to the erroneous sums from 10 to 19.

Sums greater than 15 may be identified by $C_4 = 1$, while for sums that vary from 10 to 15, we have $S_3 = 1$, either $S_2 = 1$ or $S_1 = 1$, or both.

The logic expression for the output carry, C_0^{n+1}, which permits the detection of sums greater than 9, is thus given by:

$$C_0^{n+1} = C_4 + S_3 \cdot (S_2 + S_1)$$

For a given sum $S_3S_2S_1S_0$, we have:

$$\Sigma_3\Sigma_2\Sigma_1\Sigma_0 = \begin{cases} S_3S_2S_1S_0 + 0000 & \text{if} \quad C_0^{n+1} = 0 \\ S_3S_2S_1S_0 + 0110 & \text{if} \quad C_0^{n+1} = 1 \end{cases}$$

Decimal	Binary $C_4S_3S_2S_1S_0$	BCD $C_0 + \Sigma_3\Sigma_2\Sigma_1\Sigma_0$
0	00000	$0 + 0000$
1	00001	$0 + 0001$
2	00010	$0 + 0010$
3	00011	$0 + 0011$
4	00100	$0 + 0100$
5	00101	$0 + 0101$
6	00110	$0 + 0110$
7	00111	$0 + 0111$
8	01000	$0 + 1000$
9	01001	$0 + 1001$
10	01010	$1 + 0000$
11	01011	$1 + 0001$
12	01100	$1 + 0010$
13	01101	$1 + 0011$
14	01110	$1 + 0100$
15	01111	$1 + 0101$
16	10000	$1 + 0110$
17	10001	$1 + 0111$
18	10010	$1 + 1000$
19	10011	$1 + 1001$

Table 4.20. *Function table of the BCD adder*

The correction circuit is represented in Figure 4.55.

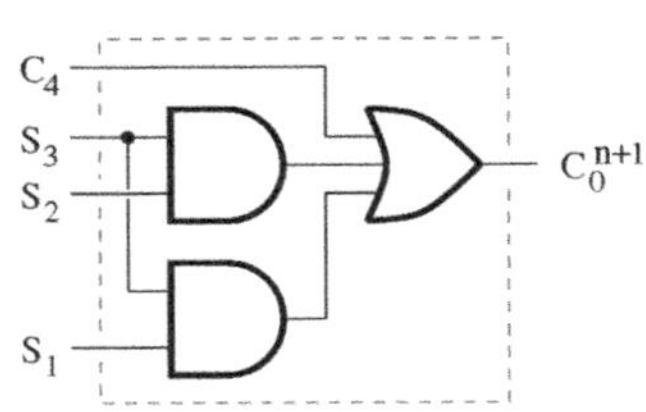

Figure 4.55. *Correction circuit*

SOLUTION 4.10.– Analysis of a logic circuit.

The analysis of the logic circuit (see Figure 4.56) can yield the following logic equations:

$$A = \overline{(X_0 \oplus Y_0) + (X_1 \oplus Y_1)} \qquad [4.69]$$

$$B = \overline{X_0} \cdot Y_0(\overline{X_1 \oplus Y_1}) + \overline{X_1}(X_1 \oplus Y_1) \qquad [4.70]$$

$$C = X_0 \cdot \overline{Y_0}(\overline{X_1 \oplus Y_1}) + X_1(X_1 \oplus Y_1) \qquad [4.71]$$

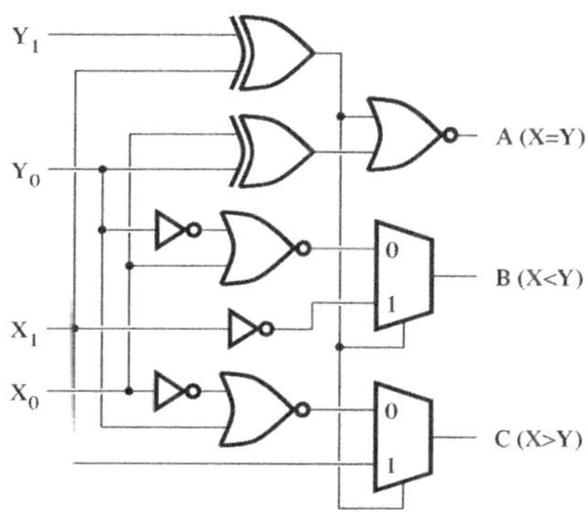

Figure 4.56. *Two-bit comparator*

This is a two-bit comparator. We can verify that $C = \overline{A + B}$ (or that two of the outputs are always the logic complement of the other output).

SOLUTION 4.11.– *Cascadable* comparator cell.

The outputs of the comparator are defined in the following manner:

$O_{A>B} = 1$ if $A > B$ or ($A = B$ and $I_{A>B} = 1$)

$O_{A=B} = 1$ if $A = B$ and $I_{A=B} = 1$

$O_{A<B} = 1$ if $A < B$ or ($A = B$ and $I_{A<B} = 1$)

Table 4.21 gives the truth table of a 1-bit *cascadable* comparator.

A	B	$O_{A>B}$	$O_{A=B}$	$O_{A<B}$
0	0	$I_{A>B}$	$I_{A=B}$	$I_{A<B}$
0	1	0	0	1
1	0	1	0	0
1	1	$I_{A>B}$	$I_{A=B}$	$I_{A<B}$

Table 4.21. *Truth table of a 1-bit cascadable comparator*

The logic expressions of the outputs obtained from the Karnaugh maps of Figure 4.57 can be written as follows:

$$\begin{aligned} O_{A>B} &= A \cdot I_{A>B} + \overline{B} \cdot I_{A>B} + A \cdot \overline{B} \\ &= (A + \overline{B}) \cdot I_{A>B} + A \cdot \overline{B} \\ O_{A=B} &= (A \cdot B + \overline{A} \cdot \overline{B}) \cdot I_{A=B} \\ &= (A + \overline{B}) \cdot (\overline{A} + B) \cdot I_{A=B} \\ O_{A<B} &= \overline{A} \cdot I_{A<B} + B \cdot I_{A<B} + \overline{A} \cdot B \\ &= (\overline{A} + B) \cdot I_{A<B} + \overline{A} \cdot B \end{aligned}$$

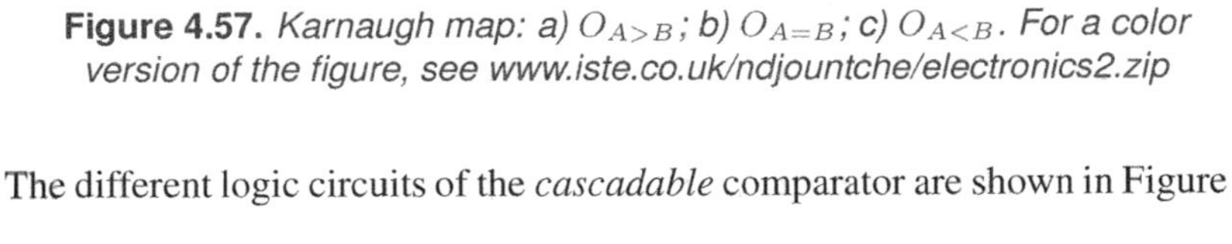

Figure 4.57. *Karnaugh map: a) $O_{A>B}$; b) $O_{A=B}$; c) $O_{A<B}$. For a color version of the figure, see www.iste.co.uk/ndjountche/electronics2.zip*

The different logic circuits of the *cascadable* comparator are shown in Figure 4.58.

SOLUTION 4.12.– *Cascadable* comparator.

An n-bit comparator is implemented by connecting 1-bit comparator cells and an output network in series.

When the bits are equal to the left of the cell i, $x_i y_i = 00$, and if A is equal to B, we then have $x_{i+1}y_{i+1} = 00$. However, when $x_i y_i = 00$, if $AB = 10$, it follows that $x_{i+1}y_{i+1} = 01$ and $A > B$, and if, on the other hand, $AB = 01$, we then have $x_{i+1}y_{i+1} = 10$ and $A < B$.

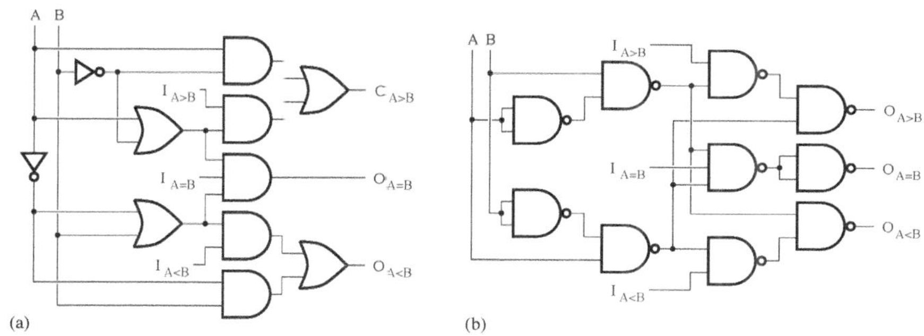

Figure 4.58. *Cascadable comparator*

When the superiority relationship, $x_i y_i = 01$, is established at the left of the cell i, for any value of A and B, we have $x_{i+1} y_{i+1} = 01$ and $A > B$.

Similarly, when the inferiority relationship $x_i y_i = 10$, is established at the left of the cell i, whatever be the values of A and B, we have $x_{i+1} y_{i+1} = 10$ and $A < B$.

Table 4.22 gives the truth table of the comparator. The logic equations for the signals x_{i+1} and y_{i+1} obtained from the Karnaugh maps of Figure 4.59 are given by:

$$x_{i+1} = x_i + \overline{A} \cdot B \cdot \overline{y_i} \qquad [4.72]$$

$$y_{i+1} = y_i + A \cdot \overline{B} \cdot \overline{x_i} \qquad [4.73]$$

For the comparator outputs, the following logic equations can be derived from the Karnaugh maps shown in Figure 4.60:

$$O_{A>B} = \overline{x_{i+1}} \cdot y_{i+1} \qquad [4.74]$$

$$O_{A=B} = \overline{x_{i+1}} \cdot \overline{y_{i+1}} \qquad [4.75]$$

$$O_{A<B} = x_{i+1} \cdot \overline{y_{i+1}} \qquad [4.76]$$

Figure 4.61 depicts the logic circuit obtained from the logic equations of the comparator cell and output section.

When both numbers A and B are coded with n bits, a comparison operation begins with the most signification bits: A_{n-1} and B_{n-1}.

x_i	y_i	A	B	x_{i+1}	y_{i+1}	$O_{A>B}$	$O_{A=B}$	$O_{A<B}$
0	0	0	0	0	0	0	1	0
0	0	0	1	1	0	0	1	0
0	0	1	0	0	1	0	1	0
0	0	1	1	0	0	0	1	0
0	1	0	0	0	1	1	0	0
0	1	0	1	0	1	1	0	0
0	1	1	0	0	1	1	0	0
0	1	1	1	0	1	1	0	0
1	0	0	0	1	0	0	0	1
1	0	0	1	1	0	0	0	1
1	0	1	0	1	0	0	0	1
1	0	1	1	1	0	0	0	1
1	1	0	0	x	x	0	0	0
1	1	0	1	x	x	0	0	0
1	1	1	0	x	x	0	0	0
1	1	1	1	x	x	0	0	0

Table 4.22. *Truth table for the comparator*

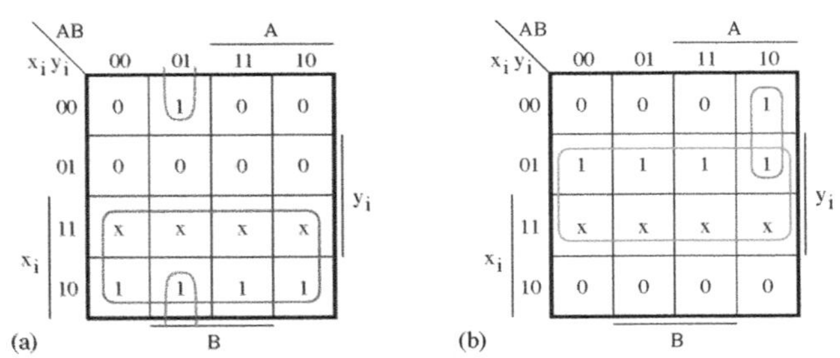

Figure 4.59. *Karnaugh maps: a)* $x_{i+1} = x_i + \overline{A} \cdot B \cdot \overline{y_i}$*; b)* $y_{i+1} = y_i + A \cdot \overline{B} \cdot \overline{x_i}$. *For a color version of the figure, see www.iste.co.uk/ndjountche/electronics2.zip*

SOLUTION 4.13.– Analysis of a logic circuit.

By analyzing the proposed logic circuit, we can obtain the following logic equations:

$$G = N \cdot V + \overline{N} \cdot \overline{V} \cdot \overline{Z} \quad [4.77]$$

$$E = Z \quad [4.78]$$

and:

$$L = N \oplus V \qquad [4.79]$$

Table 4.23 represents the truth table of this circuit, where $A = A_1 A_0$ and $B = B_1 B_0$.

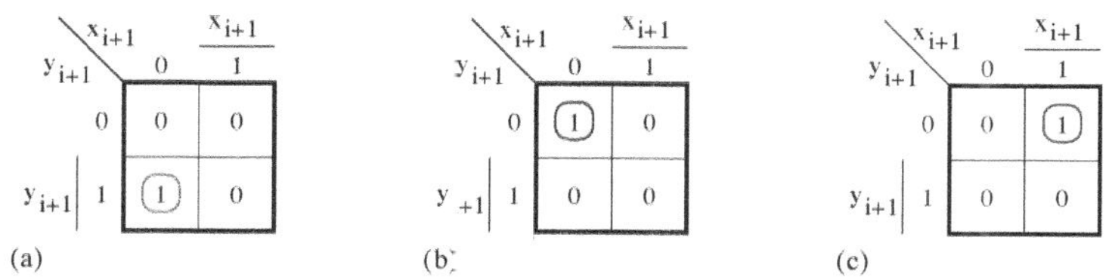

Figure 4.60. *Karnaugh maps: a)* $O_{A>B} = \overline{x_{i+1}} \cdot y_{i+1}$*; b)* $O_{A=B} = \overline{x_{i+1}} \cdot \overline{y_{i+1}}$*, (c)* $O_{A<B} = x_{i+1} \cdot \overline{y_{i+1}}$*. For a color version of the figure, see www.iste.co.uk/ndjountche/electronics2.zip*

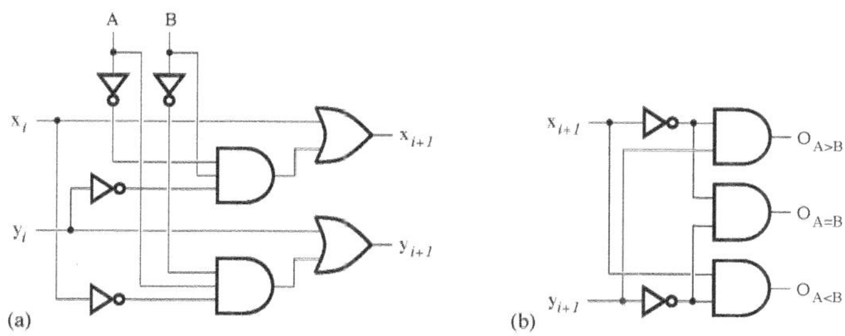

Figure 4.61. *Logic circuits: a) comparator cell; b) output section*

Noting that $G = O_{A>B}$, $E = O_{A=B}$ and $L = O_{A<B}$, we have:

$$G = \overline{E + L} \qquad [4.80]$$

Figure 4.62 shows the logic circuit of a comparator for 4-bit signed numbers.

SOLUTION 4.14.– Arithmetic and logic unit.

The analysis of a stage in the ALU yields the following logic equations:

$$F = (\overline{S_2} \cdot C_i) \oplus (A \oplus S_0) \oplus (S_1 \cdot B) + S_2 \cdot S_1 \cdot B \qquad [4.81]$$

and:

$$C_{i+1} = (\overline{S_2} \cdot C_i) \cdot [(A \oplus S_0) \oplus (S_1 \cdot B)] + \overline{S_2} \cdot (S_1 \cdot B) \cdot (A \oplus S_0) \qquad [4.82]$$

A_1	A_0	B_1	B_0	$C_2S_1S_0$	Z	N	V	G	E	L	
0	0	0	0	1 0 0	1	0	0	0	1	0	$A = B$
0	0	0	1	0 1 1	0	1	0	0	0	1	$A < B$
0	0	1	0	0 1 0	0	1	1	1	0	0	$A > B$
0	0	1	1	0 0 1	0	0	0	1	0	0	$A > B$
0	1	0	0	1 0 1	0	0	0	1	0	0	$A > B$
0	1	0	1	1 0 0	1	0	0	0	1	0	$A = B$
0	1	1	0	0 1 1	0	1	1	1	0	0	$A > B$
0	1	1	1	0 1 0	0	1	1	1	0	0	$A > B$
1	0	0	0	1 1 0	0	1	0	0	0	1	$A < B$
1	0	0	1	1 0 1	0	0	1	0	0	1	$A < B$
1	0	1	0	1 0 0	1	0	0	0	1	0	$A = B$
1	0	1	1	0 1 1	0	1	0	0	0	1	$A < B$
1	1	0	0	1 1 1	0	1	0	0	0	1	$A < B$
1	1	0	1	1 1 0	0	1	0	0	0	1	$A < B$
1	1	1	0	1 0 1	0	0	0	1	0	0	$A > B$
1	1	1	1	1 0 0	1	0	0	0	1	0	$A = B$

Table 4.23. *Truth table of the comparator*

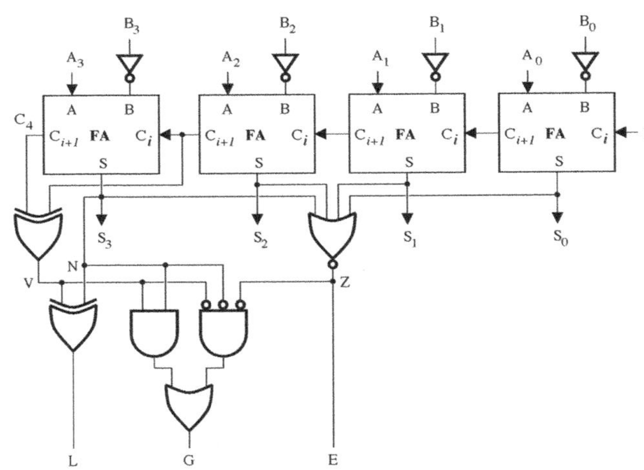

Figure 4.62. *Logic circuit of the 4-bit comparator*

The expressions for F and C_{i+1} have common terms. This helps reduce the number of logic gates in the resulting implementation.

Table 4.24 gives the function table of the ALU.

$S_2S_1S_0$	F	C_{i+1}	Operation
0 0 0	$A \oplus C_i$	$A \cdot C_i$	Transfer ($C_i = 0$) or Incrementing ($C_i = 1$) of A
0 0 1	$\overline{A} \oplus C_i$	$\overline{A} \cdot C_i$	1's complement ($C_i = 0$) or 2's complement ($C_i = 1$) of A
0 1 0	$A \oplus B \oplus C_i$	$C_i \cdot (A \oplus B) + A \cdot B$	A plus B if $C_i = 0$ or A plus B plus 1 if $C_i = 1$
0 1 1	$\overline{A} \oplus B \oplus C_i$	$C_i \cdot (\overline{A \oplus B}) + \overline{A} \cdot B$	$\overline{A}$ plus B if $C_i = 0$ or B minus A if $C_i = 1$
1 0 0	A	0	Transfer of A
1 0 1	$\overline{A}$	0	Complement of A
1 1 0	$A + B$	0	A OR B
1 1 1	$\overline{A} + B$	0	Complement of A OR B

Table 4.24. *Function table of the arithmetic and logic unit*

SOLUTION 4.15.– Programmable logic circuit.

The analysis of the programmable logic circuit can help to extract the different equivalent circuits for carrying out arithmetic functions (addition/subtraction, absolute value, minimum and maximum) as shown in Figure 4.63.

The function table of this logic circuit is given in Table 4.25.

S_2	S_1	S_0	Operation	$F(A, B)$
0	0	1	Addition	$A + B$
1	0	0	Absolute value	$\|B\|$
1	0	1	Subtraction	$A - B$
1	1	0	Minimum	$\min(A, B)$
1	1	1	Maximum	$\max(A, B)$

Table 4.25. *Function table*

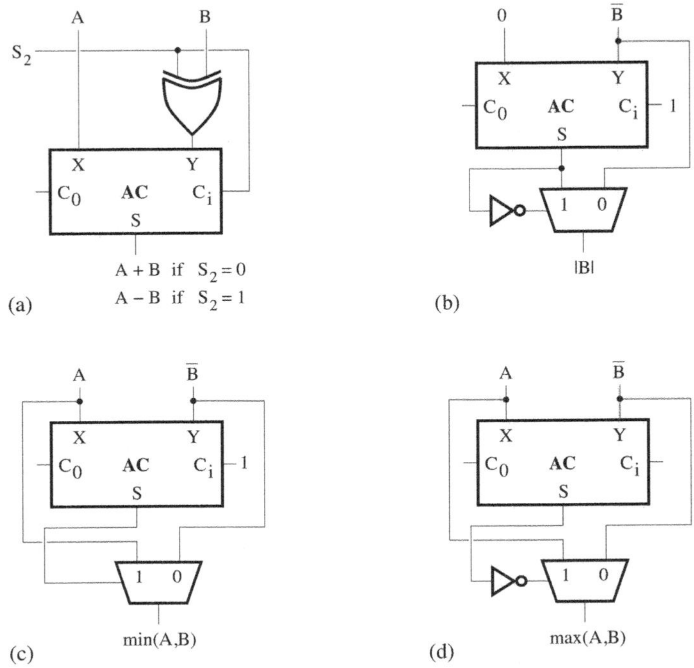

Figure 4.63. *Programmable logic circuit: a) adder/subtractor; b) absolute value; c) minimum; d) maximum*

SOLUTION 4.16.– Bidirectional counter.

The analysis of the counter can yield the following logic equations:

$$Q_0^+ = (Q_0 \oplus Dir) \cdot \overline{\text{Load}} + P_0 \cdot \text{Load} \quad [4.83]$$

$$Q_1^+ = (Q_1 \oplus C_0) \cdot \overline{\text{Load}} + P_1 \cdot \text{Load} \quad [4.84]$$

$$Q_2^+ = (Q_2 \oplus C_1) \cdot \overline{\text{Load}} + P_2 \cdot \text{Load} \quad [4.85]$$

$$Q_3^+ = (Q_3 \oplus C_2) \cdot \overline{\text{Load}} + P_3 \cdot \text{Load} \quad [4.86]$$

$$RCO = \overline{\text{Dir}} \cdot Q_3 \cdot C_2 + \text{Dir} \cdot \overline{Q_1} \cdot C_2 \quad [4.87]$$

where:

$$C_0 = \overline{\text{Dir}} \cdot Q_0 \cdot \text{Cnt} + \text{Dir} \cdot \overline{Q_0} \cdot \text{Cnt} \quad [4.88]$$

$$C_1 = \overline{\text{Dir}} \cdot Q_1 \cdot C_0 + \text{Dir} \cdot \overline{Q_1} \cdot C_0 \quad [4.89]$$

$$C_2 = \overline{\text{Dir}} \cdot Q_2 \cdot C_1 + \text{Dir} \cdot \overline{Q_2} \cdot C_1 \quad [4.90]$$

Table 4.26 presents the function table of the counter. Counting is enabled by the *Cnt* signal depending on the direction specified by the *Dir* signal. Setting the *Load* signal to 1 ensures the loading of data and the deactivation of counting, while resetting the *Load* signal to 0 enables the counter to operate normally.

It is a modulo 10 up counter.

Inputs					Operations
CK	$\overline{\text{Reset}}$	Load	Cnt	Dir	
x	0	x	x	x	Reset
x	1	0	0	x	Hold
↑	1	0	1	0	Count up
↑	1	0	1	1	Count down
↑	1	1	x	x	Load

Table 4.26. *Function table*

5

Digital Integrated Circuit Technology

5.1. Introduction

Several families of technology have been developed for manufacturing integrated circuits, the most popular being transistor–transistor logic (TTL) and complementary metal oxide semiconductor (CMOS). TTL technology uses bipolar transistors, whereas CMOS technology is based on MOS field-effect transistors.

The technology chosen has an impact on the physical and electrical characteristics of the circuit (speed, integration density, power consumption, fan-in, fan-out, etc.). Logic gates, therefore, are not ideal components. In practice, their characteristics limit their performance.

5.2. Characteristics of the technologies

With the widespread diffusion of small-scale integration, medium-scale integration and large-scale integration chips, it is essential to understand the electric characteristics of most of the major families (TTL and CMOS) of integrated circuits.

5.2.1. *Supply voltage*

The nominal value of the supply voltage for TTL components is 5 V, and for CMOS components it is presently 3.3 V and continuing to decrease with the diminishing size of the transistor.

5.2.2. *Logic levels*

The voltage levels associated with each logic state vary depending on the technology. Figure 5.1 gives a representation of the logic levels of the TTL and CMOS components.

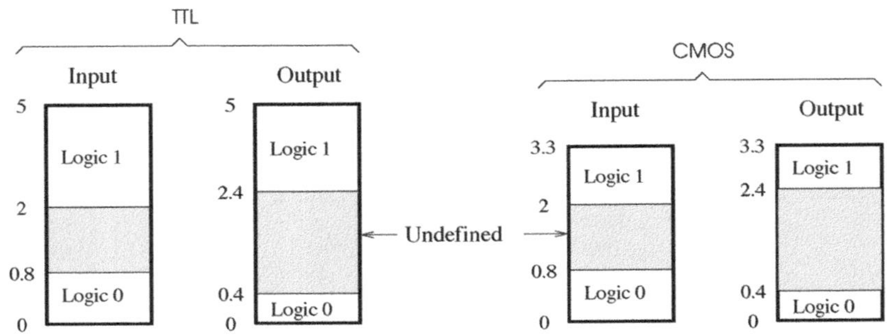

Figure 5.1. *Logic levels of TTL and CMOS components*

5.2.3. *Immunity to noise*

Undesirable signals (electromagnetic disturbances, supply voltage fluctuations, variations due to parasitic coupling or caused by the thermal agitation of charge carriers) arising in electric circuits are called "noise" and by modifying the propagation delay or the signal logic level, these may be an impediment to the proper operation of a logic circuit.

In order for its performance to remain unaffected by noise, a logic circuit must manifest a certain immunity to noise. Two values are used to indicate the noise margin in a logic circuit: high-level noise margin and low-level noise margin.

Based on input and output logic levels represented in Figure 5.2, the different noise margins can be defined as follows:

– high-level noise margin:

$$V_{NH} = V_{OH} - V_{IH} \qquad [5.1]$$

where the minimum voltage required for a high level at the input and output are designated by V_{IH} and V_{OH}, respectively;

– low-level noise margin:

$$V_{NL} = V_{IL} - V_{OL} \qquad [5.2]$$

where V_{IL} and V_{OL} designate, respectively, the maximum voltage for the input and the output.

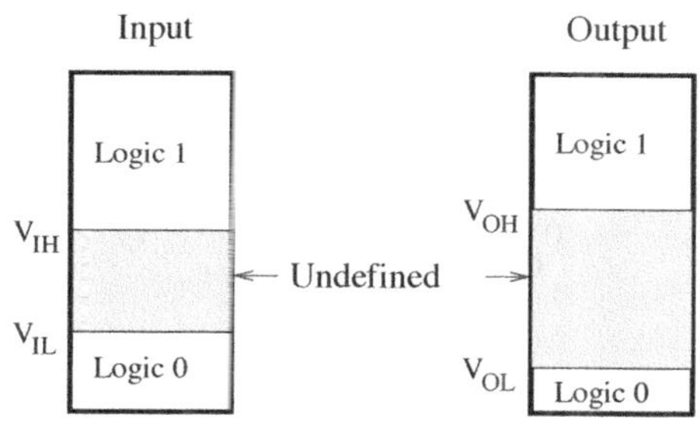

Figure 5.2. *Input and output logic levels*

5.2.4. *Propagation delay*

A logic signal that passes through a logic circuit is subject to a propagation delay. Hence, there is a difference between the instant the input changes state and the instant the output changes state. Depending on the edge of the signal, two types of propagation delays can be defined:

– t_{PLH}: propagation delay corresponding to the transition from low to high;

– t_{PHL}: propagation delay corresponding to the transition from high to low.

5.2.5. *Electric power consumption*

Electric power is defined as the product of voltage and electric current. It is constant across the range of operational frequencies for TTL technology and varies with frequency for CMOS technology, as shown in Figure 5.3.

5.2.6. *Fan-out or load factor*

When the output signal of a logic gate is connected to the inputs of other gates, a load develops on the driving gate. There is a limit to the number of inputs that an output can safely drive. This limit is called the fan-out of the gate.

In TTL technology, the increase in the output load can lead to an increase in the power consumption and a reduction in the low-level noise margin.

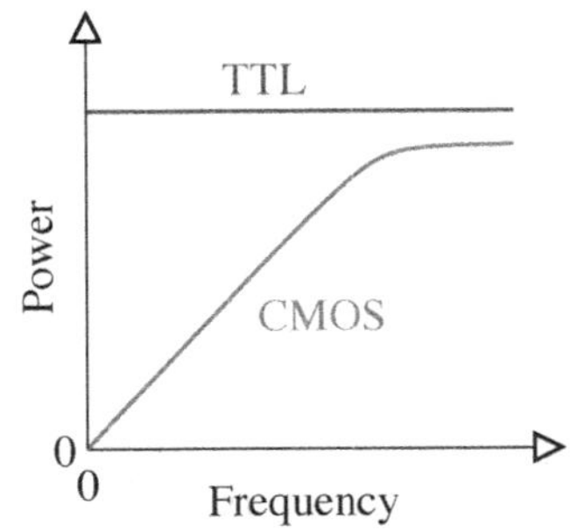

Figure 5.3. *Electric power consumption of TTL and CMOS components*

In CMOS technology, the load factor depends on the maximum operation frequency. The fewer logic gates connected to the output, the higher its maximum frequency.

5.3. TTL logic family

5.3.1. *Bipolar junction transistor*

The bipolar junction transistor (BJT) is the active component present in all TTL circuits. Figure 5.4 depicts a BJT. A bipolar transistor has three terminals: base, emitter and collector. It consists of two junctions: the base–emitter junction and the base–collector junction.

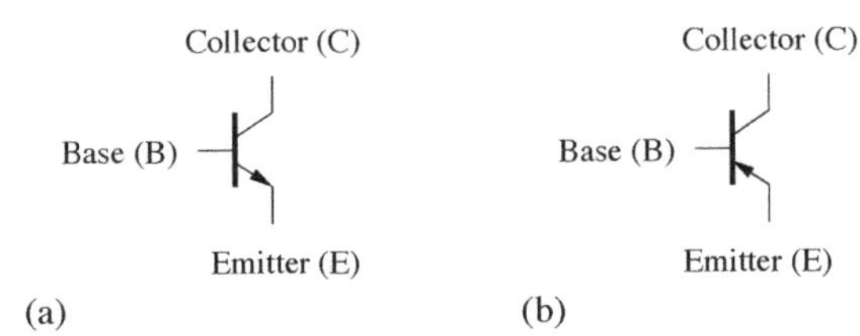

Figure 5.4. *a) npn and b) pnp bipolar junction transistors*

In general, the transistors used in logic circuits operate in commutation mode, as shown in Figure 5.5.

When the voltage applied at the base is higher than the base–emitter voltage, which is of the order of 0.7 V, and when the current intensity at the base is sufficient, the transistor begins to conduct current and then reaches saturation state. In this state, the transistor acts as closed switch between the collector and the emitter.

When the input voltage at the base is lower than 0.7 V, the transistor is in the cut-off region and then acts as an open switch between the collector and emitter.

The BJT transistor can also operate as an amplifier. In this case, it is equivalent to a non-ideal current source controlled by a current.

It must be noted that the *pnp* transistor operates with the polarization voltages inverted with respect to an *npn* transistor.

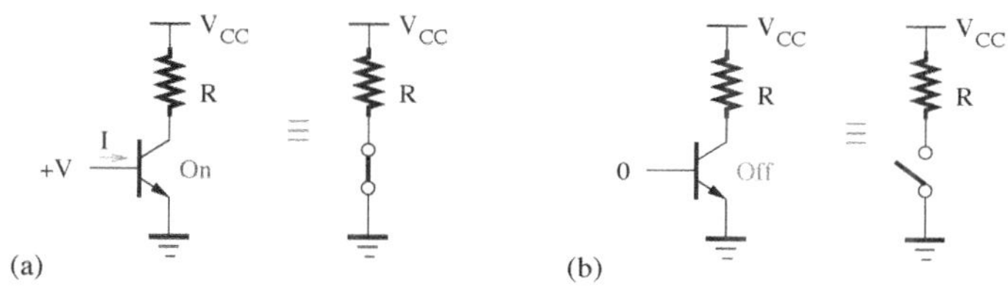

Figure 5.5. *Operation of a bipolar transistor*

5.3.2. *TTL NAND gate*

A NAND gate in the TTL logic family is represented in Figure 5.6. The transistor, Q_1, has two emitters; as a result, there are two emitter–base junctions that can set Q_1 to the conduction state. Transistor Q_2 plays the role of a phase splitter and the section, $Q_3 - D_1 - Q_4$, that constitutes the output stage is called the *totem-pole* stage.

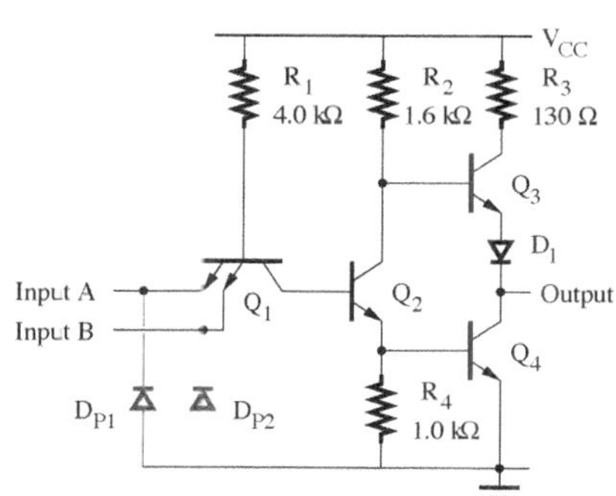

Figure 5.6. *TTL NAND gate (D_{P1} and D_{P2} are protection diodes)*

The equivalent circuits shown in Figure 5.7 are obtained by replacing the transistor Q_1 with its equivalent diode circuit. Diodes D_2 and D_3 correspond to the base–emitter junctions and D_4 corresponds to the base–collector junction.

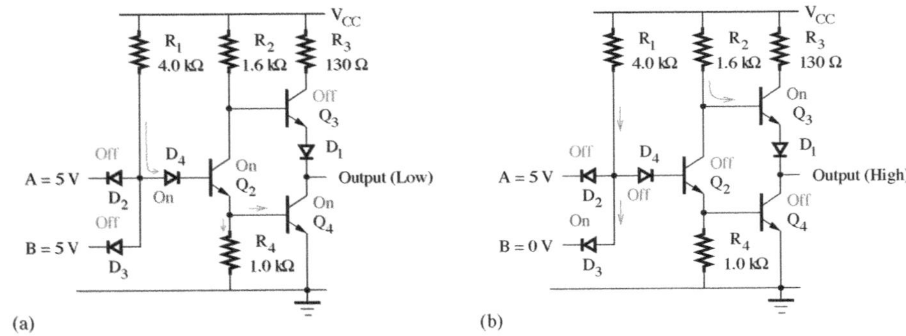

Figure 5.7. *Equivalent circuits of a TTL NAND gate. For a color version of this figure, see www.iste.co.uk/ndjountche/electronics2.zip*

For Figure 5.7(a), both inputs are at the high level. The input currents are very weak. Transistor Q_3 is turned-off and Q_4 conducts so that the output voltage is lower than 0.4 V.

The output signal provided by the equivalent circuit of Figure 5.7(b) is at the high level. In this case, the input A or B can be set to low level, or both inputs (A and B) can also be set to low level at the same time. The transistor, Q_3, conducts while Q_4 is turned-off. The output voltage is, thus, greater than 2.4 V.

NOTE 5.1.– There is another type of output in TTL integrated circuits: the open collector output. In this case, an external polarization linked to the supply voltage source, V_{CC}, must be connected to the collector of the output transistor to obtain the correct low and high logic levels.

NOTE 5.2.– In TTL technology, an unused or floating input acts as a high logic level because it corresponds to an inversely polarized base–emitter junction.

5.3.3. *Integrated TTL circuit*

Different versions of TTL circuits have been developed to meet the requirements of increasingly diversified applications:

– 74-Standard TTL;

– 74S-Schottky TTL;

– 74F-Fast TTL.

5.4. CMOS logic family

5.4.1. *MOSFET transistor*

Metal-oxide semiconductor field-effect transistors (MOSFET) can play the role of active switches in CMOS digital circuits.

The symbols for the n-channel and p-channel MOSFET transistors are given in Figure 5.8. A MOSFET transistor has the following terminals: gate, drain, source and substrate (or body). In general, the body is connected to the source.

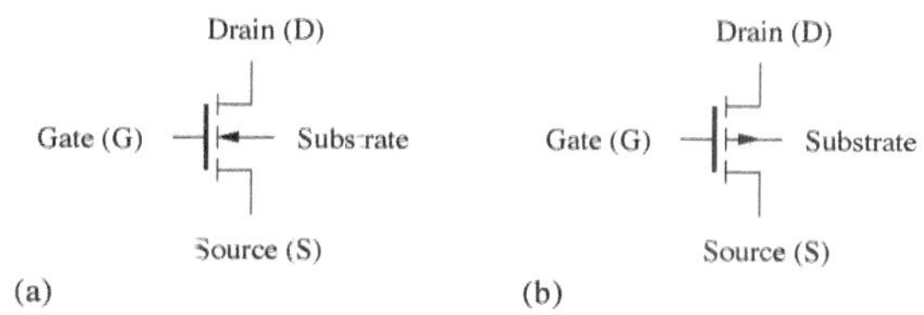

Figure 5.8. *n-channel a) and p-channel b) MOSFET transistor*

When the gate voltage of an n-channel MOSFET transistor is higher than a certain threshold voltage, the transistor conducts and ideally acts as a closed switch between the drain and source. When the gate voltage becomes equal to zero, the transistor is turned-off and acts as an open switch. In this case, the channel resistance is of the order of 10^{10} Ω. This operation is illustrated in Figure 5.9.

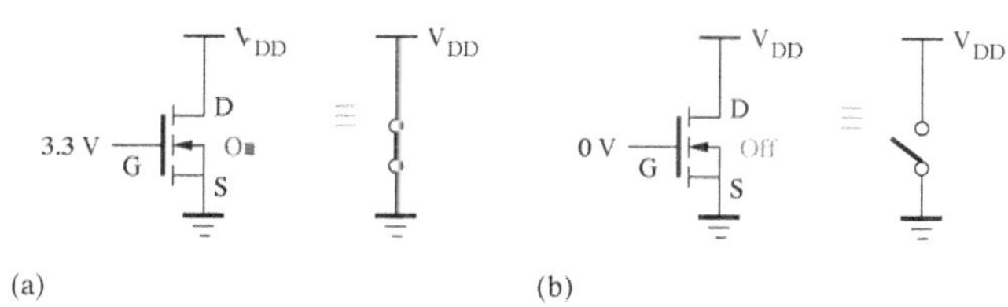

Figure 5.9. *Operation of an n-channel MOSFET transistor*

In dynamic operation, a MOSFET transistor is equivalent to a current source controlled by voltage and whose operation is perticularly affected by parasitic capacitances.

The p-channel MOSFET transistor operates with voltage polarities that are opposite to those of the n-channel MOSFET transistor, as illustrated in Figure 5.10.

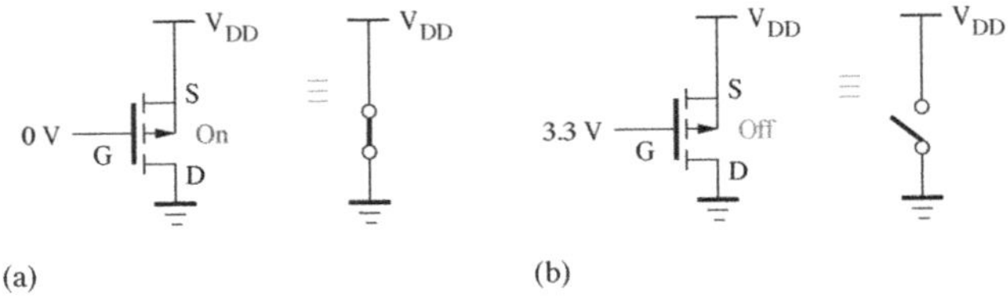

Figure 5.10. *Working of a p-channel MOSFET transistor*

5.4.2. *CMOS logic gates*

CMOS technology is characterized by high integration density and low-power consumption.

An inverter can be implemented as shown in Figure 5.11(a). It consists of two complementary (n-channel and p-channel) transistors. When the input signal assumes the high level, the transistor M_1 conducts while M_2 is turned off, and the output takes the low level. When the logic level of the input signal is low, the transistor M_1 is turned off while M_2 conducts, and the output assumes the high level. The truth table of the inverter is illustrated in Table 5.1.

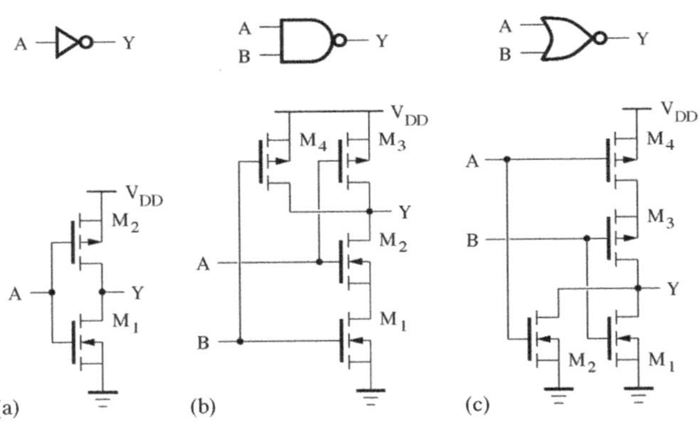

Figure 5.11. *CMOS logic gates: a) inverter; b) NAND gate; c) NOR gate*

A NAND gate can be implemented as shown in Figure 5.11(b). It consists of four transistors: the two n-channel transistors are connected in series while the two p-channel transistors are connected in parallel. The truth table given in Table 5.2 summarizes the operation of the NAND gate.

A	M_1	M_2	Y
Low	Off	On	High
High	On	Off	Low

Table 5.1. *Truth table of the inverter*

A	B	M_1	M_2	M_3	M_4	Y
Low	Low	Off	Off	On	On	High
Low	High	On	Off	On	Off	High
High	Low	Off	On	Off	On	High
High	High	On	On	Off	Off	Low

Table 5.2. *Truth table of the NAND gate*

The circuit that can used to implement a NOR gate is represented in Figure 5.11(c). It is composed of four transistors: the two n-channel transistors are connected in parallel, while the two p-channel transistors are connected in series. It is, therefore, the dual circuit of the NAND gate circuit. Table 5.3 describes the operation of the NOR gate.

A	B	M_1	M_2	M_3	M_4	Y
Low	Low	Off	Off	On	On	High
Low	High	On	Off	Off	On	Low
High	Low	Off	On	On	Off	Low
High	High	On	On	Off	Off	Low

Table 5.3. *Truth table of the NOR gate*

NOTE 5.3.– The CMOS components may be damaged by electrostatic discharge. It is, therefore, essential to take precautions when handling them (conductive foam, special bracelet connected to a high-value series resistance).

5.5. Open drain logic gates

In microprocessor systems, it is common to connect the outputs of several different logic gates to a single interconnection wire. As the use of a conventional logic gate is not possible in this case, one possible solution consists of using open drain logic gates. The expression *open drain* signifies that the output transistor drain is floating and must be connected through an external resistor to the supply voltage.

In CMOS technology, a NAND gate is composed of two n-channel transistors forming a network that is loaded by two p-channel transistors. The open drain NAND gate is implemented, as shown in Figure 5.12(a), by suppressing the load (p-channel transistors) to leave the output transistor drain floating. In practice, to implement the NAND logic function, a pull-up resistor is often connected to the output, as illustrated in Figure 5.12(b).

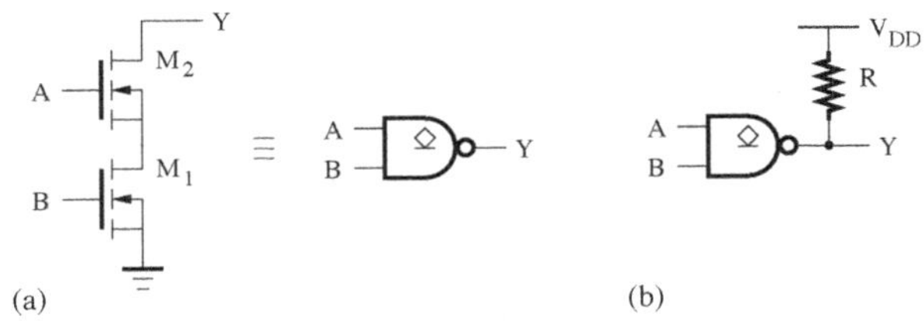

Figure 5.12. *a) Circuit and symbol of an open drain NAND gate; b) implementation of a NAND gate*

The truth table of the open drain NAND gate is given in Table 5.4, while Table 5.5 shows the truth table of the corresponding NAND gate.

A	B	M_1	M_2	Y
Low	Low	Off	Off	Open
Low	High	On	Off	Open
High	Low	Off	On	Open
High	High	On	On	High

Table 5.4. *Truth table of the open drain NAND gate*

A	B	M_1	M_2	Y
Low	Low	Off	Off	High
Low	High	On	Off	High
High	Low	Off	On	High
High	High	On	On	Low

Table 5.5. *Truth table of the NAND gate*

Several open drain gates can be connected to a pull-up resistor, as shown in Figure 5.13(a). A wired AND connection is thus established, characterized by one of the following logic equations:

$$Y = \overline{A \cdot B} \cdot \overline{C \cdot D} \cdot \overline{E \cdot F} \qquad [5.3]$$

$$= \overline{A \cdot B + C \cdot D + E \cdot F} \qquad [5.4]$$

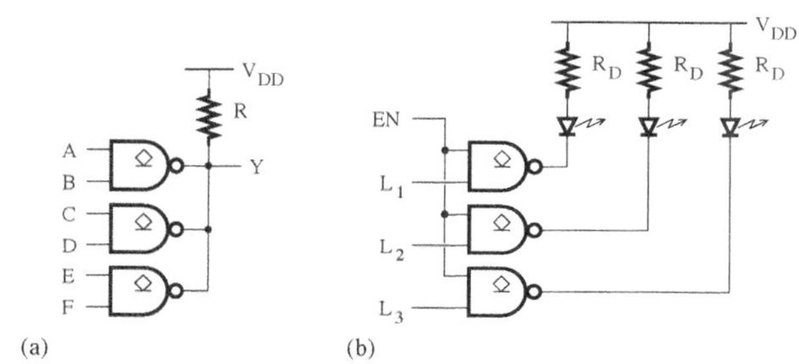

Figure 5.13. *Applications: a) wired AND connection; b) LED driver*

The open drain gate can also be used to implement a light-emitting diode (LED) driver. In the case shown in Figure 5.13(b), the enable signal is represented by EN, and each LED is switched on depending on the logic level of the corresponding signal L_k $(k = 1, 2, 3)$.

In general, the use of open drain gates offers more flexibility by allowing for the adjustment of each output signal and the advantage of reducing the number of components. However, the sizing of the pull-up resistor may affect the circuit's precision.

NOTE 5.4.– In the case of technology based on bipolar transistors, we talk about *open collectors*.

5.5.1. *Three-state buffer*

In general, a three-state buffer circuit is used to connect different components to a bus. In addition to the logic levels 0 and 1, it can take the high impedance state, z, that is used to disconnect a component.

Figure 5.14 shows the symbol and logic circuit, respectively, of a three-state buffer. When the enable signal E assumes the high state, one of the transistors, M_p or M_n, conducts while the other is turned off, and the output signal takes the same logic state as the input signal. When the logic level of the enable signal is low, the two transistors are turned off, thus enabling the output to take the high impedance state. The truth table shown in Table 5.6 gives a summary of the operation of the three-state buffer.

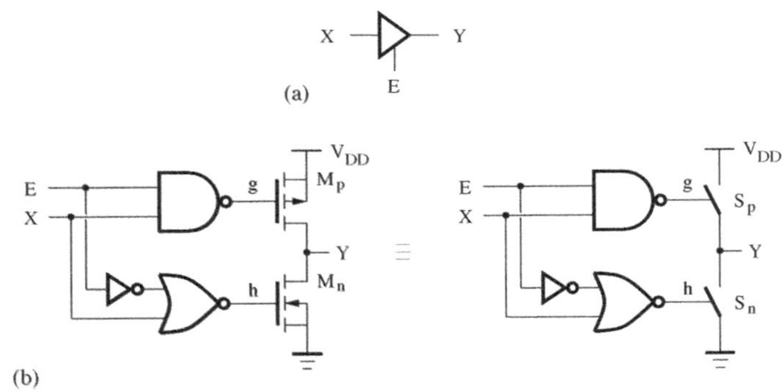

Figure 5.14. *Three-state buffer: a) symbol; b) logic circuit*

E	X	g	h	S_p	S_n	Y
Low	Low	High	Low	Open	Open	z
Low	High	High	Low	Open	Open	z
High	Low	High	High	Open	Closed	Low
High	High	Low	Low	Closed	Open	High

Table 5.6. *Truth table of the three-state buffer*

A bidirectional input/output (I/O) pin can be implemented as shown in Figure 5.15. It consists of two three-state buffers and logic gates. When the enable signal, $\overline{EN}$, is set to 0, the data can be transmitted from pin X to pin Y, and *vice versa*, depending on the logic state of the signal DIR.

5.5.2. *CMOS integrated circuit*

Several versions of CMOS circuits are available:

– the 74HC-CMOS fast series (74HCT-CMOS compatible with the TTL family);

– the advanced 74AC-CMOS series (74ACT-CMOS compatible with the TTL family);

– the low-voltage 74LV-CMOS series.

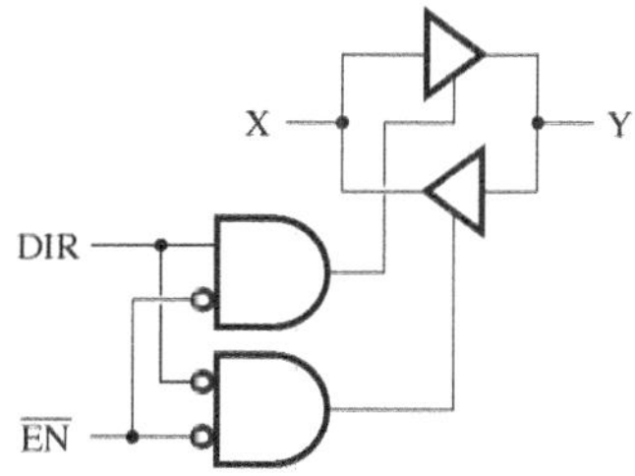

Figure 5.15. *Bidirectional input/output pin*

Among CMOS circuits offered by manufacturers, some are compatible with TTL circuits while others are not.

5.6. Other logic families

To satisfy other constraints, such as speed or reprogrammability, other technologies have been developed in addition to TTL and CMOS technologies.

– The *emitter-coupled logic* (ECL) family: The coupled logic family is a bipolar family just like the TTL family. ECL logic circuits use an input stage with differential amplifier, a polarization circuit, and an output stage with emitter-follower. The ECL family is much faster than the TTL family because the transistors do not operate in the saturation region.

– The *electrically erasable CMOS* (EECMOS) logic family: EECMOS logic is based on a combination of CMOS and NMOS technology. It is mainly used in reprogrammable modules, which are built around a floating gate MOS transistor that is loaded or unloaded by an external current.

5.7. Interfacing circuits of different technologies

The input and output levels of a logic circuit can vary depending on the component technology and the supply voltage used. The interfacing of logic gates may be further divided into two categories: interfacing circuits that can operate at the same voltage levels and interfacing of circuits that operate at different voltage levels. In the first case, the circuits are compatible and may be directly connected, while in the second case, it is necessary to convert the voltage level.

In order to interface two logic circuits, some conditions must be met:

1) the V_{OH} voltage of the driving circuit must be higher than the V_{IH} voltage of the loading circuit;

2) the V_{OL} voltage of the driving circuit must be lower than the V_{IL} voltage of the loading circuit;

3) the output voltage of the driving circuit must not exceed the input/output voltage tolerances of the loading circuit.

The interfacing of a TTL circuit to a CMOS circuit using a resistor is illustrated, as an example, in Figure 5.16(a), while Figure 5.16(b) shows a CMOS circuit connected to a TTL circuit through a three-state buffer that can shift the logic level.

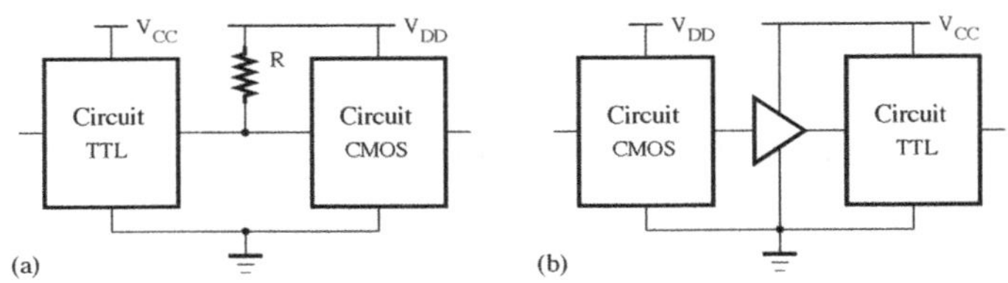

Figure 5.16. *Examples of interfacing: a) TTL-CMOS; b) CMOS-TTL*

5.8. Exercises

EXERCISE 5.1.– Answer the following questions:

a) There are four different voltage thresholds for the TTL logic family and CMOS logic family:

$\alpha-$ true $\beta-$ false

b) The input signal frequency of a component whose critical path introduces a propagation delay, t_p, must be lower than $1/t_p$:

$\alpha-$ true $\beta-$ false

c) For applications where a higher switching speed is required and a relatively high current is required for the output load, CMOS logic gates are preferred over TTL logic gates:

$\alpha-$ true $\beta-$ false

d) The ECL logic family is faster than the TTL family but consumes more energy:

$\alpha-$ true $\beta-$ false

e) Unused inputs for a logic circuit must be connected either to the supply voltage, or to the ground, or to an used input:

$\alpha-$ true $\beta-$ false

f) In general, the field-effect transistor is faster than a bipolar transistor:

$\alpha-$ true $\beta-$ false

g) Careful handing of a CMOS gate is required because of:

$\alpha-$ its fragile construction $\beta-$ its immunity to high-level noise

$\gamma-$ its sensitivity to electrostatic charge $\delta-$ its low energy consumption.

h) When the frequency of an input signal for a CMOS gate increases, the average energy consumption:

$\alpha-$ diminishes $\beta-$ increases $\gamma-$ does not change.

i) The operation of a CMOS gate is more reliable than a TTL gate in a noisy environment due to:

$\alpha-$ the narrower noise margin $\beta-$ the input capacitance

$\gamma-$ the larger noise margin $\delta-$ low energy consumption.

j) The highest currents associated with high and low input level being I_{IH} and I_{IL}, and high and low output levels being I_{OH} and I_{OL}, respectively; and with $\lfloor x \rfloor$ denoting the greatest integer that is less than or equal to x, the maximum number of loading gates that the output of a logic gate can drive, as given by the formula:

$$\min(\lfloor I_{OH}/I_{IH} \rfloor, \lfloor I_{OL}/I_{IL} \rfloor)$$

is not applicable, in practice, to the CMOS family, which are only affected by increased propagation delays caused by an increase in the load capacitance:

$\alpha-$ true $\beta-$ false

EXERCISE 5.2.– Which logic gate is implemented by each circuit shown in Figure 5.17.

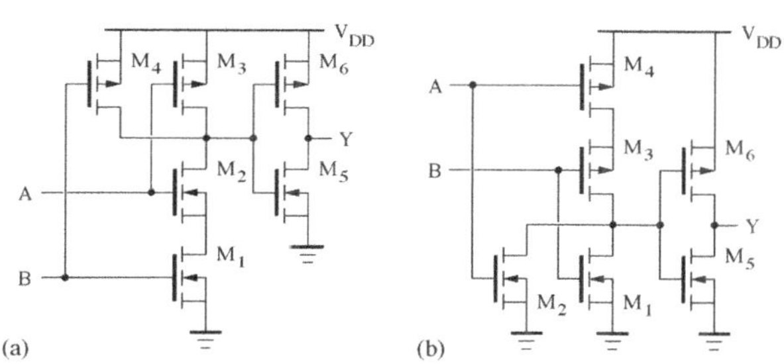

Figure 5.17. *CMOS logic gate: a) AND gate; b) OR gate*

EXERCISE 5.3.– Show that the logic circuits shown in Figures 5.18(a) and 5.18(b) implement the function of AND and OR gates, respectively.

EXERCISE 5.4.– Power dissipation.

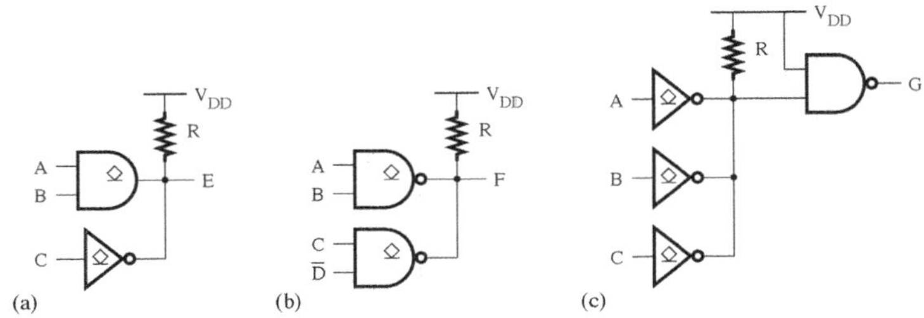

Figure 5.18. *CMOS logic circuits: a) AND gate, b) OR gate*

The electric power dissipated by a TTL component is given by:

$$P_D = V_{CC} I_{CC} \quad [5.5]$$

where V_{CC} is the supply voltage and I_{CC} is the current flowing through the circuit.

For a CMOS component, the electric power dissipated is the sum of dynamic and static contributions and can be written as follows:

$$P_D = P_{\text{dynamique}} + P_{\text{statique}} \quad [5.6]$$

$$= (C_L + C_{PD}) V_{DD}^2 f + V_{DD} I_{DD} \quad [5.7]$$

where C_L is the output load capacitance, C_{PD} is the equivalent capacitance for the gate, f is the output signal frequency, V_{DD} is the supply voltage and I_{DD} is the leakage current.

Determine the power, P_D, of this circuit, assuming that the outputs of the circuit commutate with a cyclic relationship, η, of 50%, the current, I_{CC} is of the form:

$$I_{CC} = \eta I_{CCH} + (1 - \eta) I_{CCL} = (I_{CCH} + I_{CCL})/2$$

where I_{CCH} and I_{CCL} represent the current when all the outputs are at the high and low logic levels, respectively.

What is the highest possible frequency (f) if the maximum value of the power, P_D, of the CMOS circuit is 15 mW and the output load capacitance, C_L, is 10 pF?

From the datasheet, we have:

– TTL circuit: $V_{CC} = 5$ V, $I_{CCH} = 1.5$ mA, and $I_{CCL} = 4.5$ mA;

– CMOS circuit: $C_{PD} = 22$ pF, $V_{DD} = 3$ V, and $I_{DD} = 20$ μA.

5.9. Solutions

SOLUTION 5.1.– Answers to the questions

a) α	b) α	c) β	d) α	e) α
f) β	g) γ	h) β	i) γ	j) α

SOLUTION 5.2.– The logic gate shown in Figure 5.17(a) is implemented by connecting a NAND gate and an inverter in series. The truth table of this gate is given in Table 5.7. It is an AND logic function.

A	B	M_1	M_2	M_3	M_4	M_5	M_6	Y
Low	Low	Off	Off	On	On	On	Off	Low
Low	High	On	Off	On	Off	On	Off	Low
High	Low	Off	On	Off	On	On	Off	Low
High	High	On	On	Off	Off	Off	On	High

Table 5.7. *Truth table of the AND gate*

The logic gate shown in Figure 5.17(b) is composed of a NOR gate and an inverter. It can be characterized by the truth table shown in Table 5.8 and, thus, implements the OR logic function.

A	B	M_1	M_2	M_3	M_4	M_5	M_6	Y
Low	Low	Off	Off	On	On	On	Off	Low
Low	High	On	Off	Off	On	Off	On	High
High	Low	Off	On	On	Off	Off	On	High
High	High	On	On	Off	Off	Off	On	High

Table 5.8. *Truth table of the OR gate*

SOLUTION 5.3.– For each circuit, the logic equations obtained can be written as follows:

– circuit as shown in Figure 5.18(a):

$$E = A \cdot B \cdot \overline{C} \quad [5.8]$$

– circuit as shown in Figure 5.18(b):

$$F = \overline{\overline{A} \cdot B} \cdot \overline{C \cdot \overline{D}} \quad [5.9]$$

$$= (\overline{A} + \overline{B})(\overline{C} + D) \quad [5.10]$$

– circuit as shown in Figure 5.18(c):

$$G = \overline{\overline{A} \cdot \overline{B} \cdot \overline{D}} \quad [5.11]$$

$$= A + B + C \quad [5.12]$$

SOLUTION 5.4.– Power dissipation.

The power of the TTL circuit is given by:

$$P_D = 5(1.5 + 4.5)10^{-3}/2 = 15 \times 10^{-3} \text{ W} = 15 \text{ mW}$$

The maximum frequency of the CMOS circuit is given by:

$$f = f_{\max} = \frac{P_D - V_{DD}I_{DD}}{(C_L + C_{PD})V_{DD}^2} \quad [5.13]$$

$$= \frac{10 \times 10^{-3} - 3(20 \times 10^{-6})}{(10 + 20)10^{-12} \times 3^2} = 36.8 \times 10^6 \text{ Hz} = 36.8 \text{ MHz}$$

6

Semiconductor Memory

6.1. Introduction

Semiconductor memories are digital information storage devices. They are used in all appliances including microprocessors and also in implementing programmable logic circuits.

Semiconductor memories have very low access time compared to other memories such as hard disks and optical storage discs (*Blu-ray* discs, DVD, etc.). Semiconductor memory can be classified into two types:

– volatile memory: these are memories where the integrity of information is not guaranteed unless they are electrically powered. They are readable and writable;

– non-volatile memories: these memories preserve information even in the absence of an electric power supply.

The operation of storing information in a memory corresponds to writing or programming.

6.2. Memory organization

The elementary cell of a semiconductor memory is used to store one bit of data. For non-volatile memory, this can be assimilated to one of the nodes in a network that is connected, through a switch, to either electric voltage (corresponding to the high logic level) or to the ground (or the voltage defining the low logic level); in the case of volatile memory, the elementary cell is based on a flip-flop, usually a D flip-flop.

A memory word is a set of juxtaposed n bits. In practice, n is equal to 4, 8, 16, 32, or 64. When n is equal to 8, the memory word is called an *octet*. Each row of cells represents a memory location, which is typically equal to the word size.

Each word is selected by providing a binary address. A set of wires used to identify specific memory locations constitutes the address bus.

Using a decoder, n wires are required to address 2^n memory locations. The memory capacity when expressed as the number of memory locations or words is, therefore, 2^n. It can also be expressed as the number of bits. In this case, it is multiplied by the number of bits per word.

EXAMPLE 6.1.– Determine the number of address lines for a memory with a capacity of $64K \times 8$.

A $64K \times 8$ memory is organized into words of eight bits, or octets, and has a capacity of 64 K octets.

As each word requires an address line and $1K$ equals 2^{10} or corresponds to $1,024$ bits, implementing this memory requires a total number, n, of address lines, where $n = \log(64 \times 1\ 024)/\log(2) = 16$.

The set of wires through which the information to be stored or read is transmitted is called the *data bus*.

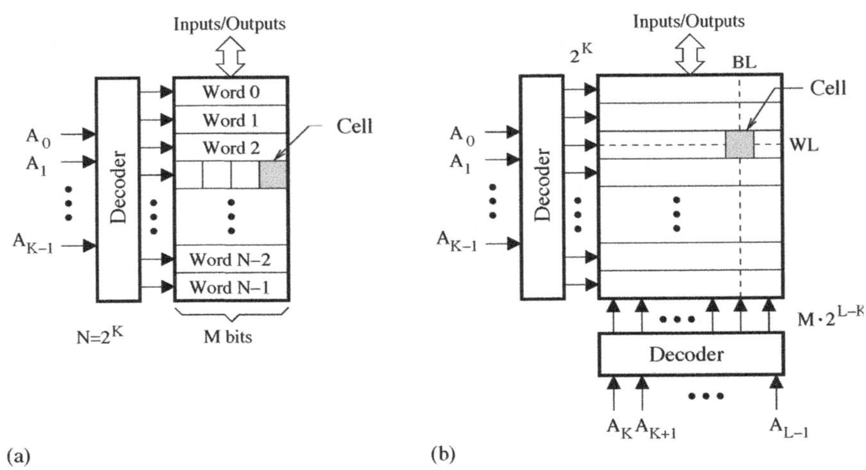

Figure 6.1. *a) Column structure for a memory; b) matrix structure for a memory*

A memory can be organized in a column structure, as shown in Figure 6.1(a). The decoder is used to select, from k-bit addresses, an address of 2^K lines, or an M-bit word. When the number of bits, K, becomes very high, a regular structure is only obtained by adopting a matrix configuration, as illustrated in Figure 6.1(b).

The address of a word is divided into a line address $(A_0, \cdots, A_{K-1})$ and a column address $(A_K, \cdots, A_{L-1})$. Thus, in order to select a word, a horizontal line or a word line (WL) and a vertical line, or bit line (BL), must be activated.

6.3. Operation of a memory

Memories are designed such that they can be connected to the same data bus. Each memory chip then possess a selected input, $\overline{CS}$ (*chip select*) or $\overline{CE}$ (*chip enable*), that can be used to avoid conflicts when the bus is in use for other purposes. Each of these control inputs can serve to connect or disconnect a memory from the bus.

– if $\overline{CS}$ takes the logic state 0, the memory is selected and connected to the data bus;

– if $\overline{CS}$ assumes the logic state 1, the data buffer is set to the high impedance state in order to disconnect the memory.

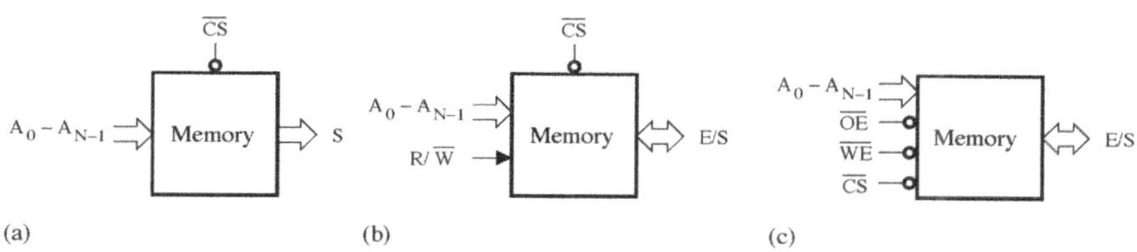

Figure 6.2. *a) Read only memory; b) and c) read/write memory*

A memory has an *output enable* input that, often symbolized as $\overline{OE}$, can be used to enable or disable the output, and a *write enable* input (or *read/write* $R/\overline{W}$ input) that can be used to select which operation is to take place, read or write. The conditions for a read operation are:

– if $\overline{OE}$ takes the logic state 0, it is possible to read data;

– if $\overline{OE}$ takes the logic state 1, the data bus is in a high-impedance state and reading data are not possible.

The necessary conditions for a write operation are as follows:

– if $\overline{WE}$ takes the logic state 0, it is possible to write data;

– if $\overline{WE}$ takes the logic state 1, it is not possible to write data.

A signal may take the low level, high level or an intermediate level that corresponds to the high impedance state (see equation [6.3]).

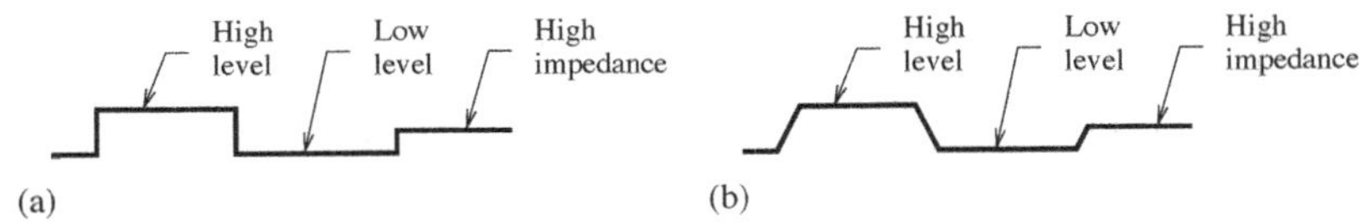

Figure 6.3. *Representation of a signal: a) ideal case; b) taking into account the rise and fall times*

Data representation on a bus is illustrated in Figure 6.4. Two states can be observed:

– a low-impedance state for which the logic level of each wire may be 0 or 1 (two horizontal lines; the hexadecimal value of the binary configuration of the wires is written between these lines);

– a high impedance state for which the logic level is not defined (a median line). In this case, the bus is disconnected.

Figure 6.4. *Data representation on a bus*

Memory timing diagrams for a read cycle and a write cycle are shown in Figures 6.5 and 6.6, respectively. In general, a transient state is seen on the bus before the data level stabilizes. The timing characteristics of a read operation can be defined as follows:

– t_{RC}: duration of read cycle;

– t_{AC}: access time with respect to select input;

– t_{AA}: access time with respect to the address bus;

– t_{OE}: data appearance delay with respect to read initialization.

The following are the timing characteristics associated with a write operation:

– t_{WC}: duration of the write cycle;

– t_{SA}: address setup time;

– t_{WD}: $\overline{WE}$-hold duration at the low level after data have been applied;

– t_{HD}: data hold time.

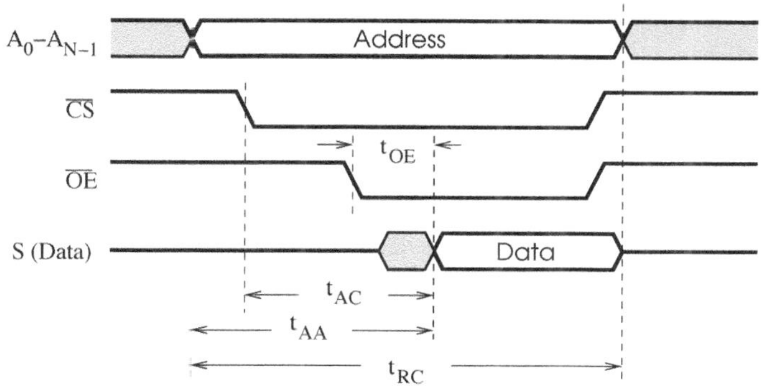

Figure 6.5. *Memory timing diagram for a read cycle*

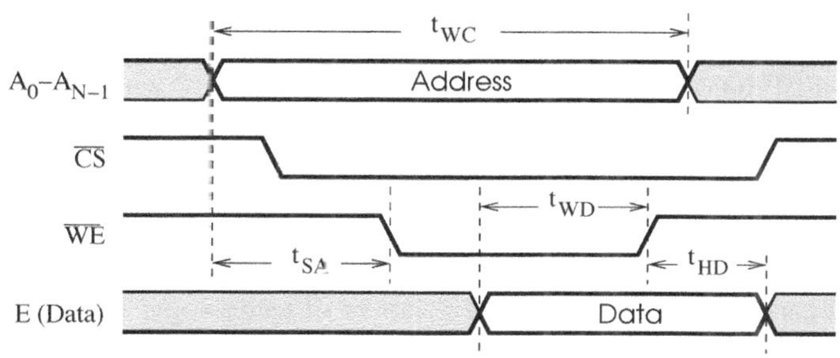

Figure 6.6. *Memory timing diagram for a write cycle*

6.4. Types of memory

Volatile and non-volatile memory can be differentiated based on the data-retention feature (or the ability to return correct stored data over a period time).

6.4.1. *Non-volatile memory*

Non-volatile memory or *read-only memory* (ROM) is used to permanently store data. The information contained in a ROM is recorded at the time of production, based on an exposure technique for photosensitive material, through a mask that reflects the desired binary configuration.

We can distinguish between many types of non-volatile memories that differ in internal structure, the number of times they can be reprogrammed and the methods for erasure:

– PROM or *programmable ROM* is a non-volatile memory that can only be programmed once. It is composed of a set of fuses that is destroyed during programming;

– EPROM or *erasable PROM* is a kind of PROM that can be programmed, erased and reprogrammed. Erasure is carried out by exposing the active part of the EPROM to ultraviolet light. After this operation, all memory cells store a logic 1. The bulk erase operation for an EPROM is non-selective (that is: it affects all the cells);

– EEPROM or *electrically erasable PROM* is functionally identical to EPROM. The erase operation consists of applying a predetermined electric pulse to the memory. The main difference between EEPROM and EPROM is that EEPROM can be erased and reprogrammed without being displaced from its support and in a selective manner. To erase an EPROM, it must be removed from its support to be handed with special equipment. EEPROM is almost 10 times slower than random-access memory (RAM) and its storage capacity is about 100 times lower than that of a RAM. They are mainly used in applications where non-volatility is a requirement;

– Flash EEPROM can be erased partially or completely with an electric pulse. Contrary to what happens with the EEPROM, where the minimum erasure concerns an octet, the partial erasure of a flash EEPROM is performed on a block of data. A flash EEPROM can be programmed or erased without being removed from its support. The flash EEPROM offers a higher storage density than EEPROM.

A ROM can be manufactured using a variety of technology: bipolar, *metal-oxide semiconductor* (MOS) or *complimentary MOS* (CMOS).

The structure of a mask ROM is illustrated in Figure 6.7. It comprises a decoder, n-channel MOS transistors, pull-up resistors and three-state buffers. The vertical line (BL) is connected to the supply voltage, V_{DD}. The mask used during the manufacture does not permit transistors to be inserted anywhere except those positions where the logic 1 must be stored. For each read operation, a single horizontal line (WL) is activated by the decoder. This enables the transistors to function as closed switches between the corresponding nodes on the vertical lines and the ground. As the outputs for the buffer are logically complemented, the transistor's presence results in the storage of logic 1. The absence of a transistor results in the storage of the logic state 0. A $2^n \times m$ ROM can be used to store m different logic functions of n variables.

As Figure 6.8 illustrates, the structure of an EPROM is similar to that of a ROM, but using floating-gate transistors.

A memory cell is programmed by applying voltage that is higher than the normal operational voltage to the transistor gate. This results in the capture by the floating gate of a portion of channel electrons that have passed through the thin oxide layer. The charge thus acquired by the gate is maintained even after the programming voltage is disconnected, and the transistor is biased in the cut-off region and then behaves as an open switch. Because the vertical line is connected to the supply voltage, V_{DD}, and the output of the three-state buffer is complemented, the programming operation corresponds to the storage of the logic state 0.

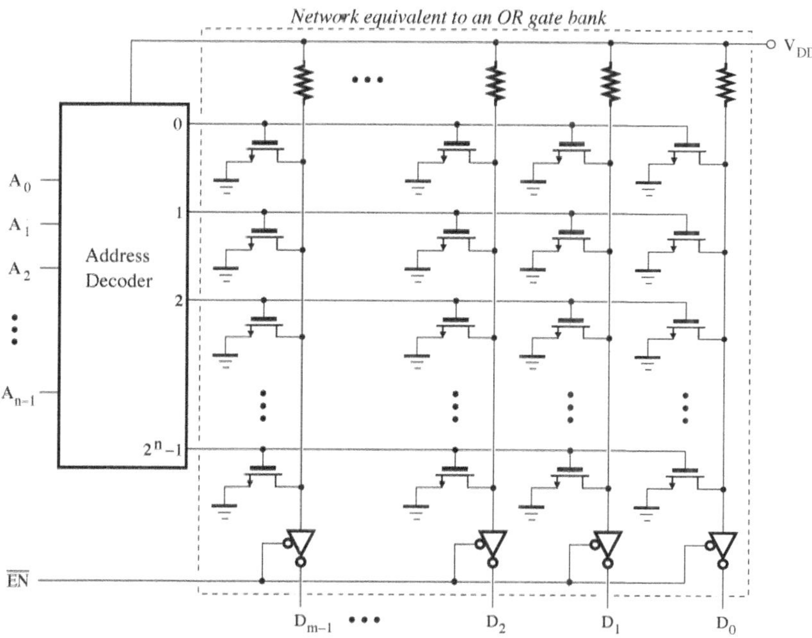

Figure 6.7. *Structure of a read-only memory (in MOS technology)*

Under normal operation, the selection of a horizontal line by the decoder results in the application of the voltage corresponding to the high logic level to the transistor gates. The transistors, which have been connected to the programming voltage beforehand, are not affected and continue to operate as open switches (or cells set to the logic state 0), while the others become equivalent to closed switches (or cells set to the logic state 1).

The erase operation consists of applying an ionizing ray (e.g. ultraviolet light), which has enough energy to free the captured electrons from the floating gate. An EPROM is encapsulated in a plastic-molded package having a window that is translucent to ultraviolet light.

In general, an EPROM cannot be erased electrically because once a sufficient number of electrons have been captured by the gate, the transistor conduction can no longer be reestablished just by reversing the voltage used during programming.

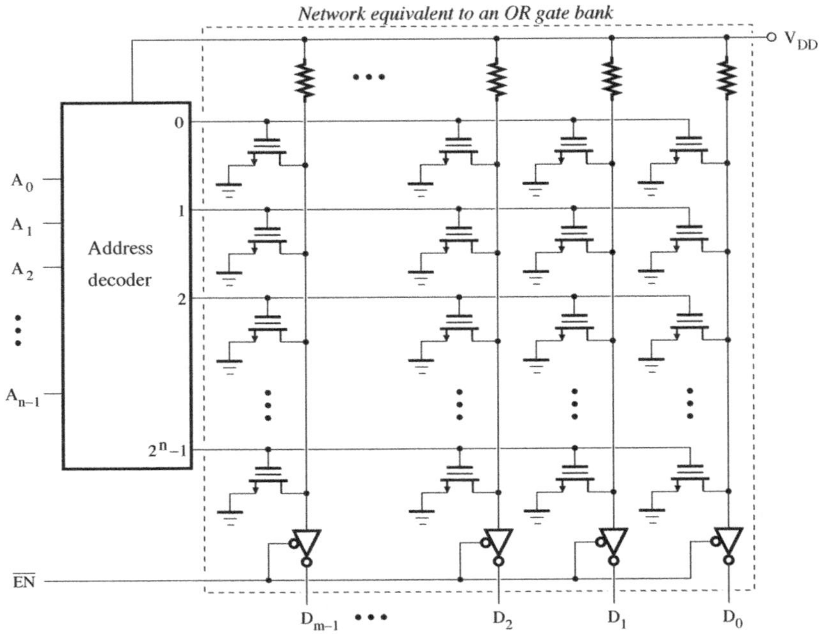

Figure 6.8. *Structure of an EPROM*

However, electrically erasable ROMs can be implemented by using cells with two transistors in series, a MOS transistor for the selection and a floating-gate tunneling oxide transistor for the storage. This approach is characterized by a lower integration density as compared to a ROM.

Non-volatile programmable memories present the disadvantage of getting damaged after some time (retention time of about 10 years) and having a capped number of write/erase cycles (10^2–10^5).

6.4.2. *Volatile memories*

A volatile memory, or RAM gives access to each of its cells for a read or write operation, with the same access time and in any order.

There are two types of volatile memory:

– static RAM or SRAM can operate at high speeds, but present the disadvantage of requiring high power consumption as its elementary cells are based on D flip-flops;

– dynamic RAM or DRAM is characterized by a large storage capacity. The basic cell is implemented by associating a transistor and a capacitor (see Figure 6.11) that can store or restore electric charge. The memorization of a bit is materialized by the presence (logic 1) or absence (logic 0) of a charge on the capacitor.

Charge stored on a capacitor tends to gradually decrease over time. This is why DRAM must be periodically refreshed to maintain memorization. As the external logic circuits must take into account these priority actions linked to refreshing, DRAM requires more complex interfacing techniques than SRAM.

The structure of a 4×4 memory with cells based on D flip-flops is represented in Figure 6.9. It can be useful for the illustration of the SRAM operation principle. The truth table is given in Table 6.1, where the memory is activated by the chip-enable signal, $\overline{CE}$, and the logic state of the write signal $\overline{WR}$ is used to choose between the write and read operations. For each combination of address bits, the decoder activates a horizontal line (or WL) thereby enabling the writing or reading of a 4-bit word through the bidirectional I/O pins.

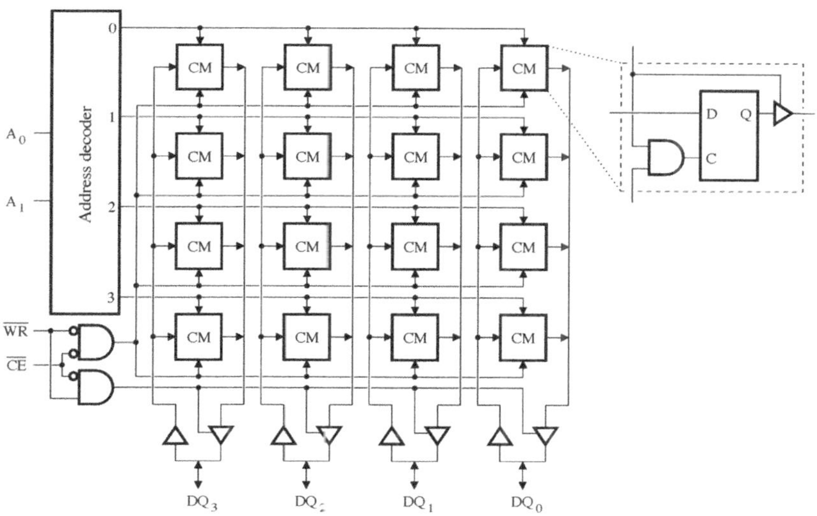

Figure 6.9. *Structure of a 4×4 memory with cells based on D flip-flops*

In general, an SRAM may be implemented as shown in Figure 6.10. The matrix structure, with a line decoder and a column decoder, offers the advantage of being more regular. Table 6.2 shows the truth table of the memory, where the control inputs are designated by $\overline{CE}$ (*chip enable*), $\overline{OE}$ (*output enable*) and $\overline{WE}$ (*write enable*).

$\overline{CE}$	$\overline{WR}$	DQ_i	Remarks
1	x	Z	Deactivate
0	0	1	Write 1
0	0	0	Write 0
0	1	Data	Read

Table 6.1. *Truth table for the 4×4 $(i = 0, 1, 2, 3)$ memory*

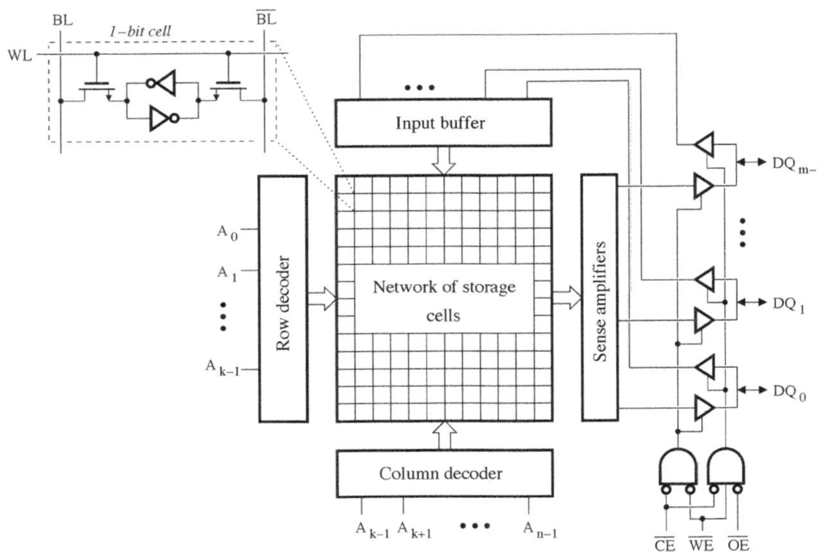

Figure 6.10. *Structure of an SRAM*

$\overline{CE}$	$\overline{OE}$	$\overline{WE}$	DQ_i	Remarks
1	x	x	Z	Memory disabled
0	1	1	Z	Outputs disabled
0	0	1	Data	Read
0	x	0	Data	Write

Table 6.2. *Truth table of the $(i = 0, 1, \cdots, m-1)$ SRAM*

The elementary cell, which can store 1-bit data, consists of two inverters that form a closed loop, each of whose outputs are connected to a transistor. Access to the cell is controlled by the logic state of the WL, while the BL and its complement $(\overline{BL})$

are used to transfer data during read and write operations. The elementary cell of an SRAM is sized so as to minimize its surface. Consequently, it operates with very low electric currents, resulting in a small voltage difference between the lines BL and $\overline{BL}$. Access to memory is accelerated by using sense amplifiers to capture and to increase this small voltage difference so that this latter can reach recognizable logic levels.

A memory with higher density, like the DRAM, can be implemented by reducing the size of the elementary cell. Figure 6.11 depicts the structure of a DRAM, which is based on an elementary cell that is composed of a MOS transistor and a capacitor, C.

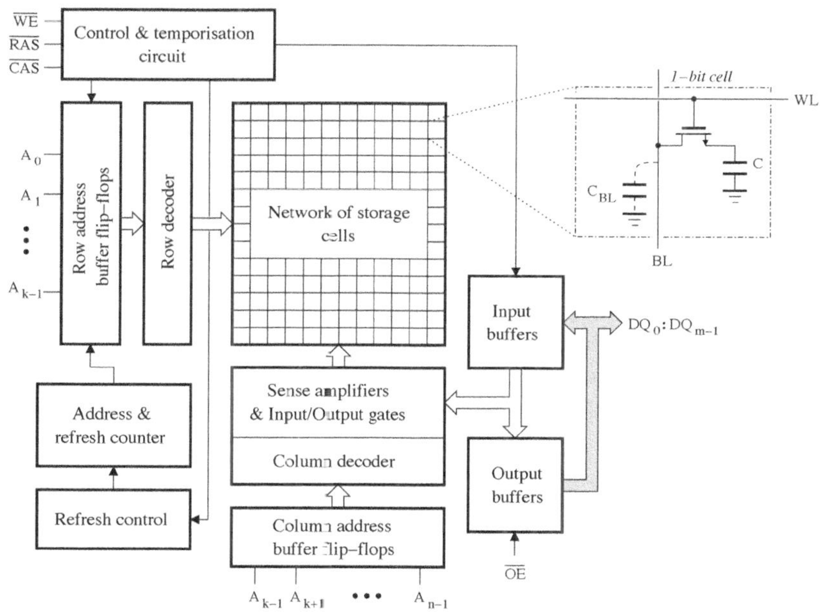

Figure 6.11. *Structure of a DRAM*

During a write operation, a data bit is placed on the line BL, and the signal applied to the line WL assumes the high logic level. Depending on the logic state (1 or 0) of the data bit, the capacitor is either charged or discharged. Before a read operation, the BL line is connected to a precharge voltage that is equal to half of the sum of high-level and low-level voltages; this enables the charging of the parasite capacitor, C_{BL}, of the BL line. As soon as the signal in the WL line takes the high level, a charge redistribution occurs between the capacitor C_{BL} and the capacitor C. This results in an increase or decrease of the voltage on the BL line, depending on whether the logic state of the initially stored data is 1 or 0. As the parasitic capacities are of the order of ten or hundred times the value of C, this variation in voltage is low and must be converted to logic levels using a sense amplifier.

The charge stored on the capacitor C is reduced by the presence of parasitic capacitances as well as by each read operation. Thus, refresh operations, which consist of rewriting the cell contents, must be periodically carried out so that the data stored in the memory is not altered. The frequency of these operations is reduced by adopting a two-dimensional architecture, where an entire line can be refreshed at a time.

To reduce the number of pins, the line and column addresses are multiplexed in most DRAMs. The decrease in access time is obtained by assigning the selection function to two control signals, $\overline{RAS}$ (*row address strobe*) and $\overline{CAS}$ (*column address strobe*). The $\overline{RAS}$ signal is used to initiate the capture of line addresses and to mark the beginning of each operation. It also triggers refresh cycles. The $\overline{CAS}$ signal is used to latch the column addresses and to start the read or write operation. Its activation is also necessary to trigger some types of refresh cycle.

The timing diagrams for a write cycle and for a read cycle in a DRAM are represented in Figure 6.12. When the $\overline{RAS}$ signal takes the logic state 0, the line address is stored on flip-flops and then replaced by the column address. When the logic state of the $\overline{CAS}$ signal becomes 0, the flip-flops are used to store the column address. Each access to the memory, therefore, requires the activation of the $\overline{RAS}$ signal, followed by that of the $\overline{CAS}$ signal. The write and read operations are also dependent on the $\overline{WE}$ and $\overline{OE}$ signals, respectively. Addresses and data are valid only when each control signal remains active for at least for a predetermined minimal time.

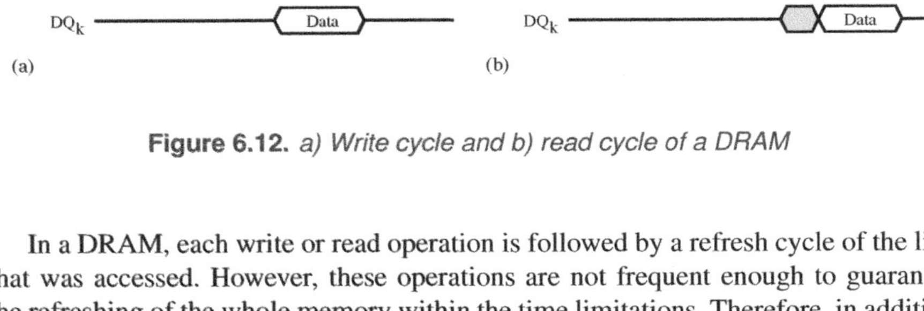

Figure 6.12. *a) Write cycle and b) read cycle of a DRAM*

In a DRAM, each write or read operation is followed by a refresh cycle of the line that was accessed. However, these operations are not frequent enough to guarantee the refreshing of the whole memory within the time limitations. Therefore, in addition to hidden refresh cycles, which are characterized by the execution of a write or read operation and a refresh cycle during a single active period of the $\overline{CAS}$ signal, it is necessary to consider other refresh options:

– the $\overline{RAS}$*-only refresh* cycle is executed when the address for the line to be refreshed is placed on the bus after activating the $\overline{RAS}$ signal;

– the $\overline{CAS}$*-before-*$\overline{RAS}$ *refresh* cycle is initiated by activating the $\overline{CAS}$ signal before the $\overline{RAS}$ signal. When this sequence is detected by the control circuit, the line whose address is specified by an internal counter is refreshed;

– the autorefresh cycle is also known as sleep mode or self refresh cycle. An on-chip oscillator is used to determine the refresh frequency and a counter helps monitor the address each time.

A conventional DRAM is controlled asynchronously and, thus, requires wait states to ensure the data synchronization, while a synchronous DRAM (SDRAM) is basically a DRAM with a synchronous interface. An SDRAM can preserve the same data transfer speed over a larger bandwidth because the inputs/outputs and control signals are synchronized with an external clock signal.

Other DRAM structures have been proposed for applications that require faster access to data:

– *Extended data out DRAM* begins the output data generation on the falling edge of the $\overline{CAS}$ signal and continues in this way until the $\overline{RAS}$ signal is deactivated or until the next falling edge of the $\overline{CAS}$ signal.

– Rambus DRAM is characterized by a higher bandwidth. It uses an interface to connect one or more memories to the same bus.

– *Double data rate SDRAM* (DDR SDRAM) enables data transfer on both the rising and falling edges of the clock signal, thus doubling the data transfer rate. It uses a prefetch buffer with a depth of two words to facilitate access to several words sharing the same line address. The increase in the input/output flow rate is obtained for later versions of the SDRAM, DDR2 and DD3, by using prefetch buffers with depths of four and eight to increase the number of words accessed by four and eight, respectively.

6.4.3. *Characteristics of the different memory types*

A semi-conductor memory uses the same technology as an integrated circuit. It is characterized by high-speed operation and low manufacturing costs. Table 6.3 summarizes the characteristics of different types of semiconductor memory.

6.5. Applications

Logic function implementations using a memory consist of storing the values for this function at well-defined addresses. These addresses are determined by properly connecting the input variables of the logic function to the address bus.

Type	Category	Erase	Erase an octet	Volatile	Use
SRAM	Read/write	Electrical	Yes	Yes	Cache
DRAM	Read/write	Electrical	Yes	Yes	Main memory
ROM	Read only	Impossible	No	No	Various applications
PROM	Read only	Impossible	No	No	Few applications
EPROM	Read mostly Write once	Ultraviolet	No	No	Prototyping
EEPROM	Read mostly Write several times	Electrical	Yes	No	Prototyping
Flash EEPROM	Read/write	Electrical	–	No	Various applications

Table 6.3. *Characteristics of different types of semiconductor memory*

When the value of certain bits of the address bus is fixed, the memory is divided into two parts:

– an accessible part;

– an inaccessible part.

EXAMPLE 6.2.– Consider a memory with an address bus of three bits $A_2A_1A_0$, where A_0 is the least significant bit.

Table 6.4 depicts the accessible and inaccessible parts (four cell zone) if the bit A_2 is fixed to 0.

Table 6.5 shows the accessible and inaccessible part (four cell zone) if the bit A_0 is set to 1.

In general, the total capacity of the accessible part is equal to the total capacity divided by 2^x, where x is the number of fixed bits.

6.5.1. *Memory organization*

We can implement a high-capacity memory using elementary memory chips. The memory to be designed contains:

– one address bus;

– one data bus;

– control signals.

A_2	A_1	A_0		
0	0	0	cell 0	Accessible part
		1	cell 1	
	1	0	cell 2	
		1	cell 3	
1	0	0	cell 4	Inaccessible part
		1	cell 5	
	1	0	cell 6	
		1	cell 7	

Table 6.4. $A_2 = 0$. *For a color version of this table, see www.iste.co.uk/ndjountche/electronics2.zip*

A_2	A_1	A_0		
0	0	0	cell 0	Inaccessible part
		1	cell 1	Accessible part
	1	0	cell 2	Inaccessible part
		1	cell 3	Accessible part
1	0	0	cell 4	Inaccessible part
		1	cell 5	Accessible part
	1	0	cell 6	Inaccessible part
		1	cell 7	Accessible part

Table 6.5. $A_0 = 1$. *For a color version of this table, see www.iste.co.uk/ndjountche/electronics2.zip*

The increase in capacity corresponds to either the increase in the number of the data bus bits (see Figure 6.13), or the increase in the number of the address bus bits (see Figure 6.14).

NOTE 6.1.– In the case where the address bus is expanded, it is often necessary to determine a logic function for the selection of elementary memories.

6.5.2. *Applications*

To meet the requirements of certain applications, elementary memories can be combined in order to obtain larger storage capacity.

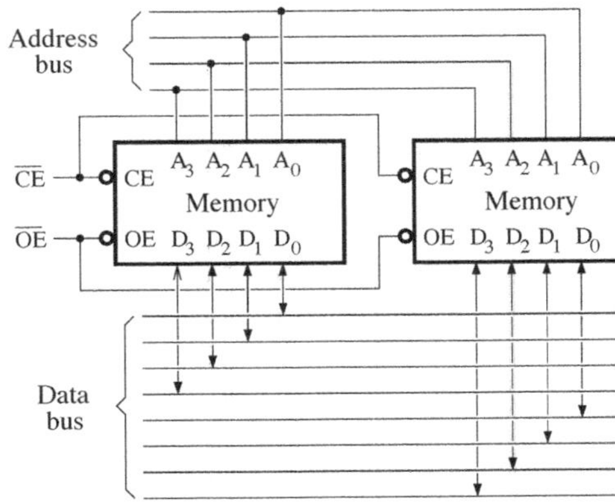

Figure 6.13. *Combining memories to expand the word length*

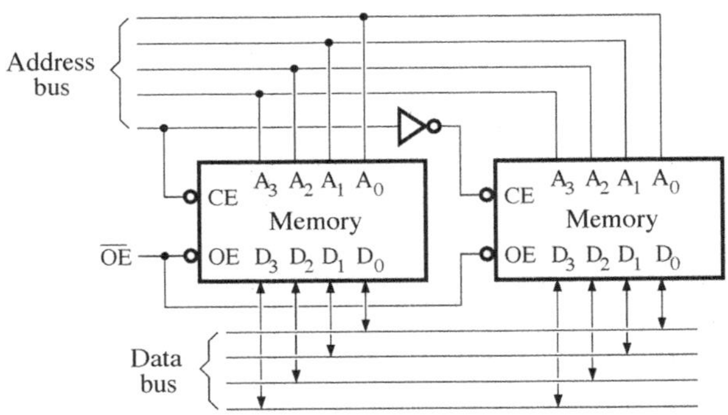

Figure 6.14. *Combining memories to expand the capacity*

6.5.2.1. *Example 1*

We wish to connect several $2\ K \times 8$ PROMs to produce a total capacity of $8\ K \times 8$.

Indicate how many PROM chips are required.

Determine the number of wires of the address bus.

Four PROM chips, each of $2\ K$ words, are required to obtain $8\ K$ words.

Because $8\ K = 8 \times 1\ 024 = 8\ 192 = 2^{13}$, the address bus must have 13 wires.

6.5.2.2. *Example 2*

We desire to implement a 4 $K \times 4$ memory, beginning with the address 0000h and using an elementary 2 $K \times 4$ memory and two elementary 1 $K \times 4$ memories.

Suggest the possible organization for this memory.

For the 4 $K \times 4$ memory, the data bus has four bits and the lowest address is 0, or:

$$0\ 0000\ 0000\ 0000 = 0000\ \text{h}$$

the highest address is $0 + 4\ K - 1 = 4\ 095$, or:

$$0\ 1111\ 1111\ 1111 = 0\text{FFFh}$$

The number of bits, n, required for the address bus is given by:

$$2^{n-1} < 4\,K - 1 \leq 2^n \qquad [6.1]$$

That is:

$$2^{n-1} < 4 \times 2^{10} - 1 \leq 2^n \qquad [6.2]$$

and finally:

$$n = 12 \qquad [6.3]$$

The address bits, therefore, are of the form: $A_{11}A_{10}A_9 \cdots A_0$.

The select logic equations will be even simpler provided that the address where each elementary memory begins is:

- either divisible by 2;
- or divisible by the capacity of the memory.

The two possible ways of organizing the 4 $K \times 4$ memory are given in Figure 6.15.

For each elementary memory, the address bits that are used only for decoding must be identified.

Starting from the bit A_0 to the least significant bits for an elementary memory, it is not necessary to take into account those bits that change from 0, in the starting address, to 1, in the final address.

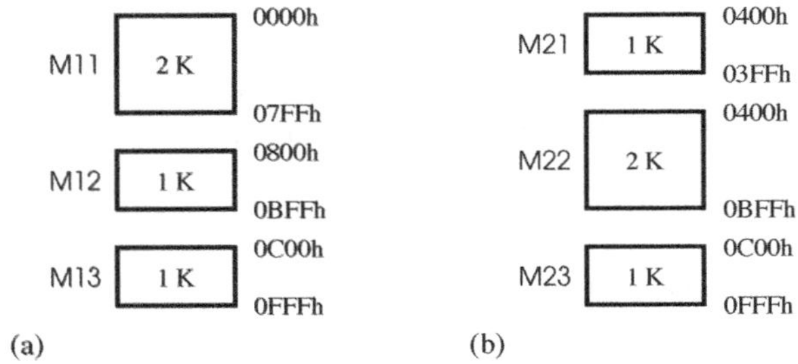

Figure 6.15. *a) Organization 1; b) organization 2*

– Organization 1 Using Table 6.6 to establish the select functions, the useful address bits are:

- A_{11} for the $2K$ memory (M11);
- A_{11} and A_{10} for the first $1K$ memory (M12);
- A_{11} and A_{10} for the second $1K$ memory (M13).

$\overline{CS}$	A_{11}	A_{10}	A_9	A_8	A_7	A_6	A_5	A_4	A_3	A_2	A_1	A_0	
0	0	0	0	0	0	0	0	0	0	0	0	0	0000h
	0	1	1	1	1	1	1	1	1	1	1	1	07FFh
0	1	0	0	0	0	0	0	0	0	0	0	0	0800h
	1	0	1	1	1	1	1	1	1	1	1	1	0BFFh
0	1	1	0	0	0	0	0	0	0	0	0	0	0C00h
	1	1	1	1	1	1	1	1	1	1	1	1	0FFFh

Table 6.6. *Address table for organization 1For a color version of this table, see www.iste.co.uk/ndjountche/electronics2.zip*

The select equations are given by:

$$\overline{CS_{11}} = \overline{\overline{\overline{CS}} \cdot \overline{A_{11}}} \qquad [6.4]$$

$$\overline{CS_{12}} = \overline{\overline{\overline{CS}} \cdot A_{11} \cdot \overline{A_{10}}} \qquad [6.5]$$

and:

$$\overline{CS_{13}} = \overline{\overline{\overline{CS}} \cdot A_{11} \cdot A_{10}} \qquad [6.6]$$

– Organization 2 Based on Table 6.7, the select functions depend on the following address bits:

- A_{11} and A_{10} for the first $1K$ memory (M21);
- A_{11} and A_{10} for the $2K$ memory(M22);
- A_{11} and A_{10} for the second $1K$ memory (M23).

$\overline{CS}$	A_{11}	A_{10}	A_9	A_8	A_7	A_6	A_5	A_4	A_3	A_2	A_1	A_0	
0	0	0	0	0	0	0	0	0	0	0	0	0	0000h
	0	0	1	1	1	1	1	1	1	1	1	1	03FFh
0	0	1	0	0	0	0	0	0	0	0	0	0	0400h
	1	0	1	1	1	1	1	1	1	1	1	1	0BFFh
0	1	1	0	0	0	0	0	0	0	0	0	0	0C00h
	1	1	1	1	1	1	1	1	1	1	1	1	0FFFh

Table 6.7. *Address table for organization 2For a color version of this table, see www.iste.co.uk/ndjountche/electronics2.zip*

The select logic equations can be written as follows:

$$\overline{CS_{21}} = \overline{\overline{\overline{CS}} \cdot \overline{A_{11}} \cdot \overline{A_{10}}} \qquad [6.7]$$

$$\overline{CS_{22}} = \overline{\overline{\overline{CS}} \cdot (\overline{A_{11}} \cdot A_{10} + A_{11} \cdot \overline{A_{10}})} \qquad [6.8]$$

and:

$$\overline{CS_{23}} = \overline{\overline{\overline{CS}} \cdot A_{11} \cdot A_{10}} \qquad [6.9]$$

The expression $\overline{CS_{22}}$ is more complex than the expression for $\overline{CS_{12}}$ because the starting address ($0400\ h$) is not divisible by the capacity of the elementary memory ($2\ K = 2 \times 2^{10} = 0800\ h$).

Figure 6.16 depicts the logic circuit of the $4\ K \times 4$ memory corresponding to each organization (1 and 2).

NOTE 6.2.– To obtain the useful address bits for the determination of the select function of each elementary memory, it is necessary to go through the address table, beginning with the least significant bit and moving toward the most significant bits, frame the bits that go from 0, in the starting address, to 1, in the final address and stop as soon as the value of the encountered bit is either 1 for the starting address or 0 for the final address. The unmarked bits will be used for the selection of each elementary memory.

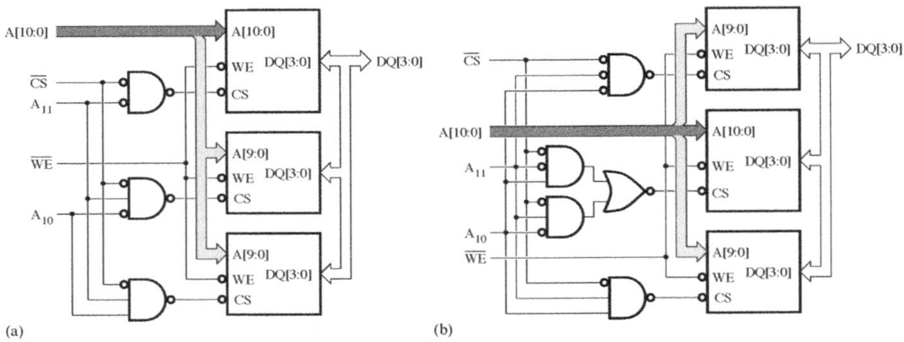

Figure 6.16. *Logic circuit of the $4K \times 4$ memory: a) organization 1; b) organization 2*

The logic expression for the select function may be obtained directly or may be simplified further by using a Karnaugh map.

A select function may be implemented using a decoder or logic gates.

6.5.2.3. *Binary-to-BCD converter*

Binary and binary-coded decimal (BCD) numbers are very frequently used in digital systems. The simplest conversion method consists of using an algorithm that allows modular synthesis based on expandable building blocks.

The conversion of a binary code to BCD is carried out based on algorithm 6.1.

Algorithm 6.1. *Conversion of a binary code to BCD*

---

[1] Shift the binary number to the left by one bit.

[2] Add 0011 each time to the bits that are found in the position of a BCD digit if the decimal value of these bits is greater than 4, before shifting them to the left by one bit.

[3] Repeat step [2] until the least significant bit of the binary code is located in the position of the least significant BCD digit. The bits thus obtained constitute the BCD code.

Convert, as an example, the 7-bit binary code, 1111011_2, to BCD.

Table 6.9 lists the different steps to carry out for the conversion. We thus have:

$$1111011_2 = 123_{BCD} \qquad [6.10]$$

The truth table shown in Table 6.8 establishes the equivalence between a 5-bit binary code and BCD number based on the conversion algorithm that recommends the addition of 3 each time there is a BCD digit greater than 4. It must be noted that the enable signal, $\overline{G}$, is an active-high signal. The logic circuit of a 5-bit binary-to-BCD (B2BCD) converter is illustrated in Figure 6.17. It is implemented using a ROM and may be considered as an expandable module.

Figures 6.18(a) and 6.18(b) show, respectively, the logic circuits for the 6-bit binary-to-BCD and 8-bit binary-to-BCD converters, obtained by cascading expandable modules so as to satisfy the shift-left requirement of the conversion algorithm.

6.5.2.4. *BCD-to-binary converter*

The steps for converting BCD to binary code are given in algorithm 6.2.

Algorithm 6.2. *Conversion of BCD to binary code*

[1] Shift the BCD number to the right by one bit.

[2] Subtract 0011 from a new consecutive 4-bit word each time its decimal value is greater than 7 (0111).

[2] Repeat steps [1] and [2] until the last bit is transferred out of the BCD number. The binary code then consists of the obtained bits.

In the case of the conversion of the BCD code 16_{BCD} or $1\ 0110_{BCD}$ to binary, the different conversion steps are described in Table 6.11.

And finally, we obtain:

$$16_{BCD} = 10000_2 \qquad [6.11]$$

To construct the truth table of the BCD-to-5-bit binary converter, as shown in Table 6.10, where the enable signal is represented by $\overline{G}$, the BCD-to-binary conversion algorithm can be interpreted as being the inverse of the binary-to-BCD conversion algorithm. Because this truth table must be implemented by a ROM with output three-state inverting buffers, the unused (or don't-care) states are set to high logic.

Figure 6.19 depicts the logic circuit of the BCD-to-5-bit binary (BCD2B) converter.

E	D	C	B	A	$\overline{G}$	Y_6	Y_5	Y_4	Y_3	Y_2	Y_1
0	0	0	0	0	0	0	0	0	0	0	0
0	0	0	0	1	0	0	0	0	0	0	1
0	0	0	1	0	0	0	0	0	0	1	0
0	0	0	1	1	0	0	0	0	0	1	1
0	0	1	0	0	0	0	0	0	1	0	0
0	0	1	0	1	0	0	0	1	0	0	0
0	0	1	1	0	0	0	0	1	0	0	1
0	0	1	1	1	0	0	0	1	0	1	0
0	1	0	0	0	0	0	0	1	0	1	1
0	1	0	0	1	0	0	0	1	1	0	0
0	1	0	1	0	0	0	1	0	0	0	0
0	1	0	1	1	0	0	1	0	0	0	1
0	1	1	0	0	0	0	1	0	0	1	0
0	1	1	0	1	0	0	1	0	0	1	1
0	1	1	1	0	0	0	1	0	1	0	0
0	1	1	1	1	0	0	1	1	0	0	0
1	0	0	0	0	0	0	1	1	0	0	1
1	0	0	0	1	0	0	1	1	0	1	0
1	0	0	1	0	0	0	1	1	0	1	1
1	0	0	1	1	0	0	1	1	1	0	0
1	0	1	0	0	0	1	0	0	0	0	0
1	0	1	0	1	0	1	0	0	0	0	1
1	0	1	1	0	0	1	0	0	0	1	0
1	0	1	1	1	0	1	0	0	0	1	1
1	1	0	0	0	0	1	0	0	1	0	0
1	1	0	0	1	0	1	0	1	0	0	0
1	1	0	1	0	0	1	0	1	0	0	1
1	1	0	1	1	0	1	0	1	0	1	0
1	1	1	0	0	0	1	0	1	0	1	1
1	1	1	0	1	0	1	0	1	1	0	0
1	1	1	1	0	0	1	1	0	0	0	0
1	1	1	1	1	0	1	1	0	0	0	1
x	x	x	x	x	1	1	1	1	1	1	1

Table 6.8. *Truth table for a 5-bit binary-to-BCD conversion*

The logic circuits of the BCD-to-6-bit binary and BCD-to-7-bit binary converters are represented in Figures 6.20(a) and 6.20(b), respectively, taking into account the shift-right operations.

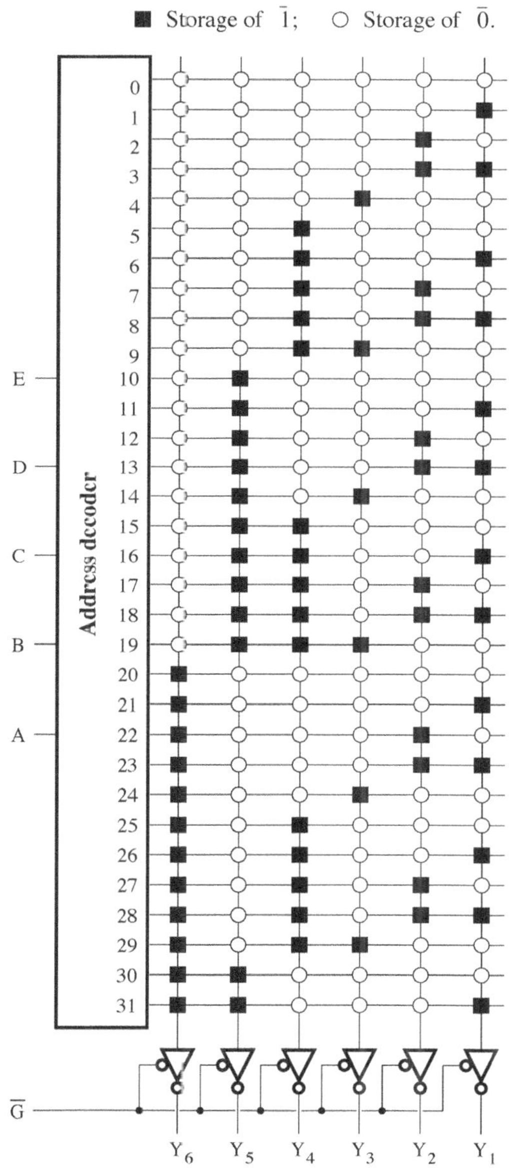

Figure 6.17. *Logic circuit of a 5-bit binary-to-BCD converter*

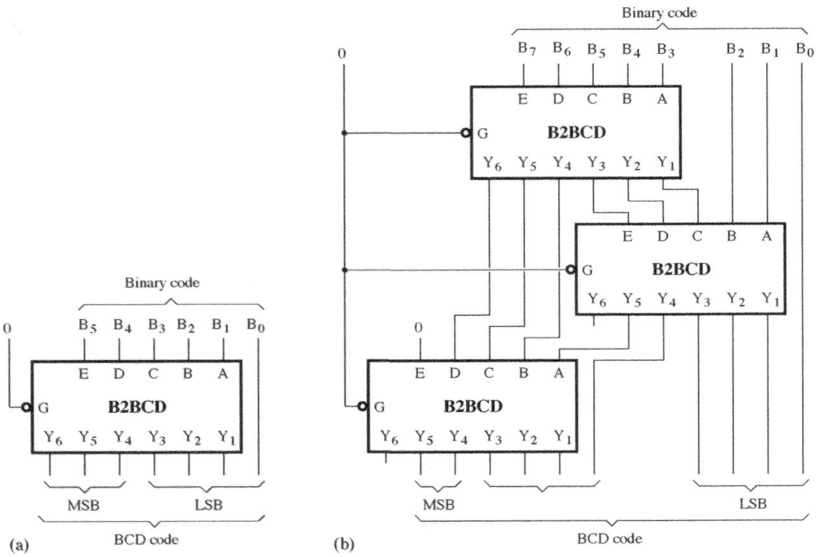

Figure 6.18. *Logic circuits: a) 6-bit binary-to-BCD and b) 8-bit binary-to-BCD converters*

Hundreds	Tens	Unit		Operation
			1111011	Store the binary code
		1	111011	Shift left
		11	11011	Shift left
		111	1011	Shift left
		1010	1011	Add 111
	1	0101	011	Shift left
	1	1000	011	Add 011 to 0101
	11	0000	11	Shift left
	110	0001	1	Shift left
	1001	0001	1	Add 011 to 110
1	0010	0011		Shift left
1	2	3		

Table 6.9. *Conversion of the code* 1111011_2 *to BCD*

6.6. Other types of memory

Several other types of memory are used to meet the requirements of increasingly different applications.

E	D	C	B	A	$\overline{G}$	Y_5	Y_4	Y_3	Y_2	Y_1
0	0	0	0	0	0	0	0	0	0	0
0	0	0	0	1	0	0	0	0	0	1
0	0	0	1	0	0	0	0	0	1	0
0	0	0	1	1	0	0	0	0	1	1
0	0	1	0	0	0	0	0	1	0	0
0	1	0	0	0	0	0	0	1	0	1
0	1	0	0	1	0	0	0	1	1	0
0	1	0	1	0	0	0	0	1	1	1
0	1	0	1	1	0	0	1	0	0	0
0	1	1	0	0	0	0	1	0	0	1
1	0	0	0	0	0	0	1	0	1	0
1	0	0	0	1	0	0	1	0	1	1
1	0	0	1	0	0	0	1	1	0	0
1	0	0	1	1	0	0	1	1	0	1
1	0	1	0	0	0	0	1	1	1	0
1	1	0	0	0	0	0	1	1	1	1
1	1	0	0	1	0	1	0	0	0	0
1	1	0	1	0	0	1	0	0	0	1
1	1	0	1	1	0	1	0	0	1	0
1	1	1	0	0	0	1	0	0	1	1
x	x	x	x	x	1	1	1	1	1	1

Table 6.10. *Truth table for a BCD-to-5-bit binary conversion*

Tens	Units		Operation
1	0110		Store the BCD code
	1011	0	Shift to right
	101	10	Shift to right
	10	110	Shift to right
	1	0110	Shift to right
		10110	Shift to right
		10000	Subtract 011 from 1011
		10000	

Table 6.11. *Conversion of 16_{BCD} to binary*

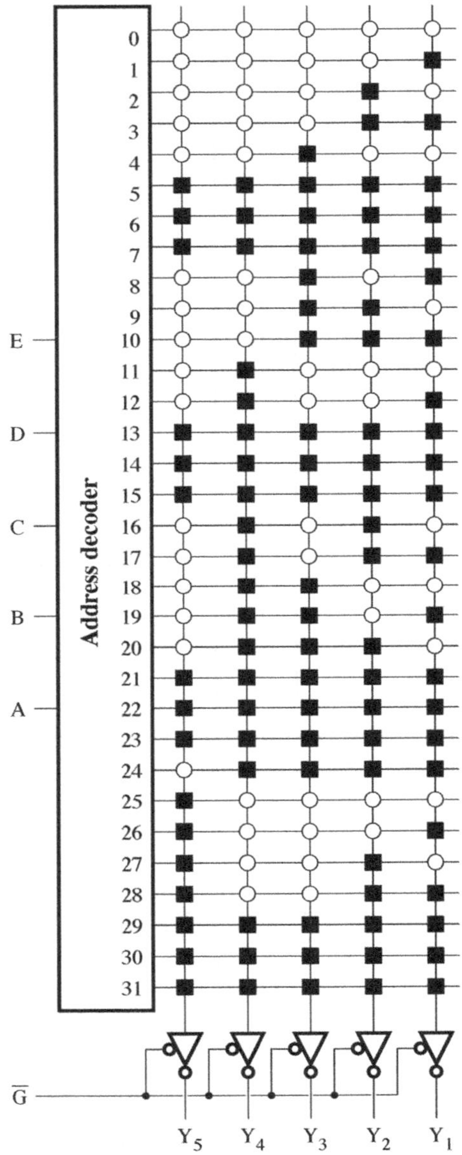

Figure 6.19. *Logic circuit of a BCD-to-5-bit binary converter*

6.6.1. *Ferromagnetic RAM*

A *ferromagnetic RAM* (FRAM) is based on a cell that is similar to that of a DRAM, but uses a capacitor with a dielectric layer made up of a ferroelectric material (perovskite crystal) to obtain non-volatility.

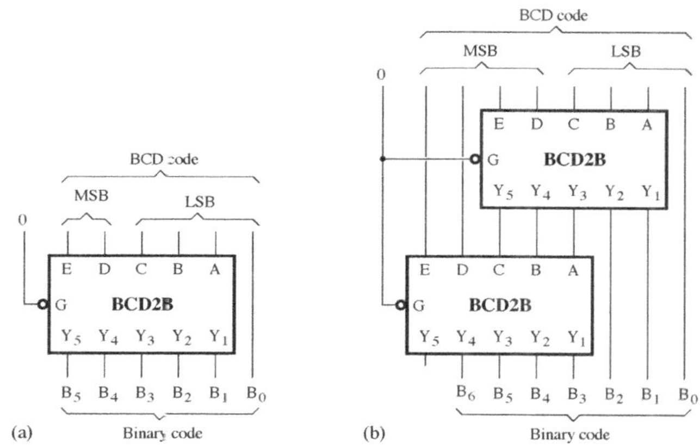

Figure 6.20. *Logic circuits for a) BCD-to-6-bit binary and b) BCD-to-7-bit binary converters*

A FRAM cell is illustrated in Figure 6.21(a). Figure 6.21(b) shows the polarization, P, that characterizes the ability of the dipole moments of the ferroelectric material to move in the direction of the electric field associated with the voltage, V, at the capacitor terminals. It presents a hysteresis cycle.

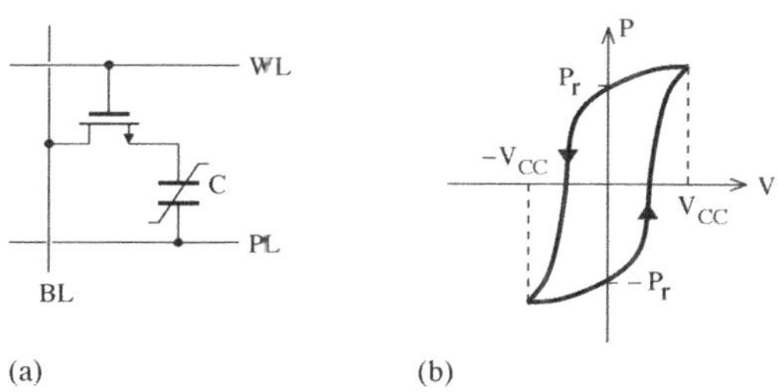

Figure 6.21. *a) FRAM cell; b) polarization characteristic*

During a write operation, the application of the voltage V_{CC} to the capacitor terminals enables the storage of the logic state 1, while the logic state 0 is stored upon the application of the voltage $-V_{CC}$.

To carry out a read operation, the BL line is precharged to zero voltage and the WL line is activated. When the *plate line* is connected to the voltage V_{CC}, a sense amplifier is used to restore and identify the corresponding logic level depending on whether the variation in the voltage applied to the BL line is increasing or decreasing. Because a read operation may cause an inversion of the polarization, it must be followed by a non-periodic refresh operation, unlike the case of the DRAM whose data retention is continually undermined by current leakage so that a periodic refresh is required.

While being characterized by low power consumption, endurance for read/write operations up to 100 trillion cycles and significantly reduced space requirement, FRAM provides access speeds up to a hundred times more faster than a conventional flash memory.

6.6.2. *Content-addressable memory*

A *content-addressable memory* (CAM), also known as associative memory, compares input data to data that is already stored, to generate a correspondence indicator and then return the address of matching data.

It is used in applications that require operations related to pattern matching on data, such as cache controllers, communication networks, image coding and data compression. The operation principle of RAM and CAM are illustrated in Figure 6.22, where the *data/command select* input, $\overline{CM}$, is used to choose between the data cycle and the command cycle.

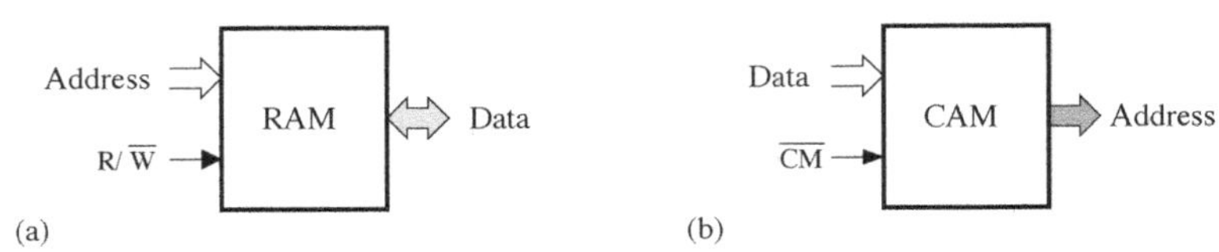

Figure 6.22. *Operation principle of a) a RAM and b) a CAM*

The structure of a CAM, as shown in Figure 6.23, generally comprises a network of storage cells with validity bits, a decoder for line addresses, a comparison register, a mask register, sense amplifiers for reading, output data register, a correspondence detector, a priority encoder, an address encoder and a control unit connected to write-enable ($\overline{W}$), chip enable ($\overline{E}$) and data/command select ($\overline{CM}$) signals. The comparison registers may be used as a buffer for data to be stored in memory. After the address decoder selects a WL line, data are written in the corresponding cells through the BL and $\overline{BL}$ lines. The validity bits are used to indicate the characteristic (valid, empty, skip or RAM) of the word stored in cells associated with the corresponding WL line. For data to be read, the signal in the BL line of the selected cell is converted to a logic level by the sense amplifier before being transferred to the output register.

The content of the memory is accessible in a random manner or through association by comparison. A correspondence search may be carried out when all the

WL lines are deactivated. Data are placed in the comparison registers and are compared simultaneously to the valid inputs of the memory. Any bit of the comparison register may be excluded or not from the comparison operation depending on the logic level (0 or 1) stored in the mask register. The correspondence detector generates the status (no correspondence, single correspondence, multiple correspondences) based on the result of the comparison operation. In the case of the single correspondence or multiple correspondence, the address with the highest priority is identified by the priority encoder and can then be generated by the address encoder.

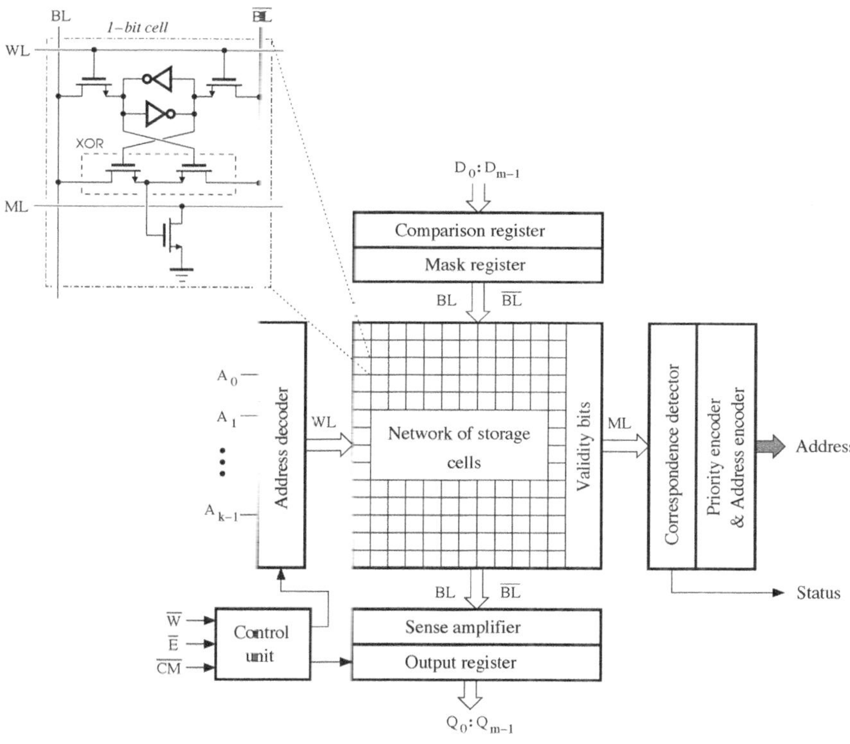

Figure 6.23. *Simplified structure of a CAM*

6.6.3. *Sequential access memory*

Sequential access memories are used as a synchronization buffer between components operating at different speeds or operating irregularly. Otherwise, the slowest component would determine the operating speeds of all other components involved in the data transfer.

First in first out (FIFO) and *last in first out* (LIFO) memories are classes of sequential access memories. The choice between these two structures depends on the application.

6.6.3.1. *FIFO memory*

A FIFO memory is organized such that data are read in the same order in which they were written. This memory can be implemented using shift registers or a RAM.

A 4×5 register-based FIFO memory is represented in Figure 6.24, where data D_i $(i = 0, 1, 2, 3, 4)$ are applied to the identical sections composed of storage flip-flops. Each new 5-bit word is transferred from one flip-flop stage to another to be stored in the last available position. Reading a word causes a shift by one position toward the output of the remaining words. The clock signal generator successively produces pulses at the nodes C_1, C_2, C_3 and C_4, when a word is written. It also provides information on the status of each word, specifying whether a flip-flop already contains valid data ($\overline{FULL}$) or whether it is empty ($\overline{EMPTY}$).

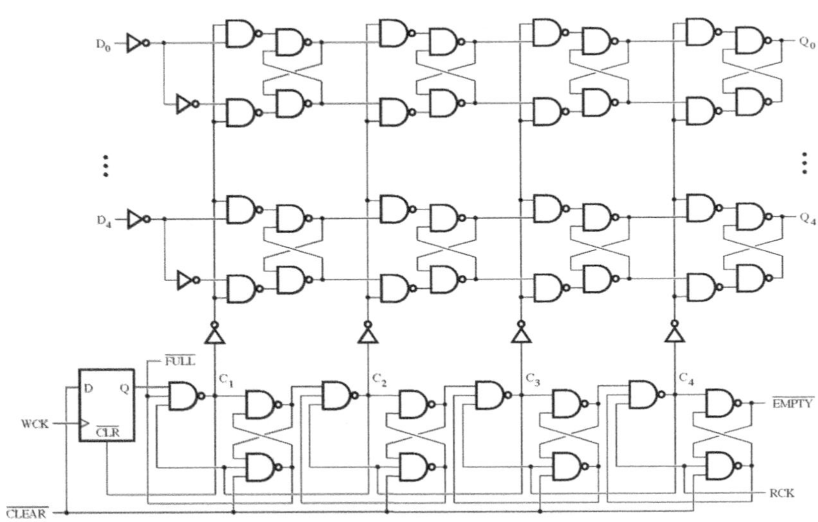

Figure 6.24. *Structure of a 4×5 register-based FIFO memory*

The time required for a word to be moved from the input to the output is called the fall-through time.

NOTE 6.3.– Comparing the operation of a conventional register and a FIFO register. Tables 6.12 and 6.13 are drawn up to illustrate the difference between the operation of a conventional shift register and the operation of a FIFO shift register (or register-based FIFO memory). The sequence 0110 is applied, one bit at a time, to the serial input, D_i, of each four-bit capacity register (B_1, B_2, B_3 and B_4) to be transferred toward the output Q_i.

Input D_i	B_1 x	B_2 x	B_3 x	B_4 x	Output Q_i
0	0	x	x	x	→
1	1	0	x	x	→
1	1	1	0	x	→
0	0	1	1	0	→

Table 6.12. *Shift register (x represents a don't-care state)*

Input D_i	B_1 –	B_2 –	B_3 –	B_4 –	Output Q_i
0	0	0	0	0	→
1	1	1	1	0	→
1	1	1	1	0	→
0	0	1	1	0	→

Table 6.13. *FIFO shift register (– indicates an empty position)*

In contrast to a register-based FIFO memory, the fall-through time for a RAM-based FIFO memory is not dependent on the number of words that can be stored. Figure 6.25 depicts the structure of a RAM-based FIFO memory where the reset signal is designated by $\overline{RS}$ and the output enable signal is represented by $\overline{OE}$. Write and read operations are controlled by the clock signals $WCLK$ and $RCLK$ and the enable signals $\overline{WEN}$ and $\overline{REN}$, respectively. Write addresses are generally produced by the write pointer and read addresses by the read pointer. Pointers can be implemented like binary counters. The status signals $\overline{EF}$, $\overline{FF}$, $\overline{PAE}$ and $\overline{PAF}$ (empty, full, almost empty, almost full) are produced by the flag logic circuit. The statuses $\overline{EF}$ and $\overline{FF}$ cannot be redefined while the statuses $\overline{PAE}$ and $\overline{PAF}$ are programmable using the offset register, initialized by the load signal $\overline{LD}$.

6.6.3.2. *LIFO memory*

A LIFO memory is based on the following operation principle: the last bit to be written is the first bit to be read. The LIFO memory can be used for the storage of data that are to be retrieved in reverse order.

A LIFO memory is often called a *stack*. Most microprocessors use a stack to save status flag bits and the contents of certain registers, in case of interruptions.

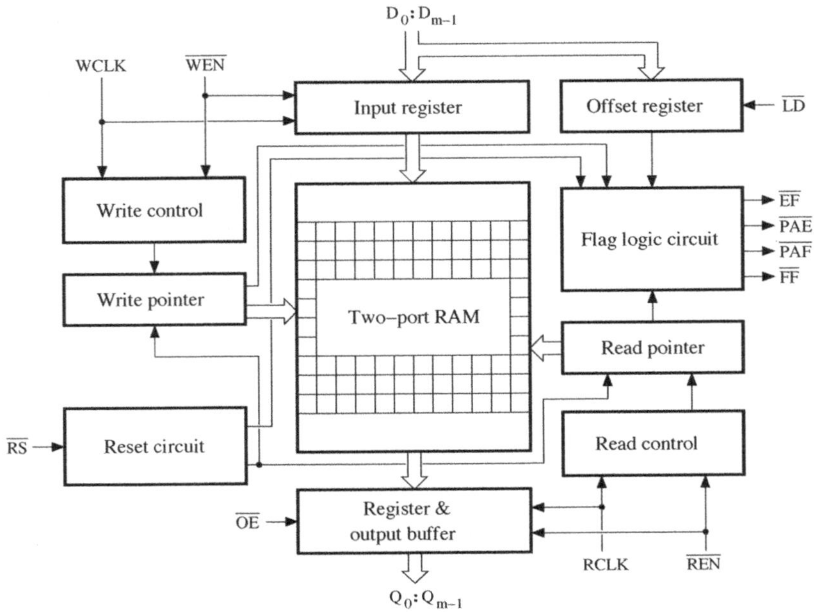

Figure 6.25. *Structure of a RAM-based FIFO memory*

The structure of a LIFO memory is similar to that of a FIFO memory and it may be based on shift registers or RAM. In principle, data stored in a register-based LIFO memory can be moved from one position to the other at each clock pulse. However, for a RAM-based LIFO memory, it is not the data but the access positions that move, controlled by counters that are called write or read pointers.

6.7. Exercises

EXERCISE 6.1.– A 64 $K \times 8$ memory comprises an EPROM located from the address 0000 h, and three RAMs located from the addresses 4000 h, 8000 h and C000 h, respectively. The capacity of each of the elementary memories is 16 $K \times 8$.

Draw up an address decoding table for this memory.

Propose a functional block diagram for this memory.

EXERCISE 6.2.– Implementing a 4 $K \times 8$ RAM requires the use of four elementary 1 $K \times 8$ memories, as shown in Figure 6.26, where the number of bits used for the addressing is 16.

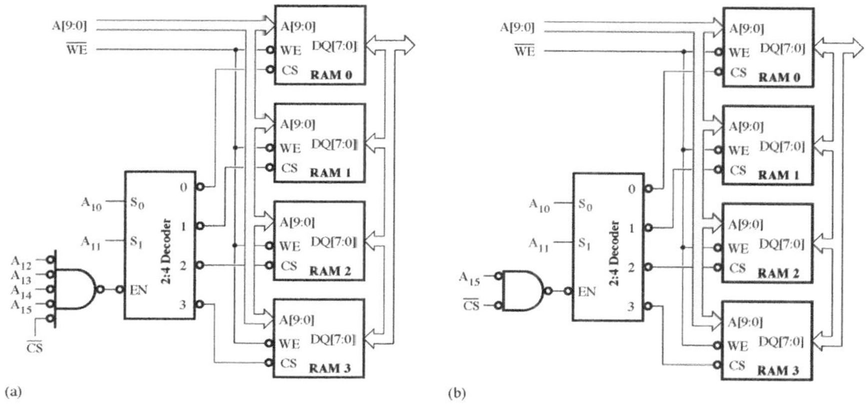

Figure 6.26. *Functional block diagram of the memory with a) full decoding and b) partial decoding*

Determine the range of addresses for each of the elementary memories shown in Figure 6.26(a).

To reduce the complexity of the address decoding in some applications, a memory with partial decoding instead of one with full decoding can be adopted.

What is the disadvantage of the partial decoding used for the memory shown in Figure 6.26(b)?

Can A_{15} be replaced by A_{14} in this memory if the range of addresses from 8C00h to 8CFFh is assigned to input/output devices?

EXERCISE 6.3.– Consider the implementation of a RAM composed of two elementary $16\ K \times 8$ memories and which is based on the map depicted in Figure 6.27, where the address space is $64\ K$:

– Determine the different address decoding functions.

– Propose a functional block diagram for this memory.

EXERCISE 6.4.– An EPROM has a capacity of $32\ K$ octets and can store 8-bit words.

a) Determine the number of bits of the address bus and the data bus for this memory;

b) Give the symbol of this memory;

c) Determine the number of accessible zones, their starting addresses and final addresses for the following cases:

– $A_{14} = 0$;

– $A_{13} = 1$.

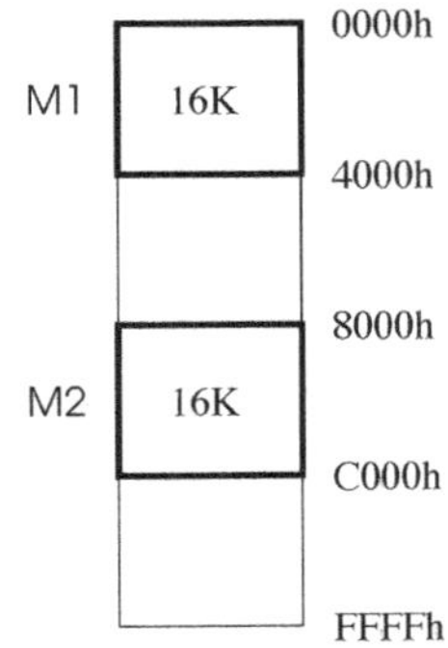

Figure 6.27. *Memory map*

EXERCISE 6.5.– A 10 $K \times 8$ memory is realized using elementary 4 $K \times 8$ and 2 $K \times 8$ memories.

Determine the address decoding equations for the different elementary memories when the memory system to be constituted begins at:

a) 0000 h;

b) 0800 h.

EXERCISE 6.6.– A microcontroller system with a 16-bit address bus $((A_{15} - A_0))$, an 8-bit data bus $((D_7 - D_0))$, a $\overline{WR}$ signal for writing and a $\overline{RD}$ signal for reading should be connected to the following read-only memories:

– a 16 $K \times 8$ memory located from 2000 h;

– an 8 $K \times 8$ memory located from 8000 h;

– a 16 $K \times 8$ memory located just after the 8 $K \times 8$ memory.

Determine the address decoding equations for these memories.

EXERCISE 6.7.– A microcontroller system has a 16-bit address bus $(A_{15}A_{14} \cdots A_0)$, an 8-bit data bus $(D_7D_6 \cdots D_0)$, and control signals $\overline{RD}$ (active-low) for reading and $\overline{WR}$ (active on rising edge) for writing.

Implement the address decoding functions for the following read-only memories:

– M1 is a 8 $K \times 8$ memory located in the address space that covers the addresses 0000h – 1FFFh;

– M2 is a 4 $K \times 8$ memory located in the address space covering the addresses 2000h – 2FFFh;

– M3 is a 16 $K \times 8$ memory located in the address space covering the addresses 3000h – 6FFFh.

EXERCISE 6.8.– A programmable circuit for the control of a play of lights (or *chaser*), can be realized as shown in Figure 6.28, using logic gates (XOR, inverter), two parallel-in parallel-out shift registers, a modulo 16 binary counter, and a 256 × 8 EPROM. The four LEDs to be controlled are L0, L1, L2, and L3. CK represents the clock signal.

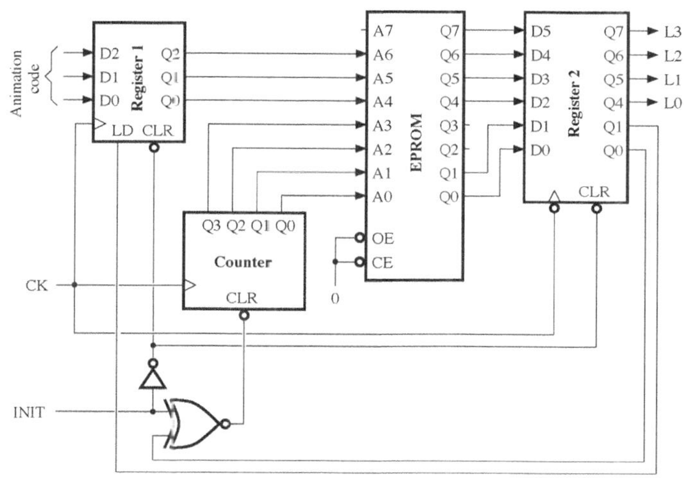

Figure 6.28. *Block diagram of a chaser*

An animated sequence can have up to sixteen steps and is selected based on a 3-bit code applied to the inputs of register 1. Each step in the sequence is stored as a word in the memory. An animation sequence is initiated by an INIT signal pulse and begins with a reset of register 2 and the counter, followed by the loading of register 1. It ends after the diode switching-on operations by the reset of the counter:

a) What is the maximum number of animation sequences possible?

b) Representing the EPROM memory addresses as hexadecimal numbers, determine the content of the memory for the sequence 0 whose animation code is

0 and which is characterized by the successive switching on of diodes from left to right, right to left and, finally, the switching on of all diodes;

c) Same question for sequence 1 which switches on the even-numbered diodes and then odd-numbered diodes, alternately;

d) Same question for sequence 2 that switches on the diodes based on the binary representation of the numbers from 1 to 15, diode L3 being considered as the most significant bit and diode L0 as the least significant bit.

6.8. Solutions

SOLUTION 6.1.– Organization of a $64K \times 8$ memory.

The address space allocated to each elementary memory ranges from 0 to $2^{14} - 1$. We thus have:

– EPROM:

Starting address: 0000 0000 0000 0000 = 0000h
Final address: 0011 1111 1111 1111 = 3FFFh

– RAM 1:

Starting address: 0100 0000 0000 0000 = 4000h
Final address: 0111 1111 1111 1111 = 7FFFh

– RAM 2:

Starting address: 1000 0000 0000 0000 = 8000h
Final address: 1011 1111 1111 1111 = BFFFh

– RAM 3:

Starting address: 1100 0000 0000 0000 = C000h
Final address: 1111 1111 1111 1111 = FFFFh

Table 6.14 depicts the address decoding table for the memory system. Only the logic states of the signals $\overline{CS}$, A_{15} and A_{14} are useful in the determination of the address decoding functions for each of the elementary functions.

In addition to the condition, $\overline{CS} = 0$, the address decoding function takes the form $A_{15}A_{14} = 00$ for the EPROM, $A_{15}A_{14} = 01$ for RAM 1, $A_{15}A_{14} = 10$ for RAM 2 and $A_{15}A_{14} = 11$, for RAM 3. The address decoding function can, therefore, be implemented using a 2 : 4 decoder with an enable signal.

The functional block diagram for the $64K \times 8$ memory system is shown in Figure 6.29.

$\overline{CS}$	A_{15}	A_{14}	A_{13}	A_{12}	A_{11}	A_{10}	A_9	A_8	A_7	A_6	A_5	A_4	A_3	A_2	A_1	A_0	
0	0	0	0	0	0	0	0	0	0	0	0	0	0	0	0	0	0000h
	0	0	1	1	1	1	1	1	1	1	1	1	1	1	1	1	3FFFh
0	0	1	0	0	0	0	0	0	0	0	0	0	0	0	0	0	4000h
	0	1	1	1	1	1	1	1	1	1	1	1	1	1	1	1	7FFFh
0	1	0	0	0	0	0	0	0	0	0	0	0	0	0	0	0	8000h
	1	0	1	1	1	1	1	1	1	1	1	1	1	1	1	1	BFFFh
0	1	1	0	0	0	0	0	0	0	0	0	0	0	0	0	0	C000h
	1	1	1	1	1	1	1	1	1	1	1	1	1	1	1	1	FFFFh

Table 6.14. *Address decoding table. For a color version of this table, see www.iste.co.uk/ndjountche/electronics2.zip*

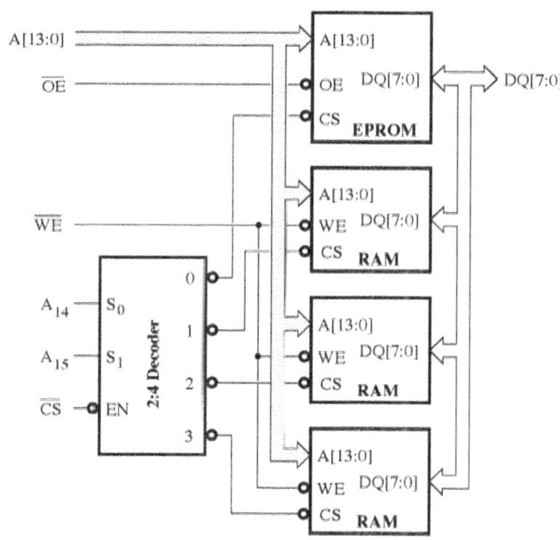

Figure 6.29. *Functional block diagram of the memory*

SOLUTION 6.2.– The analysis of the memory system based on the full decoding yields, for the elementary memories, the following addresses:

– RAM 0

$$A_{15}A_{14}A_{13}A_{12}A_{11}A_{10}A_9A_8A_7A_6A_5A_4A_3A_2A_1A_0$$

0 0 0 0 0 0 x x x x x x x x x x

– RAM 1

$$A_{15}A_{14}A_{13}A_{12}A_{11}A_{10}A_9A_8A_7A_6A_5A_4A_3A_2A_1A_0$$
$$0\ 0\ 0\ 0\ 0\ 1\ x\ x\ x\ x\ x\ x\ x\ x\ x\ x$$

– RAM 2

$$A_{15}A_{14}A_{13}A_{12}A_{11}A_{10}A_9A_8A_7A_6A_5A_4A_3A_2A_1A_0$$
$$0\ 0\ 0\ 0\ 1\ 0\ x\ x\ x\ x\ x\ x\ x\ x\ x\ x$$

– RAM 3

$$A_{15}A_{14}A_{13}A_{12}A_{11}A_{10}A_9A_8A_7A_6A_5A_4A_3A_2A_1A_0$$
$$0\ 0\ 0\ 0\ 1\ 1\ x\ x\ x\ x\ x\ x\ x\ x\ x\ x$$

where the don't-care state, x, is assigned to the address bits that are not used for decoding.

Each elementary memory has a capacity of $1K = 2^{10}$ and thus requires ten address lines. For all elementary memories, the address space occupies $4K = 2^2 \times 2^{10} = 2^{12}$, corresponding to 12 address lines.

$\overline{CS}$	A_{15}	A_{14}	A_{13}	A_{12}	A_{11}	A_{10}	A_9	A_8	A_7	A_6	A_5	A_4	A_3	A_2	A_1	A_0	
0	0	0	0	0	0	0	0	0	0	0	0	0	0	0	0	0	0000h
	0	0	0	0	0	0	1	1	1	1	1	1	1	1	1	1	03FFh
0	0	0	0	0	0	1	0	0	0	0	0	0	0	0	0	0	0400h
	0	0	0	0	0	1	1	1	1	1	1	1	1	1	1	1	07FFh
0	1	0	0	0	1	0	0	0	0	0	0	0	0	0	0	0	0800h
	1	0	0	0	1	0	1	1	1	1	1	1	1	1	1	1	0BFFh
0	0	0	0	0	1	1	0	0	0	0	0	0	0	0	0	0	0C00h
	0	0	0	0	1	1	1	1	1	1	1	1	1	1	1	1	0FFFh

Table 6.15. *Address table. For a color version of this table, see www.iste.co.uk/ndjountche/electronics2.zip*

The combination of logic states for the bits A_{10} and A_{11} differs from one elementary memory to another, while the logic state for each of the bits A_{15}, A_{14}, A_{13}, and A_{12} is unchanged. The address decoding table for the memory system can be drawn up as shown in Table 6.15. We thus have:

– RAM 0:

Starting address: 0000h

Final address: 03FFh

– RAM 1:

Starting address: 0400h
Final address: 07FFh

– RAM 2: Starting address: 0800h
Final address: 0BFFh

– RAM 3:

Starting address: 0C00h
Final address: 0FFFh

In the case of the memory system based on partial decoding, the addresses take the following forms:

– RAM 0

$$\begin{array}{cccccccccccccccc} A_{15} & A_{14} & A_{13} & A_{12} & A_{11} & A_{10} & A_9 & A_8 & A_7 & A_6 & A_5 & A_4 & A_3 & A_2 & A_1 & A_0 \\ 0 & x & x & x & 0 & 0 & x & x & x & x & x & x & x & x & x & x \end{array}$$

– RAM 1

$$\begin{array}{cccccccccccccccc} A_{15} & A_{14} & A_{13} & A_{12} & A_{11} & A_{10} & A_9 & A_8 & A_7 & A_6 & A_5 & A_4 & A_3 & A_2 & A_1 & A_0 \\ 0 & x & x & x & 0 & 1 & x & x & x & x & x & x & x & x & x & x \end{array}$$

– RAM 2

$$\begin{array}{cccccccccccccccc} A_{15} & A_{14} & A_{13} & A_{12} & A_{11} & A_{10} & A_9 & A_8 & A_7 & A_6 & A_5 & A_4 & A_3 & A_2 & A_1 & A_0 \\ 0 & x & x & x & 1 & 0 & x & x & x & x & x & x & x & x & x & x \end{array}$$

– RAM 3

$$\begin{array}{cccccccccccccccc} A_{15} & A_{14} & A_{13} & A_{12} & A_{11} & A_{10} & A_9 & A_8 & A_7 & A_6 & A_5 & A_4 & A_3 & A_2 & A_1 & A_0 \\ 0 & x & x & x & 1 & 1 & x & x & x & x & x & x & x & x & x & x \end{array}$$

where the don't-care state, x, is assigned to the address bits that are not used for decoding.

Allowing only the change in logic state of each of the bits A_{14}, A_{13} and A_{12}, there are $2^3 = 8$ possible combinations that can be generated to access the same elementary memory.

The partial decoding helps reduce the complexity of the address decoding functions, but more than one range of addresses can access the same elementary memory. This is not acceptable unless these ranges do not include address already assigned to other memories or other input/output devices.

No, because RAM 3 can be selected using the address that belongs to the range from 8C00h to 8CFFh.

SOLUTION 6.3.– The characteristics of an elementary memory and the address space translate into:

$$16\,K = 16 \times 2^{10} = 2^4 \times 2^{10} = 2^{14}$$

and

$$64\,K = 64 \times 2^{10} = 2^6 \times 2^{10} = 2^{16}$$

A $16K$ memory system requires fourteen address lines and for a $64K$ address space, 16 address lines are required.

The address decoding table is given in Table 6.16. The condition to select one of the elementary memories is $CS \cdot \overline{A_{15}} \cdot \overline{A_{14}}$, and for the other it is $CS \cdot A_{15} \cdot \overline{A_{14}}$. Figure 6.30 depicts the functional block diagram of the memory system.

$\overline{CS}$	A_{15}	A_{14}	A_{13}	A_{12}	A_{11}	A_{10}	A_9	A_8	A_7	A_6	A_5	A_4	A_3	A_2	A_1	A_0	
0	0	0	0	0	0	0	0	0	0	0	0	0	0	0	0	0	0000h
	0	0	1	1	1	1	1	1	1	1	1	1	1	1	1	1	3FFFh
0	1	0	0	0	0	0	0	0	0	0	0	0	0	0	0	0	8000h
	1	0	1	1	1	1	1	1	1	1	1	1	1	1	1	1	BFFFh

Table 6.16. *Address decoding table. For a color version of this table, see www.iste.co.uk/ndjountche/electronics2.zip*

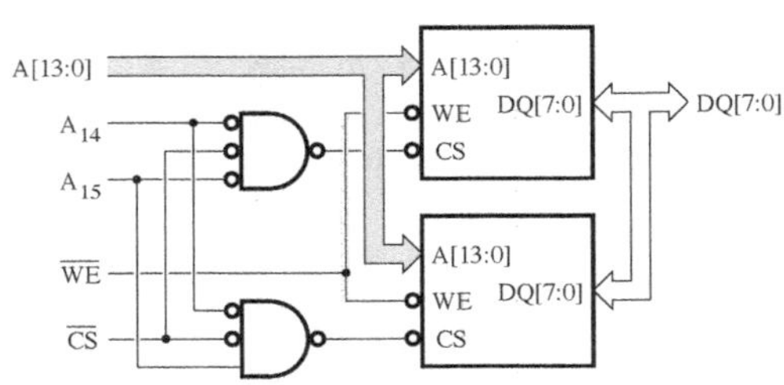

Figure 6.30. *Functional block diagram for the memory system*

SOLUTION 6.4.– The EPROM has a capacity of $32K \times 8$.

a) The number of wires of the address bus is 15, because:

$$32\,K = 32 \times 2^{10} = 2^5 \times 2^{10} = 2^{15}$$

b) The symbol for this EPROM is given in Figure 6.31.

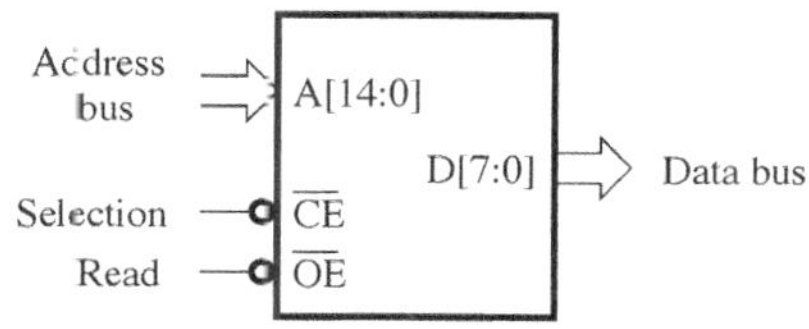

Figure 6.31. *Symbol for the EPROM*

c) Number of accessible zones.

– When $A_{14} = 0$

According to the address decoding table shown in Table 6.17, 0000h to 3FFFh constitutes an accessible zone.

	A_{14}	A_{13}	A_{12}	A_{11}	A_{10}	A_9	A_8	A_7	A_6	...	A_0	
Start	0	0	0	0	0	0	0	0	0	...	0	0000h
End		1	1	1	1	1	1	1	1	...	1	03FFh

Table 6.17. *Address decoding table* $(A_{14} = 0)$

– When $A_{13} = 1$

As given in the address table shown in Table 6.18, the split bit is A_{13} and there is a single unfixed bit, A_{14}.

	A_{14}	A_{13}	A_{12}	A_{11}	A_{10}	A_9	A_8	A_7	A_6	...	A_0	
Zone 0	0	1	0	0	0	0	0	0	0	...	0	2000h
		1	1	1	1	1	1	1	1	...	1	3FFFh
Zone 1	1	1	0	0	0	0	0	0	0	...	0	6000h
		1	1	1	1	1	1	1	1	...	1	7FFFh

Table 6.18. *Address decoding table* $(A_{13} = 1)$

The number of zones, therefore, is $2^1 = 2$.

The accessible zones depend on the logic state of A_{14}:

$A_{14} = 0$, zone 0: 2000h – 3FFFh

$A_{14} = 1$, zone 1: 6000h – 7FFFh.

SOLUTION 6.5.– Organization of a 10 $K \times 8$ memory system.

The capacity of the memory system is 10 $K \times 8$, and we have:

$$10\,K = 10 \times 2^{10} = 5 \times 2^{11}$$

a) Starting address 0000h.

The final address is 0000h $+ 5 \times 2^{11} - 1$.

The number, n, of address bits required, may be determined using the equation:

$$2^{n-1} < 5 \times 2^{11} - 1 < 2^n$$

Thus:

$$n = 14$$

The address bits are: $A_{13}A_{12} \cdots A_0$.

Two $4K$ elementary memories and one $2K$ elementary memory are required:

– M11 4 K memory: 0000h–0FFFh;

– M12 4 K memory: 1000h–1FFFh;

– M13 2 K memory: 2000h–27FFh.

To obtain simple address decoding functions in each case, the starting address is a power of 2 and is divisible by the capacity of the elementary memory (especially, 4, 096 and 2, 048).

	$\overline{CS}$	A_{13}	A_{12}	A_{11}	A_{10}	A_9	A_8	A_7	A_6	A_5	...	A_0	
M11	0	0	0	0	0	0	0	0	0	0	...	0	0000h
		0	0	1	1	1	1	1	1	1	...	1	0FFFh
M12	0	0	1	0	0	0	0	0	0	0	...	0	1000h
		0	1	1	1	1	1	1	1	1	...	1	1FFFh
M13	0	1	0	0	0	0	0	0	0	0	...	0	2000h
		1	0	0	1	1	1	1	1	1	...	1	27FFh

Table 6.19. *Address decoding table. For a color version of this table, see www.iste.co.uk/ndjountche/electronics2.zip*

Table 6.19 gives the address decoding table that can be used to obtain the following logic equations:

$$\overline{CS_{11}} = \overline{\overline{\overline{CS}} \cdot \overline{A_{13}} \cdot \overline{A_{12}}} \quad [6.12]$$

$$\overline{CS_{12}} = \overline{\overline{\overline{CS}} \cdot \overline{A_{13}} \cdot A_{12}} \quad [6.13]$$

and:

$$\overline{CS_{23}} = \overline{\overline{\overline{CS}} \cdot A_{13} \cdot \overline{A_{12}} \cdot \overline{A_{11}}} \quad [6.14]$$

b) Starting address 0800h.

The final address is given by:

$$0800\text{h} + 5 \times 2^{11} - 1 = 2^{11} + 5 \times 2^{11} - 1 = 6 \times 2^{11} - 1$$

The required number, n, of address bits, must verify the relationship:

$$2^{n-1} < 6 \times 2^{11} - 1 < 2^{n}$$

Thus:

$$n = 14$$

The address bits are $A_{13}A_{12} \cdots A_0$.

Two elementary $4K$ memories and one elementary $2K$ memory are needed:

– M21 2K memory: 0800h–0FFFh ;

– M22 4K memory: 1000h–1FFFh ;

– M23 4K memory: 2000h–2FFFh.

Table 6.20 gives the address decoding table. It can be used to obtain the following address decoding equations:

$$\overline{CS_{21}} = \overline{\overline{\overline{CS}} \cdot \overline{A_{13}} \cdot \overline{A_{12}} \cdot A_{11}} \quad [6.15]$$

$$\overline{CS_{22}} = \overline{\overline{\overline{CS}} \cdot \overline{A_{13}} \cdot A_{12}} \quad [6.16]$$

	$\overline{CS}$	A_{13}	A_{12}	A_{11}	A_{10}	A_9	A_8	A_7	A_6	A_5	...	A_0	
M21	0	0	0	1	0	0	0	0	0	0	...	0	0800h
		0	0	1	1	1	1	1	1	1	...	1	0FFFh
M22	0	0	1	0	0	0	0	0	0	0	...	0	1000h
		0	1	1	1	1	1	1	1	1	...	1	1FFFh
M23	0	1	0	0	0	0	0	0	0	0	...	0	2000h
		1	0	1	1	1	1	1	1	1	...	1	2FFFh

Table 6.20. *Address decoding table. For a color version of this table, see www.iste.co.uk/ndjountche/electronics2.zip*

and:

$$\overline{CS_{23}} = \overline{\overline{\overline{CS}} \cdot A_{13} \cdot \overline{A_{12}}} \qquad [6.17]$$

SOLUTION 6.6.– For the different elementary memories, we have:

$$16\,K = 16 \times 2^{10} = 2^4 \times 2^{10} = 2^{14}, \text{ or 4000h (in hexadecimal)}$$

$$8\,K = 8 \times 2^{10} = 2^3 \times 2^{10} = 2^{13}, \text{ or 2000h (in hexadecimal)}$$

$$4\,K = 4 \times 2^{10} = 2^2 \times 2^{10} = 2^{12}, \text{ or 1000h (in hexadecimal)}$$

Because the last address of a memory is equal to its word capacity minus one, we arrive at:

– M1 16 K memory: 2000h + 4000h − 1 = 5FFFh;

– M2 8 K memory: 8000h + 2000h − 1 = 9FFFh;

– M3 4 K memory: it is implanted just after the 8K memory. The starting address is, therefore, A000h and the last address is A000h + 1000h − 1 = AFFFh.

We assume that $\overline{RD}$ is connected to the $\overline{OE}$ inputs of the memories M1, M2 and M3.

The $\overline{RD}$ and $\overline{WR}$ signals are mutually exclusive. We do not need to consider them in the address decoding equations.

	A_{15}	A_{14}	A_{13}	A_{12}	A_{11}	A_{10}	A_9	A_8	...	A_0	
M1	0	0	1	0	0	0	0	0	...	0	2000h
	0	1	0	1	1	1	1	1	...	1	5FFFh
M2	1	0	0	0	0	0	0	0	...	0	8000h
	1	0	0	1	1	1	1	1	...	1	9FFFh
M3	1	0	1	0	0	0	0	0	...	0	A000h
	1	0	1	0	1	1	1	1	...	1	AFFFh

Table 6.21. *Address decoding table. For a color version of this table, see www.iste.co.uk/ndjountche/electronics2.zip*

The equations for memory selection can be determined using the address decoding table constructed as shown in Table 6.21. They are given by:

$$\overline{CS_{M1}} = \overline{\overline{A_{15}} \cdot \overline{A_{14}} \cdot A_{13} + \overline{A_{15}} \cdot A_{14} \cdot \overline{A_{13}}} \qquad [6.18]$$

$$\overline{CS_{M2}} = \overline{A_{15} \cdot \overline{A_{14}} \cdot \overline{A_{13}}} \qquad [6.19]$$

and:

$$\overline{CS_{M3}} = \overline{A_{15} \cdot \overline{A_{14}} \cdot A_{13} \cdot \overline{A_{12}}} \qquad [6.20]$$

For the memory M1, the logic combination of the starting address and that of the final address must be taken into account in the address decoding equation as they are different.

SOLUTION 6.7.– Organization of $28K \times 8$ ROM.

Once programmed during manufacture, a non-volatile memory can only be read.

The microcontroller has an address bus of sixteen bits ($A_{15}A_{14}\cdots A_0$, and the $\overline{RD}$ signal of the microcontroller is assumed to be connected to the $\overline{OE_1}$, $\overline{OE_2}$, and $\overline{OE_3}$ inputs of the different elementary memories.

The number, n, of bits that are required to address a memory block or the space containing a memory block, is given by the following equation:

$$2^{n-1} < \text{address-space capacity} \leq 2^n \qquad [6.21]$$

– For the memory M1, there is an address space of 8K and:

$$n = \log(8 \times 2^{10}) / \log(2) = 13$$

The highest address is $8 \times 2^{10} - 1 = 8\ 191$ or 1FFFh (in hexadecimal).

– For the memory M2, the address space is 12 K (or 8 K + 4 K) and:

$$n = \log(12 \times 2^{10}) / \log(2) = 13,58 \simeq 14$$

The highest address is $12 \times 2^{10} - 1 = 12\ 287$ or 2FFFh (in hexadecimal).

– For the memory M2, the address space is 28 K (or 8 K + 4 K + 16 K) and:

$$n = \log(28 \times 2^{10}) / \log(2) = 14,80 \simeq 15$$

the highest address is $28 \times 2^{10} - 1 = 28\ 671$ or 6FFFh (in hexadecimal).

To determine the logic equations for memory selection, the address decoding table can be drawn up as shown in Table 6.22. The logic combinations for the address bits, on which the address decoding functions are dependent, are identical for the memories M1 and M2 but are different for M3. The selection of M3 is also determined by the intermediate combinations that exists between the logic combinations associated with the starting address and the final address. The address decoding equations are, thus, written as follows:

$$\overline{CS_{M1}} = \overline{\overline{A_{15}} \cdot \overline{A_{14}} \cdot \overline{A_{13}}} \qquad [6.22]$$

$$\overline{CS_{M2}} = \overline{\overline{A_{15}} \cdot \overline{A_{14}} \cdot A_{13} \cdot \overline{A_{12}}} \qquad [6.23]$$

and:

$$\overline{CS_{M3}} = \overline{\overline{A_{15}} \cdot \left[\overline{A_{14}} \cdot A_{13} \cdot A_{12} + A_{14} \left(\overline{A_{13}} \cdot \overline{A_{12}} + \overline{A_{13}} \cdot A_{12} + \cdot A_{13} \cdot \overline{A_{12}}\right)\right]} \qquad [6.24]$$

The diagram showing the wiring of the memories to the microprocessor is given in Figure 6.32. The $\overline{RD}$ signal of the microcontroller is applied to the $\overline{OE}$ input of each elementary memory.

SOLUTION 6.8.– Light set (chaser).

An INIT signal pulse resets the registers and counter, and the content of the memory at the address 00h determines the initial conditions. The start of the count cycle brings about the loading of the animation sequence code. For each animation sequence, the bits from A6 to A4 remain unchanged while only the bits from A3 to

A0 can be modified on each rising edge of the clock signal. The counter is reset at the end of one animation sequence and the logic state for the bits A3, A2 and A0 becomes 0:

	A_{15}	A_{14}	A_{13}	A_{12}	A_{11}	A_{10}	A_9	A_8	...	A_0	
M1	0	0	0	0	0	0	0	0	...	0	0000h
	0	0	0	1	1	1	1	1	...	1	1FFFh
M2	0	0	1	0	0	0	0	0	...	0	2000h
	0	0	1	0	1	1	1	1	...	1	2FFFh
M3	0	0	1	1	0	0	0	0	...	0	3000h
	0	1	1	0	1	1	1	1	...	1	6FFFh

0	**0**	**1**	**1**
0	1	0	0
0	1	0	1
0	**1**	**1**	**0**

Table 6.22. *Address decoding table. For a color version of this table, see www.iste.co.uk/ndjountche/electronics2.zip*

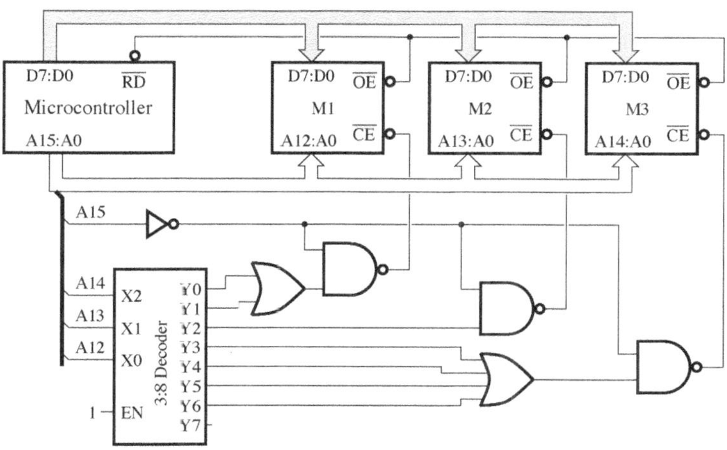

Figure 6.32. *Block diagram showing the wiring of the memories to the microcontroller*

a) The selection of the animation code is done based on a 3-bit code. There are, therefore, $2^3 = 8$ possible animation sequences.

The address for the EPROM changes at each rising edge of the clock signal.

b) The memory content for the sequence code 0 is given in Table 6.23.

c) Table 6.24 gives the memory content for the sequence code 1.

d) The memory content for the sequence code 2 is given in Table 6.25.

Address	Q_7	Q_6	Q_5	Q_4	Q_3	Q_2	Q_1	Q_0	
00h	0	0	0	0	0	0	1	0	Load code
01h	1	0	0	0	0	0	0	0	L0 On
02h	0	1	0	0	0	0	0	0	L1 On
03h	0	0	1	0	0	0	0	0	L2 On
04h	0	0	0	1	0	0	0	0	L3 On
05h	0	0	1	0	0	0	0	0	L2 On
06h	0	1	0	0	0	0	0	0	L1 On
07h	1	0	0	0	0	0	0	0	L0 On
08h	1	1	1	1	0	0	0	0	L0, L1, L2, and L3 On
09h	0	0	0	0	0	0	0	1	Counter reset

Table 6.23. *Memory content (sequence code 0)*

Address	Q_7	Q_6	Q_5	Q_4	Q_3	Q_2	Q_1	Q_0	
00h	0	0	0	0	0	0	1	0	Load code
11h	0	1	0	1	0	0	0	0	L0 and L2 On
12h	1	0	1	0	0	0	0	0	L1 and L3 On
13h	0	1	0	1	0	0	0	0	L0 and L2 On
14h	1	0	1	0	0	0	0	0	L1 and L3 On
15h	0	1	0	1	0	0	0	0	L0 and L2 On
16h	1	0	1	0	0	0	0	0	L1 and L3 On
17h	0	0	0	0	0	0	0	1	Counter reset

Table 6.24. *Memory content (sequence code 1)*

Address	Q_7	Q_6	Q_5	Q_4	Q_3	Q_2	Q_1	Q_0	
00h	0	0	0	0	0	0	1	0	Load code
21h	0	0	0	1	0	0	0	0	L0 On
22h	0	0	1	0	0	0	0	0	L1 On
23h	0	0	1	1	0	0	0	0	L1 and L0 On
24h	0	1	0	0	0	0	0	0	L2 On
25h	0	1	0	1	0	0	0	0	L2 and L0 On
26h	0	1	1	0	0	0	0	0	L2 and L1 On
27h	0	1	1	1	0	0	0	0	L2, L1, and L0 On
28h	1	0	0	0	0	0	0	0	L3 On
29h	1	0	0	1	0	0	0	0	L3 and L0 On
2Ah	1	0	1	0	0	0	0	0	L3 and L1 On
2Bh	1	0	1	1	0	0	0	0	L3, L1, and L0 On
2Ch	1	1	0	0	0	0	0	0	L3 and L2 On
2Dh	1	1	0	1	0	0	0	0	L3, L2, and L0 On
2Eh	1	1	1	0	0	0	0	0	L3, L2, and L1 On
2Fh	1	1	1	1	0	0	0	0	L0, L1, L2, and L3 On
20h	0	0	0	0	0	0	0	1	Counter reset

Table 6.25. *Memory content (sequence code 2)*

7

Programmable Logic Circuits

7.1. General overview

Given the increasing density and size of logical and digital circuits, implementation based on programmable logic circuits seems the most appropriate and economical way of meeting the needs of different applications with a low production volume, rather than an approach that uses discrete components.

A *read-only memory* (ROM) is composed of an address decoder and a network of components connected through switches for the storage of binary words. Once data is stored in a ROM, it can be read whenever but can no longer be modified under normal operating conditions.

One solution to implement logic functions is using a ROM where the output logic states given in the truth table can be stored. Thus, whenever the input bits are applied to the address lines, the logic state of the corresponding line is transferred to the output of the ROM. A *programmable read-only memory* (PROM) is implemented by adding fuses or antifuses in series with the switches of a ROM. It can be programmed once after manufacture, in an irreversible manner, using a burner.

In addition to the PROM circuit, there are other architectures for programmable circuits: *programmable array logic* (PAL), *programmable logic array* (PLA), *complex programmable logic device* (CPLD) and *field programmable gate array* (FPGA). However, the generic denomination, *programmable logic device* (PLD), is generally only used for architectures introduced to implement two-level circuits such as PAL, PLA or other versions of similar circuits.

7.2. Programmable logic device

In general, a $n \times m$ programmable circuit has n inputs and m outputs. To implement a sum of p products, p AND gates with $2n$ inputs and m OR gates with p inputs are required.

A programmable circuit may be represented completely, as shown in Figure 7.1, or in a simplified form, which is more appropriate when a large number of logic gates are involved, as shown in Figure 7.2.

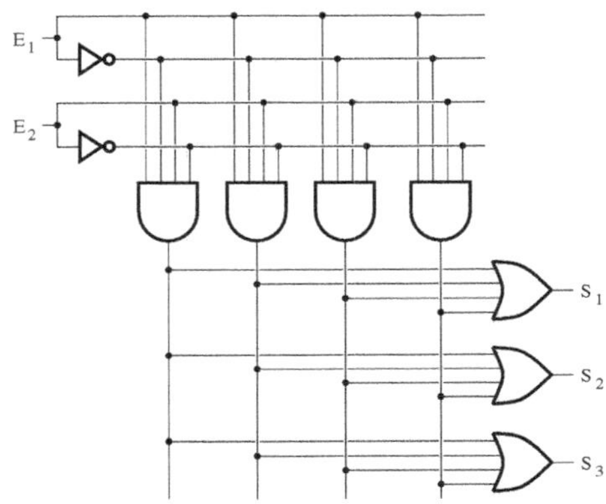

Figure 7.1. *Complete representation of a 2×3 programmable circuit*

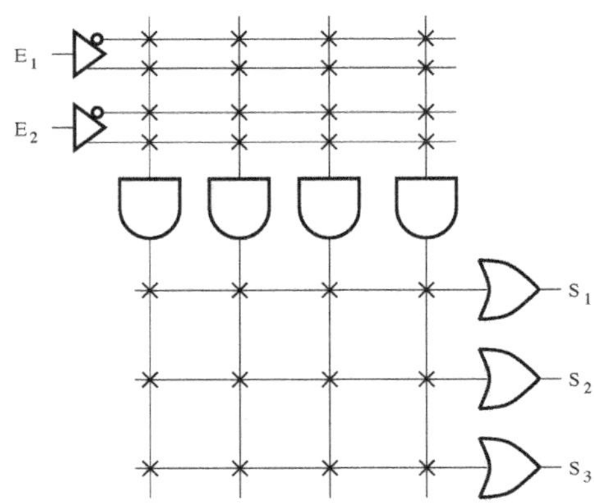

Figure 7.2. *Simplified representation of a 2×3 programmable circuit*

A PROM consists of a bank of prewired AND gates, which carry out the role of the decoder, and a bank of programmable OR gates.

A PAL comprises a bank of programmable AND gates and a bank of prewired OR gates.

A PLA is made up of a bank of programmable AND gates and a bank of programmable OR gates.

The following functions are to be implemented using a PROM, PAL or PLA:

$$X = \overline{A} + \overline{B} \cdot \overline{C} + B \cdot C \qquad [7.1]$$

$$Y = A + \overline{B} + C \qquad [7.2]$$

The logic equations for X and Y, respectively, can be expressed in the following forms:

$$\begin{aligned} X &= \overline{A} + \overline{B} \cdot \overline{C} + B \cdot C \\ &= A \cdot B \cdot C + A \cdot \overline{B} \cdot \overline{C} + \overline{A} \cdot B \cdot C + \overline{A} \cdot B \cdot \overline{C} + \overline{A} \cdot \overline{B} \cdot C + \overline{A} \cdot \overline{B} \cdot \overline{C} \end{aligned} \qquad [7\ 3]$$

and:

$$\begin{aligned} Y &= A + \overline{B} + C \\ &= A \cdot B \cdot C + A \cdot B \cdot \overline{C} + A \cdot \overline{B} \cdot C + A \cdot \overline{B} \cdot \overline{C} \\ &\quad + \overline{A} \cdot B \cdot C + + \overline{A} \cdot \overline{B} \cdot C + \overline{A} \cdot \overline{B} \cdot \overline{C} \end{aligned} \qquad [7.4]$$

– PROM-based implementation:

The initial PROM is represented in Figure 7.3. To implement the logic functions X and Y, the circuit must be configured as shown in Figure 7.4.

NOTE 7.1.– The size of a ROM with n inputs and m outputs is given by $2^n \times m$.

EXAMPLE 7.1.– A memory with three inputs and two outputs has a capacity of $2^3 \times 2 = 16$ bits, or two octets.

A memory with eight inputs and four outputs has a capacity of $2^8 \times 4 = 1\ 024$ bits or 1 kbit (kilobit).

– PAL-based implementation:

Figure 7.5 depicts the initial PAL while the circuit for a PAL programmed to implement the functions X and Y is given in Figure 7.6.

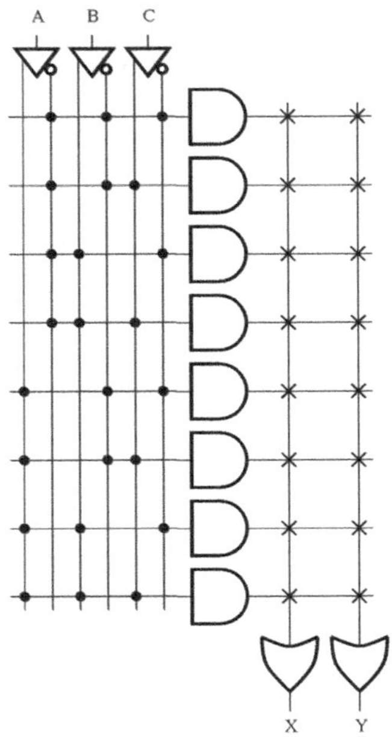

Figure 7.3. *Initial PROM (× represents a programmable connection and • symbolizes a fixed connection)*

– PLA-based implementation:

By blowing some of the fuses in the initial PLA, shown in Figure 7.7, we can obtain the programmed circuit of Figure 7.8.

Even though they offer less flexibility, PALs are adopted in practice because they are faster and less expensive. The different versions of the PAL supplied by manufacturers differ in the configuration of output pins:

– PAL with programmable output polarity (high or low). An example is given in Figure 7.9. The outputs can function as inputs or intermediate variables and can be fed back to the inputs;

– PAL with outputs stored on D flip-flops. Figure 7.10 shows the structure for this type of PAL, where the flip-flops are synchronized with the clock signal CK and the $\overline{\text{OE}}$ (*output enable*) signal can be used to deactivate the output. D flip-flops may be combined to implement state registers in synchronous finite-state machines;

– PAL with versatile outputs. This type of PAL consists of cells similar to the one shown in Figure 7.11. The inputs are connected to a programmable bank of AND gates that are connected to an OR gate. A D flip-flop is used for the storage and synchronization of data and the output configuration is determined by the multiplexer selection bits. Each cell can be configured such that the output is determined by the AND/OR bank (combinational logic) or by the flip-flop (sequential logic), with either active-high or active-low signal. Based on a cell configuration, the feedback signal can be determined by the flip-flop or any signal applied to the output pin (which is considered as an input/output node) and it may be active-high or active low. The different configurations for the feedback and output signals are given in Table 7.1.

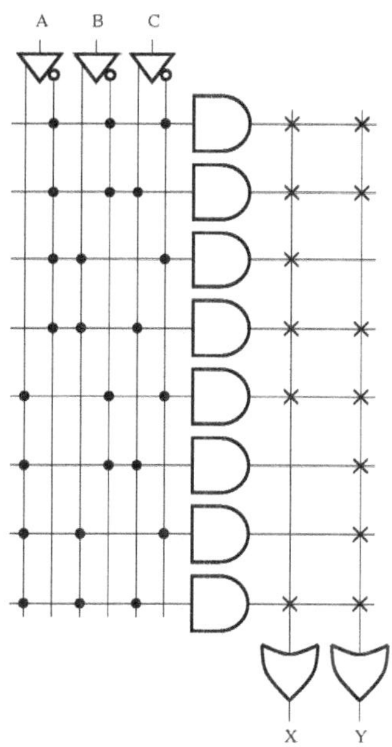

Figure 7.4. *Programmed PROM (× represents a programmable connection and • symbolizes a fixed connection)*

Fuses		Configuration		
S_1	S_0	Feedback	Output	Polarity
0	0	Flip-flop	Flip-flop	Active-low
0	1	Flip-flop	Flip-flop	Active-high
1	0	Input/output	AND/OR bank	Active-low
1	1	Input/output	AND/OR bank	Active-high

Table 7.1. *Configurations for the feedback and output signals (0: fuse intact; 1: fuse blown)*

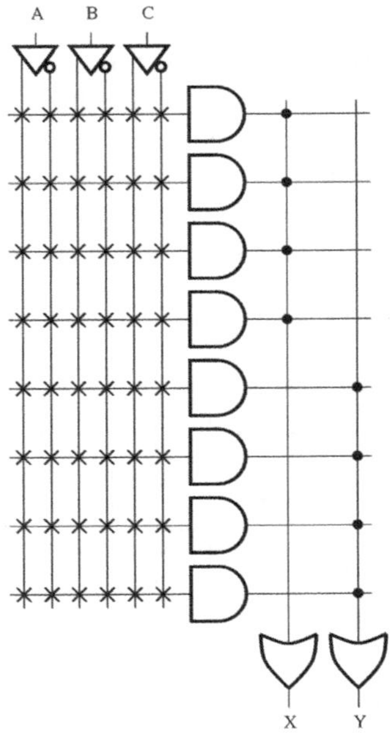

Figure 7.5. *Initial PAL*

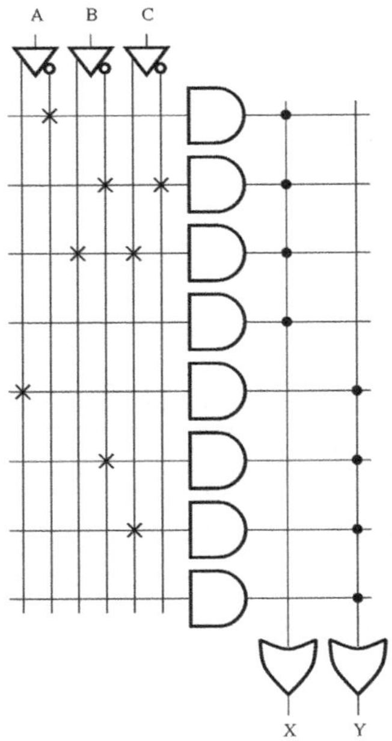

Figure 7.6. *Programmed PAL*

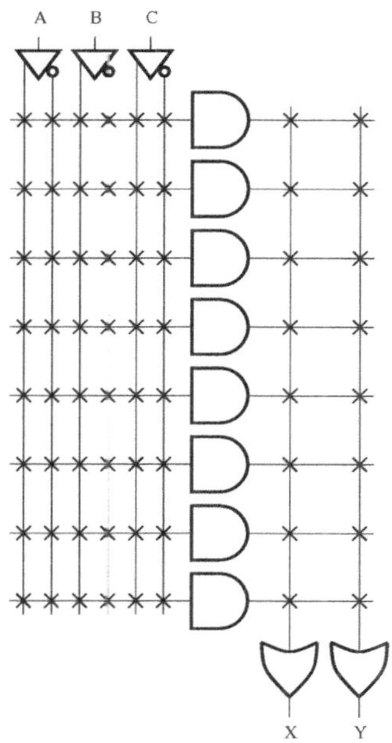

Figure 7.7. *Initial PLA*

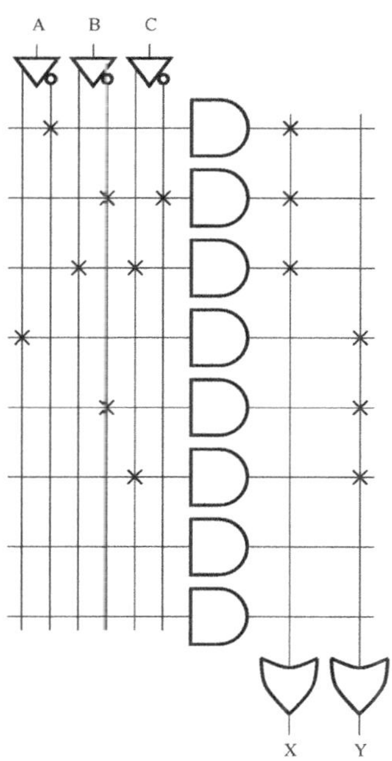

Figure 7.8. *Programmed PLA*

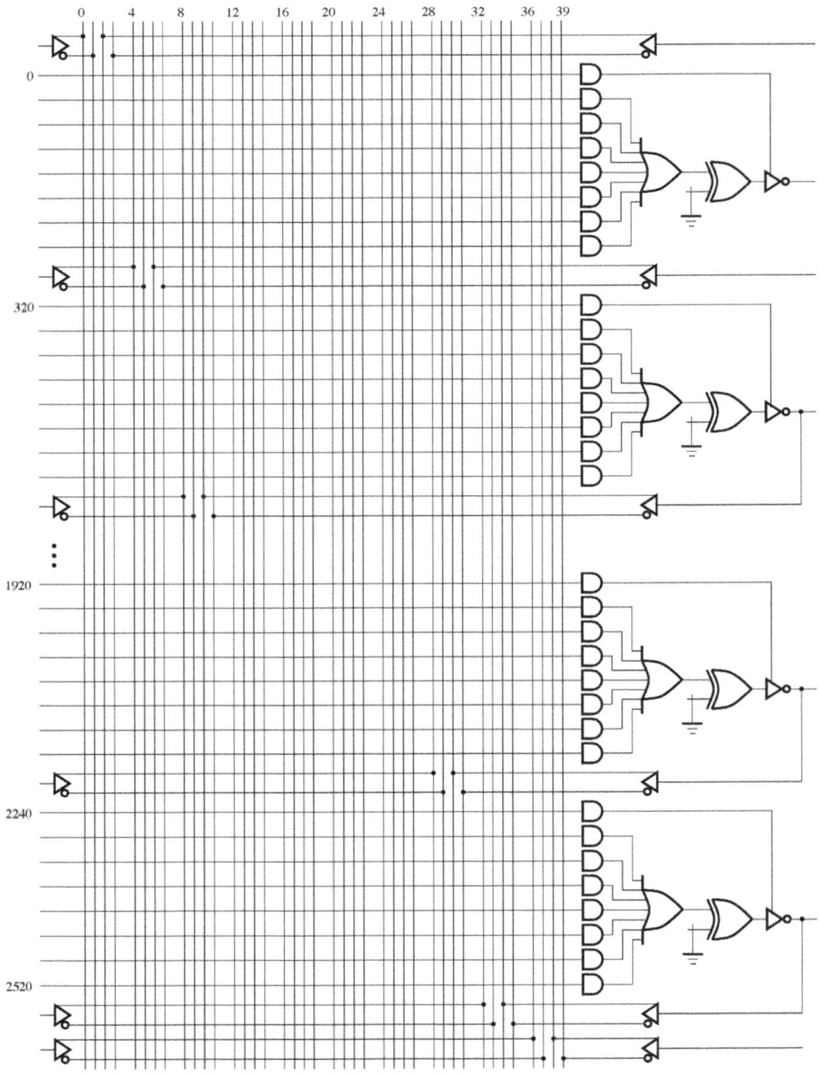

Figure 7.9. *Structure of an integrated PAL (PAL20P8)*

Generally speaking, the number of inputs of a PLD varies between 16 and 32 and the number of outputs between 8 and 16. A PLD can have a bank of programmable AND gates, a bank of prewired OR gates, D flip-flops triggered by a single clock and a fuse to protect against read operations.

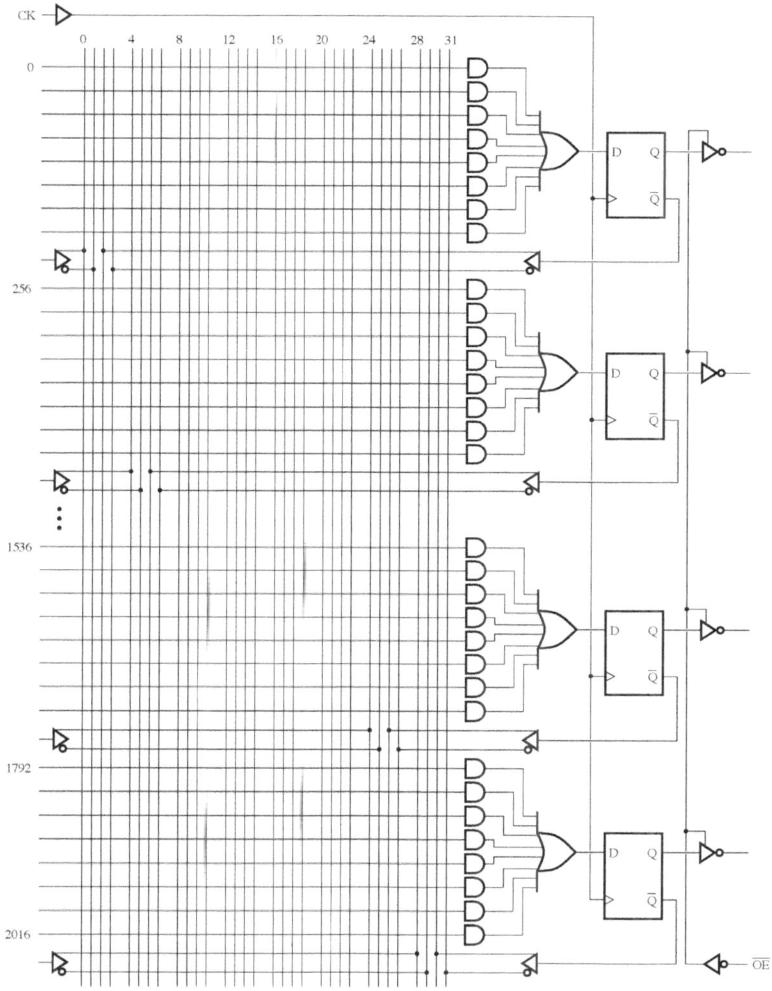

Figure 7.10. *Structure of an integrated PAL (PAL16R8)*

PLDs have fused-based interconnects (bipolar technology) or antifuse-based interconnects (*complementary metal-oxide semiconductor* [CMOS] technology) as shown in Figure 7.12. A fuse-based interconnect, implemented by combining a metal or polysilicon with a diode, is initially closed, while an antifuse-based interconnect, built around a MOS capacitor, is initially open. PLD circuits can only be programmed once, through an irreversible blowing of fuses or antifuses. This can prove a hindrance if we wish to correct any errors that could have occurred during the programming. Other structures, such as *erasable PLD* (EPLD) and *electrically erasable PLD* (EEPLD), have been proposed to offset this concern. EPLDs can be erased upon exposure to UV rays while EEPLD circuits can be erased electrically. It must be noted that the name *generic array logic* (GAL) has been trademarked by the company *Lattice Semiconductor* and is given to PALs which can be electrically erased. EPLDs and EEPLDs are based on erasable and reprogrammable memory technology and make use of the advantages of CMOS technology.

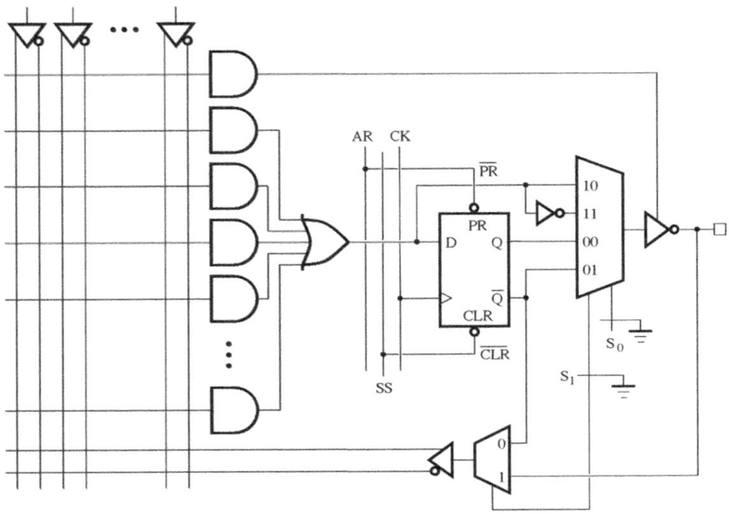

Figure 7.11. *Structure of a PLD cell (PAL22V10)*

Figure 7.12. *Programmable a) fuse-based and b) antifuse-based interconnects*

The major problem with PLDs is related to the difficulty of sharing the logic product terms between different cells.

7.3. Applications

When a memory circuit (or a programmable circuit) is used to implement combinational logic functions:

– the inputs and outputs of the functions to be implemented must be identified;

– the inputs must be connected to the address bus bits;

– the memory outputs must be considered as the outputs of the function;

– the content of the memory must be defined and programmed.

7.3.1. *Implementation of logic functions*

Implement the following functions using a PAL:

$$F(A,B,C,D) = \sum m(0,1,5,6,9,12,14) \qquad [7.5]$$

$$G(A,B,C,D) = \sum m(0,1,5,9,10,14,15) \qquad [7.6]$$

$$H(A,B,C,D) = \sum m(0,1,2,5,6,9,10) \qquad [7.7]$$

Figures 7.13–7.15 depict the Karnaugh maps that can be used to obtain the minimal forms of the logic equations for the functions F, G and H as follows:

$$F(A,B,C,D) = \overline{A}\cdot\overline{B}\cdot\overline{C} + \overline{A}\cdot\overline{C}\cdot D + \overline{B}\cdot\overline{C}\cdot D + A\cdot B\cdot\overline{D} + B\cdot C\cdot\overline{D} \qquad [7.8]$$

$$G(A,B,C,D) = \overline{A}\cdot\overline{B}\cdot\overline{C} + \overline{A}\cdot\overline{C}\cdot D + \overline{B}\cdot\overline{C}\cdot D + A\cdot B\cdot C + A\cdot C\cdot\overline{D} \qquad [7.9]$$

and:

$$G(A,B,C,D) = \overline{A}\cdot\overline{B}\cdot\overline{C} + \overline{A}\cdot\overline{C}\cdot D + \overline{B}\cdot\overline{C}\cdot D + \overline{A}\cdot C\cdot\overline{D} + \overline{B}\cdot C\cdot\overline{D} \qquad [7.10]$$

For the implementation of the functions F, G and H using a PAL with six inputs and four outputs, we assume that:

$$T = \overline{A}\cdot\overline{B}\cdot\overline{C} + \overline{A}\cdot\overline{C}\cdot D + \overline{B}\cdot\overline{C}\cdot D \qquad [7.11]$$

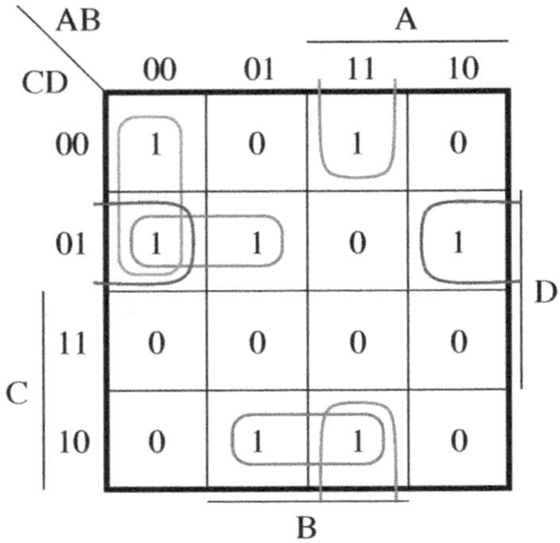

Figure 7.13. *Output F. For a color version of this figure, see www.iste.co.uk/ndjountche/electronics2.zip*

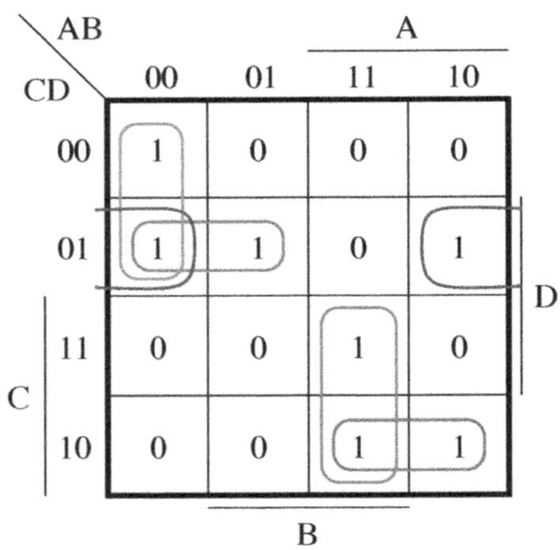

Figure 7.14. *Output G. For a color version of this figure, see www.iste.co.uk/ndjountche/electronics2.zip*

This can then lead to:

$$F(A,B,C,D) = T + A \cdot B \cdot \overline{D} + B \cdot C \cdot \overline{D} \qquad [7.12]$$

$$G(A,B,C,D) = T + A \cdot B \cdot C + A \cdot C \cdot \overline{D} \qquad [7.13]$$

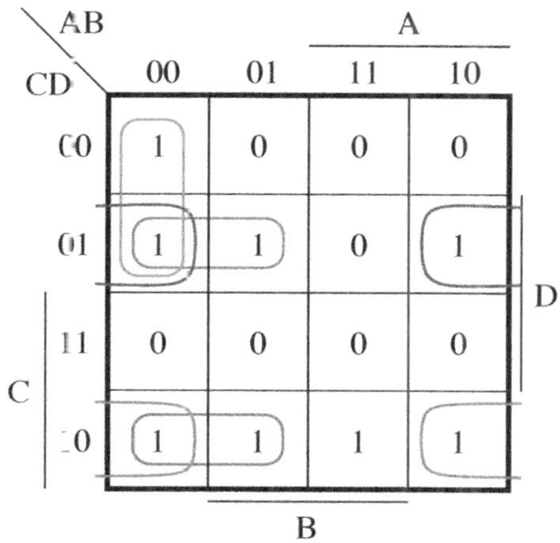

Figure 7.15. *Output H. For a color version of this figure, see www.iste.co.uk/ndjountche/electronics2.zip*

and:

$$G(A, B, C, D) = T + \overline{A} \cdot C \cdot \overline{D} + \overline{B} \cdot C \cdot \overline{D} \qquad [7.14]$$

The PAL thus obtained is represented in Figure 7.16.

7.3.2. *Two-bit adder*

Use a PLA circuit to implement an adder for two 2-bit numbers, $A = A_1A_0$ and $B = B_1B_0$, that performs the following operation:

$$\begin{array}{cccc} & & A_1 & A_0 \\ + & & B_1 & B_0 \\ \hline & C & S_1 & S_0 \end{array}$$

where the sum is represented by $S = S_1S_0$ and C designates the carry.

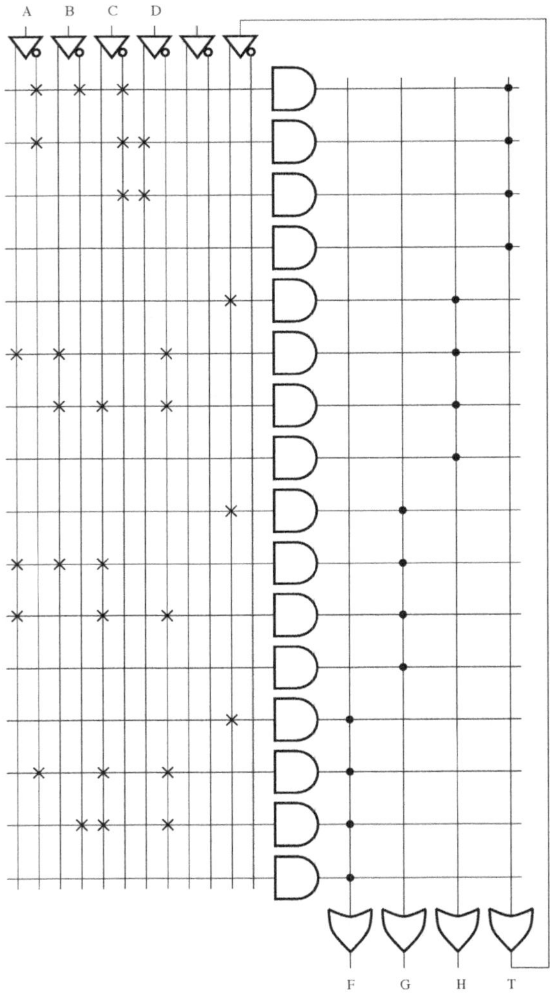

Figure 7.16. *Programmed PAL*

The truth table of the 2-bit adder may be drawn up as shown in Table 7.2. The Karnaugh maps shown in Figures 7.17–7.19 help in obtaining the minimal forms for the output logic expressions. We thus have:

$$C = A_1 \cdot B_1 + A_1 \cdot A_0 \cdot B_0 + A_0 \cdot B_1 \cdot B_0 \quad [7.15]$$

$$S_1 = A_1 \cdot \overline{B_1} \cdot \overline{B_0} + A_1 \cdot \overline{A_0} \cdot \overline{B_1} + \overline{A_1} \cdot \overline{A_0} \cdot B_1 + \overline{A_1} \cdot B_1 \cdot \overline{B_0} + \overline{A_1} \cdot A_0 \cdot \overline{B_1} \cdot B_0 + A_1 \cdot A_0 \cdot B_1 \cdot B_0 \quad [7.16]$$

and:

$$S_0 = \overline{A_0} \cdot B_0 + A_0 \cdot \overline{B_0} \qquad [7.17]$$

Inputs				Outputs		
A_1	A_0	B_1	B_0	C	S_1	S_0
0	0	0	0	0	0	0
0	0	0	1	0	0	1
0	0	1	0	0	1	0
0	0	1	1	0	1	1
0	1	0	0	0	0	1
0	1	0	1	0	1	0
0	1	1	0	0	1	1
0	1	1	1	1	0	0
1	0	0	0	0	1	0
1	0	0	1	0	1	1
1	0	1	0	1	0	0
1	0	1	1	1	0	1
1	1	0	0	0	1	1
1	1	0	1	1	0	0
1	1	1	0	1	0	1
1	1	1	1	1	1	0

Table 7.2. *Truth table*

The implementation of the 2-bit adder based on the PLA is illustrated in Figure 7.21. It is characterized by a number of products terms that is equal to 11.

One approach that can be used to reduce the number of different product terms and, consequently, reduce the size (or number of columns) of the PLA, consists of using decoders. In this case, the output logic expressions must be decomposed as follows:

$$C = (A_1 + B_1)(A_1 + \overline{B_1})(\overline{A_1} + B_1) + (A_1 + B_1)(A_0 + B_0)(A_0 + \overline{B_0})(\overline{A_0} + B_0) \qquad [7.18]$$

$$S_1 = (A_1 + \overline{B_1})(\overline{A_1} + B_1)(A_0 + B_0)(A_0 + \overline{B_0})(\overline{A_0} + B_0) + (A_1 + B_1)(\overline{A_1} + \overline{B_1})(\overline{A_0} + \overline{B_0}) \qquad [7.19]$$

and:

$$S_0 = (A_0 + B_0)(\overline{A_0} + \overline{B_0}) \qquad [7.20]$$

Using a 2 : 4 decoder, as shown in Figure 7.20 ,which can generate *maxterms*, the 2-bit adder can be implemented as shown in Figure 7.22(a). The number of columns in the PLA has been reduced to 5.

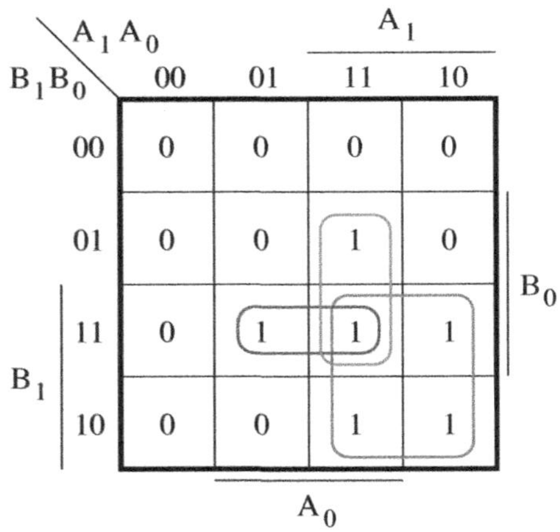

Figure 7.17. *Output C. For a color version of this figure, see www.iste.co.uk/ndjountche/electronics2.zip*

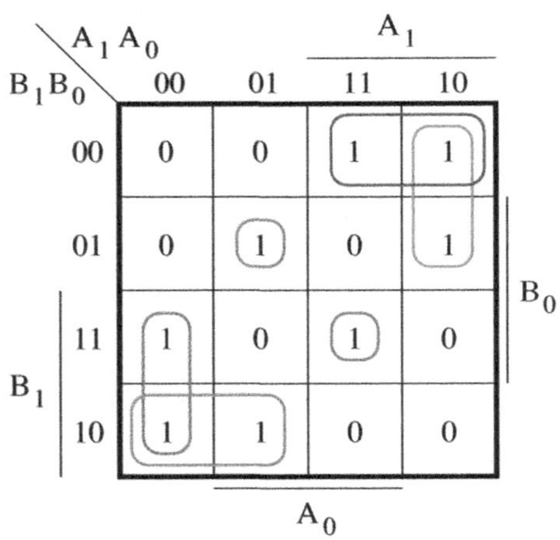

Figure 7.18. *Output S_1. For a color version of this figure, see www.iste.co.uk/ndjountche/electronics2.zip*

Another configuration used to implement the 2-bit adder can be obtained by decomposing the logical complement of the carry, $\overline{C}$, instead of the carry, C, itself.

This is useful in cases where the PLA output has programmable inverters, or in applications where the complement of the carry, $\overline{C}$, is of interest. The logic expression for $\overline{C}$ is given by:

$$\overline{C} = \overline{A_1 \cdot B_1 + A_1 \cdot A_0 \cdot B_0 + A_0 \cdot B_1 \cdot B_0} \qquad [7.21]$$

$$= \overline{A_1} \cdot \overline{A_0} + \overline{A_0} \cdot \overline{B_1} - \overline{A_1} \cdot \overline{B_0} + \overline{B_1} \cdot \overline{B_0} + \overline{A_1} \cdot \overline{B_1} \qquad [7.22]$$

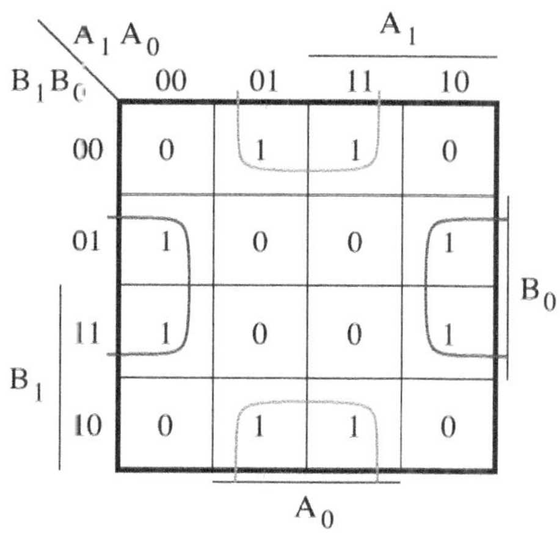

Figure 7.19. *Output S_0. For a color version of this figure, see www.iste.co.uk/ndjountche/electronics2.zip*

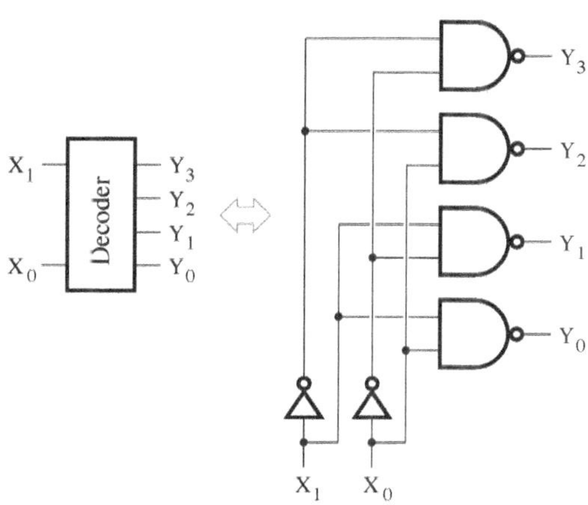

Figure 7.20. *Logic circuit of the 2 : 4 decoder*

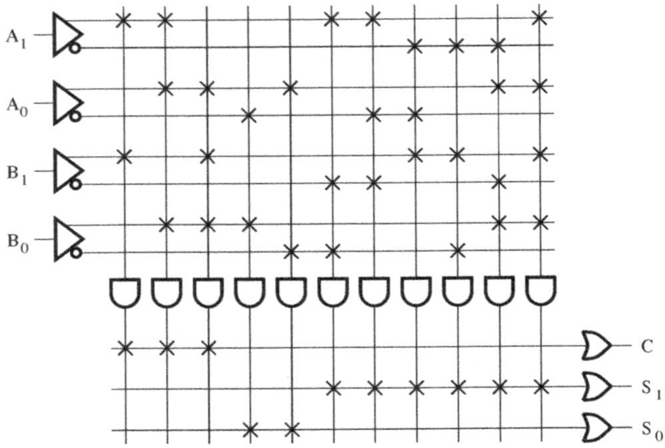

Figure 7.21. *Implementation of the 2-bit adder with a PLA*

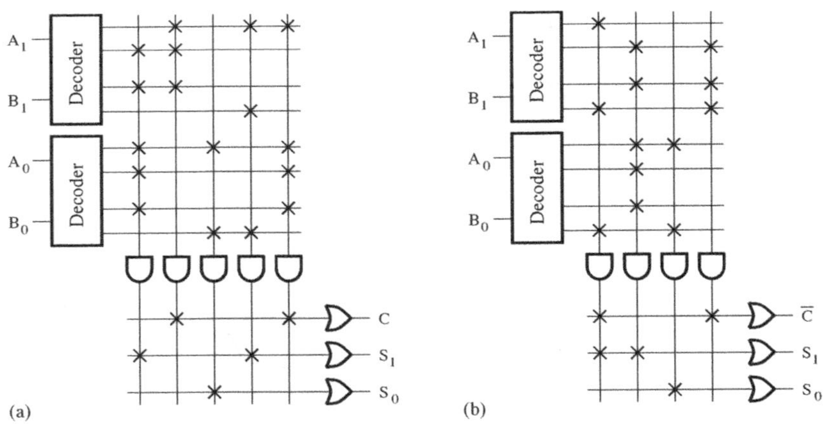

Figure 7.22. *Implementation of the two-bit adder with a PLA and 2 : 4 decoders*

Because:

$$\overline{A_1} \cdot \overline{A_0} = \overline{A_1} \cdot \overline{A_0} \cdot B_1 + \overline{A_1} \cdot \overline{A_0} \cdot \overline{B_1} \quad [7.23]$$

$$\overline{A_0} \cdot \overline{B_1} = A_1 \cdot \overline{A_0} \cdot \overline{B_1} + \overline{A_1} \cdot \overline{A_0} \cdot \overline{B_1} \quad [7.24]$$

$$\overline{A_1} \cdot \overline{B_0} = \overline{A_1} \cdot B_1 \cdot \overline{B_0} + \overline{A_1} \cdot \overline{B_1} \cdot \overline{B_0} \quad [7.25]$$

$$\overline{B_1} \cdot \overline{B_0} = A_1 \cdot \overline{B_1} \cdot \overline{B_0} + \overline{A_1} \cdot \overline{B_1} \cdot \overline{B_0} \quad [7.26]$$

it follows that:

$$\overline{C} = (\overline{A_1} \cdot B_1 + A_1 \cdot \overline{B_1})(\overline{A_0} + \overline{B_0}) + \overline{A_1} \cdot \overline{B_1}(1 + \overline{A_0} + \overline{B_0}) \quad [7.27]$$
$$= (A_1 + B_1)(\overline{A_1} + \overline{B_1})(\overline{A_0} + \overline{B_0}) + (A_1 + \overline{B_1})(\overline{A_1} + B_1)(\overline{A_1} + \overline{B_1}) \quad [7.28]$$

Figure 7.22(b) shows the resulting logic circuit. The PLA does not require more than four columns.

By allowing for the sharing of the product terms, the PLA is more flexible but slower than the PAL, which is characterized by a large number of fixed interconnects with simpler structures.

7.3.3. *Binary-to-BCD and BCD-to-binary converters*

Binary-to-BCD and BCD-to-binary converters can be implemented using a ROM. The conversion algorithm is based on shift, addition and subtraction operations and is executed assuming that the number to be converted is initially stored in a register. It can be used to deduce the truth table that establishes the correspondence between the codes and that is stored in a memory of suitable size for the implementation of the converter. By cascading converters for smaller word-length codes, it is possible to increase the size of the codes that can be converted.

7.4. Programmable logic circuits (CPLD and FPGA)

A CPLD has some programmable functional blocks (AND/OR network and macrocells) whose inputs and outputs can be connected by a reconfigurable interconnect network, as well as input/output blocks for interfacing with external components.

A FPGA is based on a matrix network comprising several programmable functional blocks (configurable logic blocks) which are interconnected using reconfigurable interconnects. This network is surrounded by input/output blocks.

The number of logic blocks (or macrocells), which is only of the order of a few hundreds for a CPLD, can generally go up to $1,000,000$ for an FPGA. A macrocell may be based on an AND/OR bank (CPLD) or on a look-up table (LUT) (CPLD and FPGA). The ratio of flip-flops to logic resources is higher in an FPGA than in a CPLD.

A CPLD is configured from an EEPROM and is, thus, considered non-volatile. On the other hand, the FPGA configuration, determined by a RAM, is considered volatile and must be updated every time it is powered on.

A CPLD is ideal for control applications as it offers very predictable timing characteristics, while an FPGA circuit, having a large number of functionalities and registers, would generally seem appropriate for datapath implementations.

7.4.1. *Principle and technology*

The architecture of a CPLD or FPGA is based on a certain number of logic blocks that are connected by a network of programmable interconnects.

For some CPLDs, each logic block is made up of an AND bank that generates product terms, a product term allocator and macrocells. But for other CPLDs and FPGAs, each logic block comprises a look-up table, a carry propagation and checking chain and macrocells, as shown in Figure 7.23. In the latter case, the RAM-based logic block is configurable.

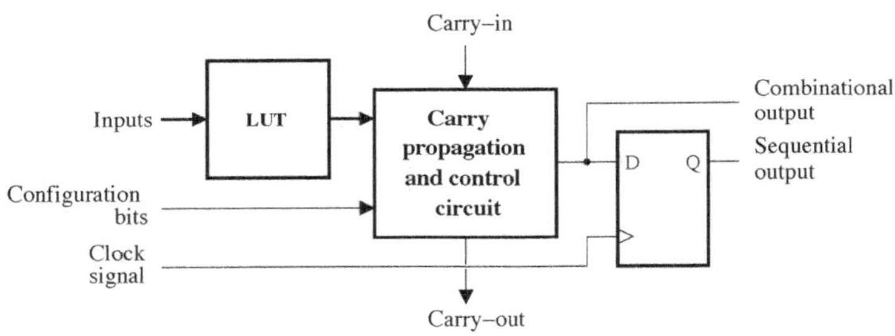

Figure 7.23. *Diagram showing the operation principle of a configurable logic block*

Consider the logic function, F, that can be written as follows:

$$F(A,B,C) = A \cdot B \cdot C + \overline{A} \cdot \overline{B} + \overline{A} \cdot \overline{C} + \overline{B} \cdot \overline{C} \quad [7.29]$$

$$= \overline{A} \oplus B \cdot C + \overline{B} \cdot \overline{C} \quad [7.30]$$

The implementation of this function using a 3-input LUT is illustrated in Figure 7.24. The LUT has eight storage cells, each of which corresponds to the output value found in each row of the truth table for the function F. Its output F takes the logic state of one of the storage cells depending on the combination of the three variables A, B and C, applied to the select inputs.

Figure 7.25 depicts the logic circuit of a 3-input LUT. Reconfigurable circuits, in general, are implemented using LUTs with three to eight inputs and one output. A

LUT allows for the implementation of any logic function which has a number of input variables that is less than or equal to the number of its inputs.

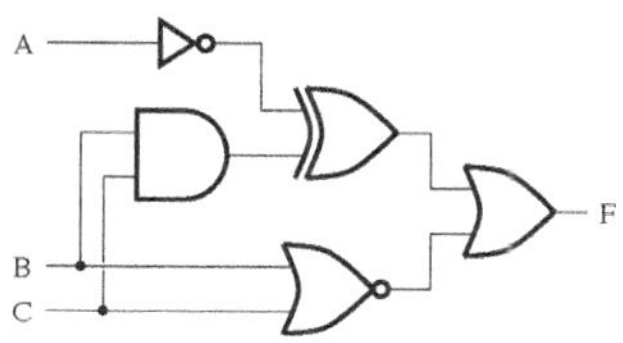

(a)

A	B	C	F(A,B,C)
0	0	0	1 = F(0,0,0)
0	0	1	1 = F(0,0,1)
0	1	0	1 = F(0,1,0)
0	1	1	0 = F(0,1,1)
1	0	0	1 = F(1,0,0)
1	0	1	0 = F(1,0,1)
1	1	0	0 = F(1,1,0)
1	1	1	1 = F(1,1,1)

(b)

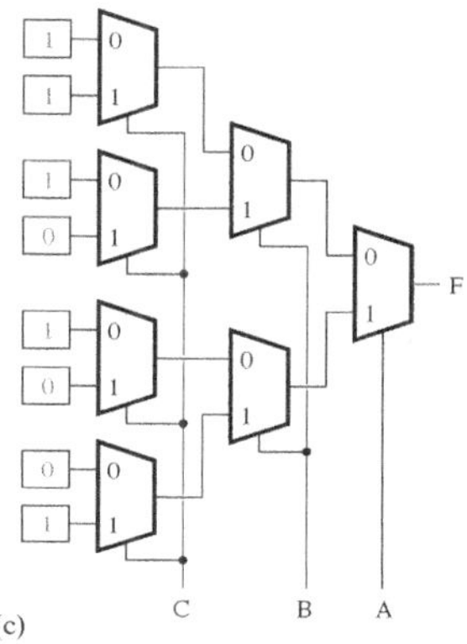

(c)

Figure 7.24. *Implementation of a function F based on a 3-input LUT. For a color version of this figure, see www.iste.co.uk/ndjountche/electronics2.zip*

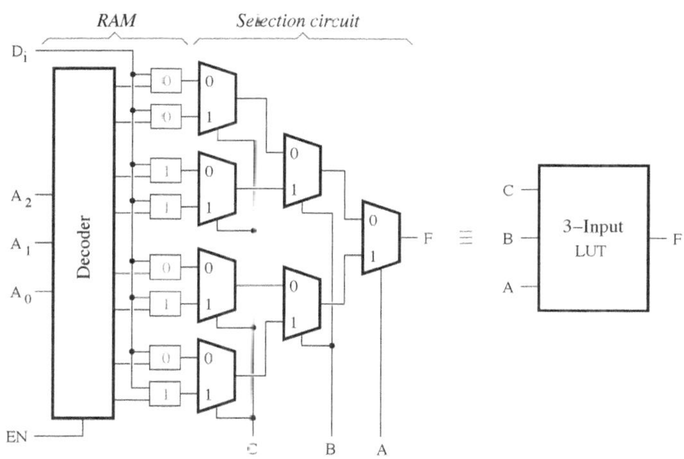

Figure 7.25. *Three-input LUT. For a color version of this figure, see www.iste.co.uk/ndjountche/electronics2.zip*

Any Boolean function can be represented in the form of a Shannon decomposition. In the specific case of the logic function $F(A, B, C)$, Shannon decomposition can be used to successively obtain the following expressions:

$$F(A,B,C) = \overline{A} \cdot F(0,B,C) + A \cdot F(1,B,C) \qquad [7.31]$$

$$= \overline{A} \cdot \overline{B} \cdot F(0,0,C) + \overline{A} \cdot B \cdot F(0,1,C) + A \cdot \overline{B} \cdot F(1,0,C) + A \cdot B \cdot F(1,1,C) \qquad [7.32]$$

$$= \overline{A} \cdot \overline{B} \cdot \overline{C} \cdot F(0,0,0) + \overline{A} \cdot \overline{B} \cdot C \cdot F(0,0,1) + \overline{A} \cdot B \cdot \overline{C} \cdot F(0,1,0) + \overline{A} \cdot B \cdot C \cdot F(0,1,1) + A \cdot \overline{B} \cdot \overline{C} \cdot F(1,0,0) + A \cdot \overline{B} \cdot C \cdot F(1,0,1) + A \cdot B \cdot \overline{C} \cdot F(1,1,0) + A \cdot B \cdot C \cdot F(1,1,1) \qquad [7.33]$$

Thus, to implement this 3-variable function based on a 3-input LUT, the inputs must be used as selection addresses for the 8×1 bit memory in which the output column of the truth table is stored.

To implement Boolean functions with more than three variables, two 3-input LUTs must be combined.

A 3-input LUT can also be implemented by associating two 2-input LUTs, as shown in Figure 7.26.

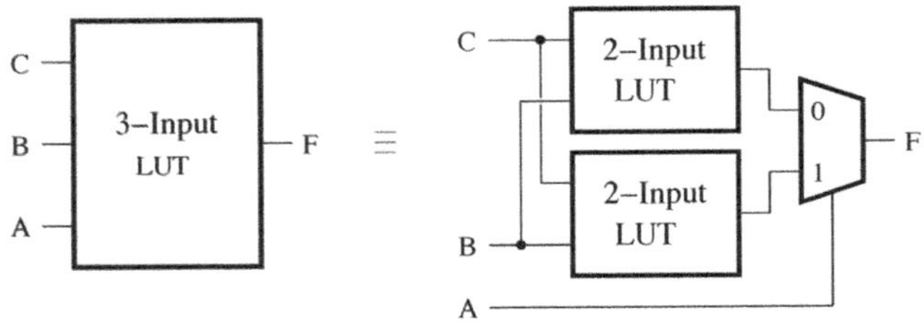

Figure 7.26. *Implementation of a 3-input LUT based on a 2-input LUT*

A K-input LUT usually allows for the implementation of any boolean function of K variables, especially 2^{2^K} different functions.

NOTE 7.2.– The implementation of the Boolean functions can be carried out using approaches other than those using LUT based circuits. The circuit shown in Figure 7.29(a), which comprises a 2-bit decoder, a network of programmable interconnects, OR gates and a 2 : 1 multiplexer can be used to implement the logic function $F(A, B, C)$ based on the decomposition as per Shannon's expansion theorem, as given by:

$$F(A, B, C) = \overline{C} \cdot F(A, B, 0) + C \cdot F(A, B, 1) \qquad [7.34]$$

On the other hand, if the function $F(A, B, C)$ is decomposed in the following manner:

$$F(A, B, C) = \overline{A} \cdot \overline{B} \cdot F(0, 0, C) + \overline{A} \cdot B \cdot F(0, 1, C) + A \cdot \overline{B} \cdot F(1, 0, C) + A \cdot B \cdot F(1, 1, C) \qquad [7.35]$$

it can be implemented by the circuit represented in Figure 7.29(b), which is based on a 4 : 1 multiplexer.

Connections can be established using a network of programmable interconnects between the input/output pins and the inputs and outputs of the different logic blocks of a reconfigurable circuit. Figure 7.27 shows the different types of interconnects used in the reconfigurable circuits. Each point of interconnection between two conductors can be implemented using a switch whose state (open or closed) is programmable based on a configuration bit saved in a memory.

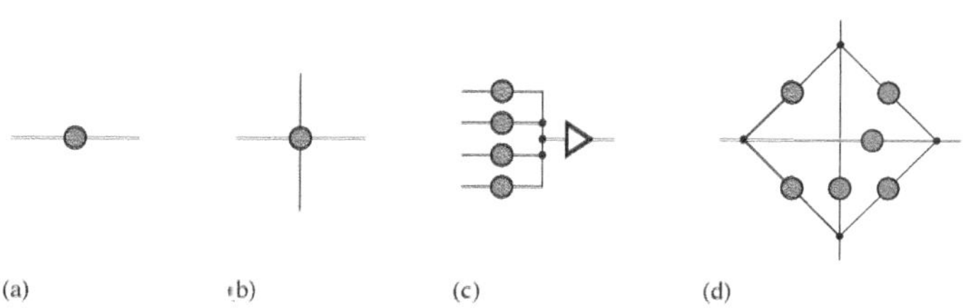

Figure 7.27. *Types of interconnects used in programable logic circuits*

A reprogrammable interconnect may be based on an EEPROM, Flash or SRAM, as shown in Figure 7.28.

Following the principle of an EEPROM, a reprogrammable interconnect uses a floating-gate avalanche-injection MOS (FAMOS) transistor. A charge induced onto the floating gate by the positive programming voltage remains there after the voltage has been disconnected. However, the charge stored on the floating gate can be removed by applying a negative voltage.

In the case of a Flash-based reprogrammable interconnect, the floating gate transistor, T_C, plays the role of a switch, while the floating-gate tunneling oxide

(FLOTOX) transistor, T_F, enables the write and erase operations. It must be noted that the FLOTOX transistor is a version of the FAMOS transistor, where the thickness of one portion of the dielectric layer is reduced to facilitate electrical erasure.

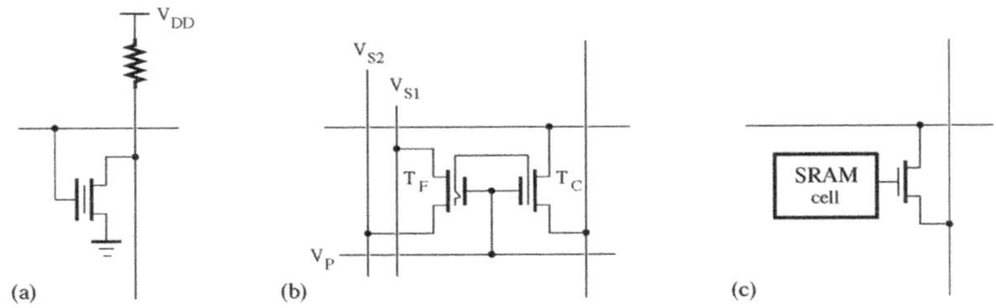

Figure 7.28. *Reprogrammable interconnects: a) EEPROM; b) Flash; c) SRAM*

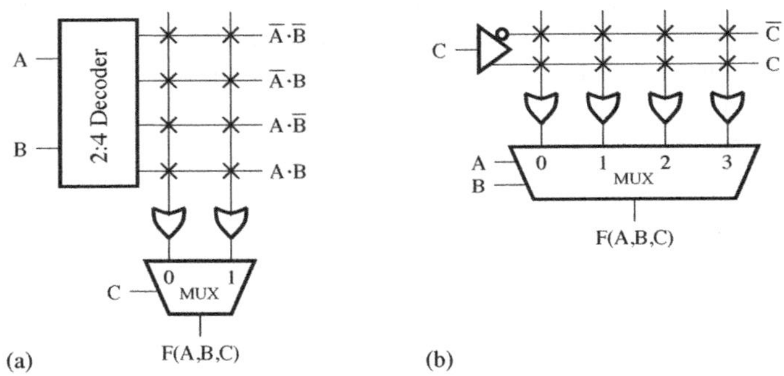

Figure 7.29. *Programmable circuits that are appropriate for the implementation of 3-variable functions*

Another implementation approach used for reprogrammable interconnects consists of using an SRAM to store the configuration bit that is required to control a MOS transistor operating as a switch.

7.4.2. *CPLD*

A CPLD contains at least two PLDs. It is made up of:

– a block of dedicated inputs;

– input/output blocks;

– macrocell blocks;

– a programmable interconnect network.

A macrocell can consist of a D flip-flop, multiplexers and three-state gates.

The programmable interconnect network is implemented either using a matrix of switches or using multiplexers. The switch matrix can help connect a given node to any output node, while the use of multiplexers is limited by the size and operating speed of the circuit, and it can generally be used only to connect a given node to a specific output node.

The structure of a CPLD, as illustrated in Figure 7.30, helps to better optimize the use of resources (bank of AND gates, bank of OR gates, macrocell, input/output blocks) than that of a PLD. The flip-flops of a CPLD can be configured as D, JK, T or RS flip-flops. The clock signal for each flip-flop can be individual or global (or can be used to control any component of the chip).

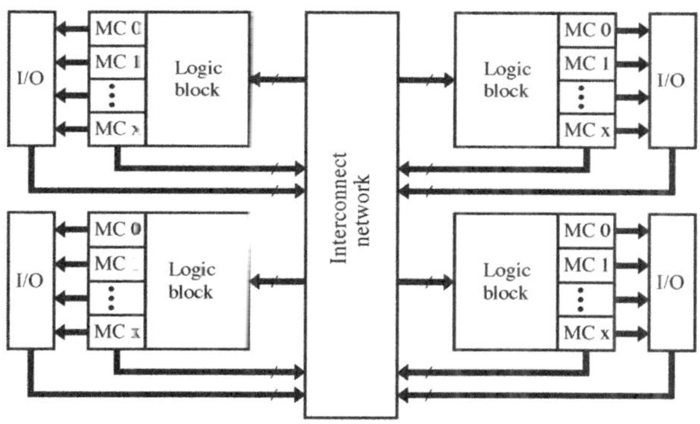

Figure 7.30. *Architecture of a CPLD*

The *Xilinx XC9500* CPLD is made up of several functional blocks and input/output blocks which are completely interconnected by a switch matrix. Each functional block comprises 18 macrocells. The structure of a macrocell is represented in Figure 7.31. It is associated with a product term allocator, a detailed diagram of which is given in Figure 7.32.

Each functional block is implemented using a sum-of-products representation. The AND gates network generates 90 product terms from 36 inputs, which correspond to 72 signals and their complements. Any number out of these 90 product terms can be assigned to each macrocell by the product term allocator.

Each macrocell may be configured to implement combinational functions or shift-register type functions. Five logic product terms from the AND bank can be used as the inputs for the OR and XOR gates to implement combinational functions, or as control inputs (clock signal, set, reset, output enable). The product term allocator offers the possibility to choose how the five product terms reserved for a macrocell are used.

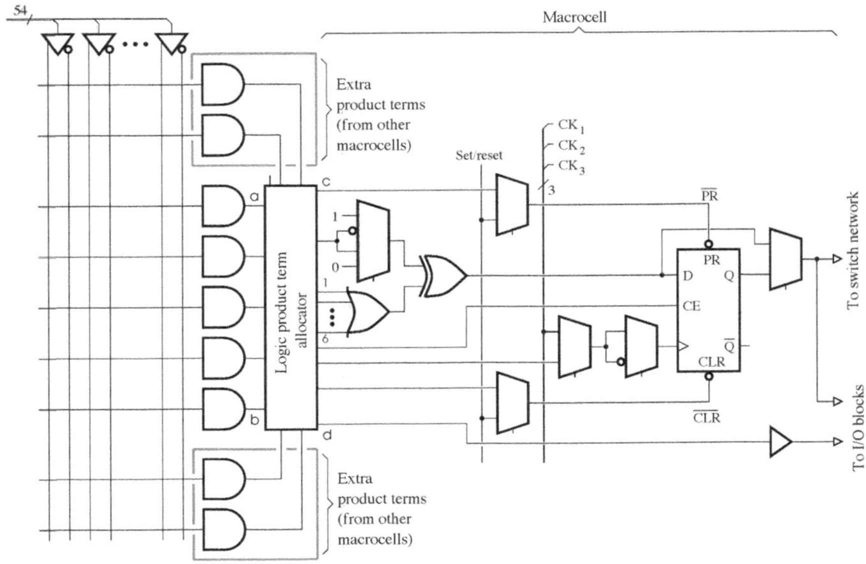

Figure 7.31. *Structure of a CPLD macrocell (Xilinx XC9500)*

In a functional block, the product terms that are not used by the macrocells can be reassigned to other macrocells in order to increase their capacity. All 90 product terms can be made available for each macrocell, but with an additional delay.

7.4.3. *FPGA*

An FGPA is based on a regular network and thus helps achieves a higher integration level than a CPLD. It has a large number of macrocells and offers great interconnect flexibility (see Figure 7.33).

The architecture of a CPLD is based on elementary logic blocks, called macrocells. These have a mainly combinational structure and realize a logic OR of the product terms from the AND gates. A flip-flop at the output can be used to implement sequential functions. This structure is characterized by a minimal input-output latency period. It can serve for the implementation of an AND gate with a high number of inputs (of the order of 136 inputs for the *Lattice* ispXPLD) and is, therefore, useful in the synthesis of address decoders for a memory or logic operations on a data bus.

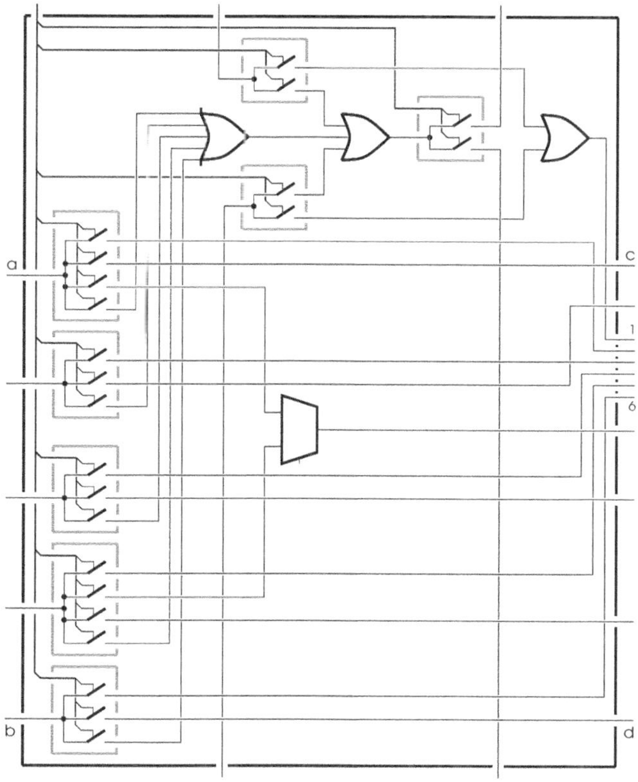

Figure 7.32. *Logic product term allocator*

FPGAs are implemented from a large number of more compact LUT-based logic blocks. But the routing of these logic blocks is more complex in FPGAs, which offer fewer input/output logic resources than CPLDs.

Each connection node can be implemented using two techniques, thus defining two classes of FPGAs:

– antifuse-based FPGA. The connection node is of ROM type. In this case, the modification of a connection node is irreversible. The advantage of an antifuse-based FPGA is the large number of connection nodes due to the reduced fuse surface;

– SRAM-based FPGA (reprogrammable FPGA). The connection node is made up of a set of transistors whose switching is controlled by a configuration register. An SRAM-based FPGA has a logic mechanism for autocharging the configuration registers from the EEPROM after powering on, or on demand.

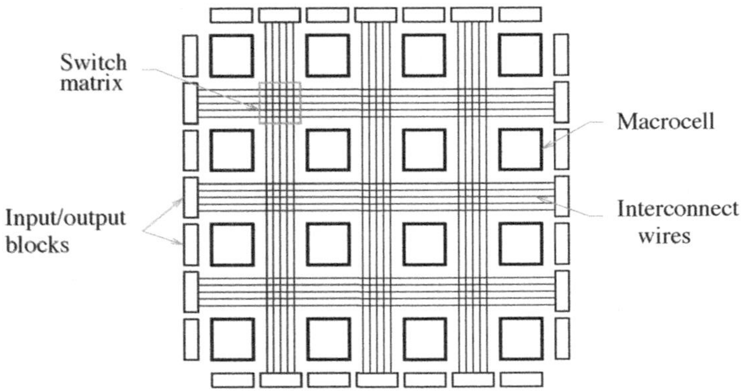

Figure 7.33. *Architecture of an FPGA*

The structure of a configurable logic block in an *Altera Stratix V* FPGA is given in Figure 7.34. It contains a combinational logic section including LUTs, two adders, flip-flops and different control circuits. A logic block can be used to implement any function with more than six inputs, some seven-inputs functions and certain combinations of two independent functions.

In general, the choice of the complexity degree for the logic block is dictated by the tradeoffs between conflicting factors. The less complex a logic block, the greater the number of resources needed for its routing; but as the complexity of the logic block increases it can result in reduced efficiency in the capacity utilization of the available logic circuits.

During normal operation, the different configuration bits of an FPGA are stored on an SRAM, which is volatile. They must, therefore, be saved on a non-volatile on-chip memory, especially a Flash or EEPROM, in order to be used during the configuration period which follows every power-on.

The architecture of the logic block varies depending on the manufacturer. The Virtex-5 FPGA from Xilinx, for example, uses two different types of logic blocks. One to implement logic, arithmetic and ROM functions, the other to store data on distributed RAM and to shift data using registers.

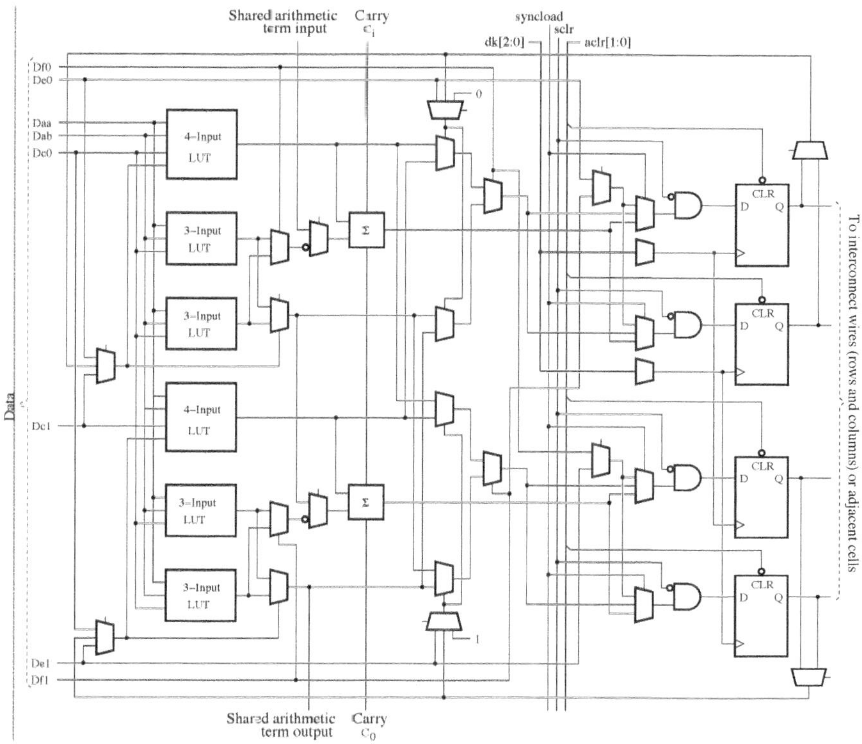

Figure 7.34. *Structure of a configurable logic block (or macrocell) in an FPGA (Altera Stratix V)*

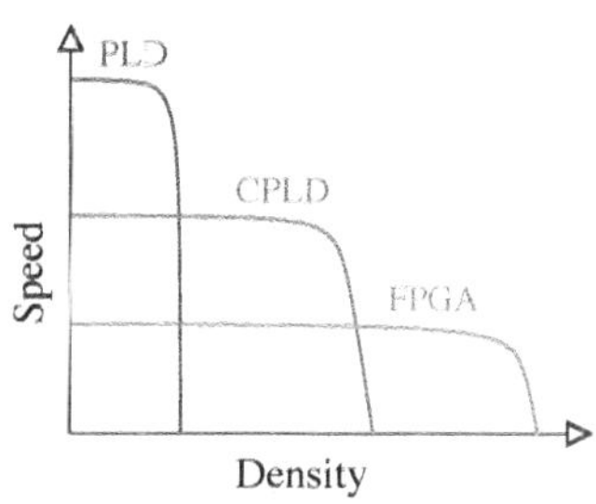

Figure 7.35. *Comparison of programmable logic circuits*

Input/output blocks of an FPGA play the role of a communication interface with external components. To ensure flexibility and versatility, they are configurable via the control bits, thus allowing for the connection of different types of components and unidirectional (input or output) or bidirectional (input/output) data transfer. Additionally, in order to meet electrical specifications, a polarization circuit, made up of resistors and transistors, is associated with each input/output pin.

The structure of a configurable input/output block is represented in Figure 7.36. It is made up of three-state buffer circuits, multiplexers and flip-flops that can be used as such or combined with internal flip-flops to form input/output registers.

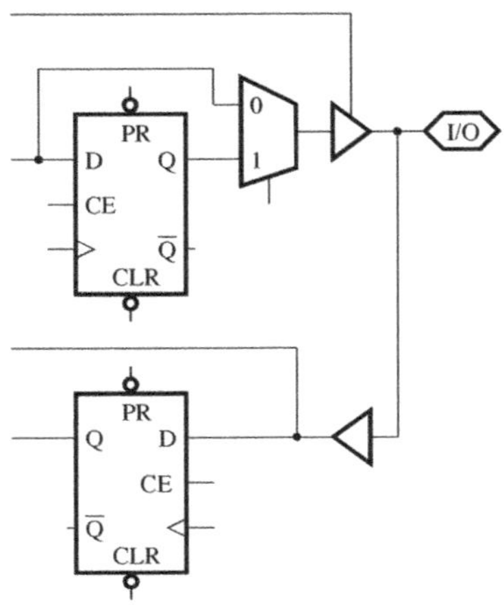

Figure 7.36. *Configurable input/output blocks*

The necessity of implementing high-speed data transfer translates to an increase in the complexity of the input/output blocks. As a result, the input/output boxes have registers instead of flip-flops, alignment and synchronization circuits, parallel-to-serial and serial-to-parallel converters, encoders and decoders.

7.5. References

– Xilinx website: *www.xilinx.com*;

– Altera website: *www.altera.com*;

– Microsemi website: *www.microsemi.com*;

– Lattice semiconductor website: *www.latticesemi.com*

7.6. Exercises

EXERCISE 7.1.– Consider the following logic functions:

$$F(A,B,C,D) = A \cdot \overline{C} + B \cdot \overline{C} \cdot D \quad [7.36]$$

$$G(A,B,C,D) = A \cdot \overline{B} \cdot \overline{C} + \overline{B} \cdot D + \overline{A} \cdot \overline{C} \cdot D \quad [7.37]$$

– Implement these functions using three $2 : 4$ decoders (see Figure 7.37 and Table 7.3) and two OR gates that can take up to eight inputs. The decoder is enabled when $\overline{EN1} = 0$ and $EN2 = 1$.

– Implement these functions using:

a) a programmable 4-input PROM circuit;

b) a programmable 4-input PAL circuit.

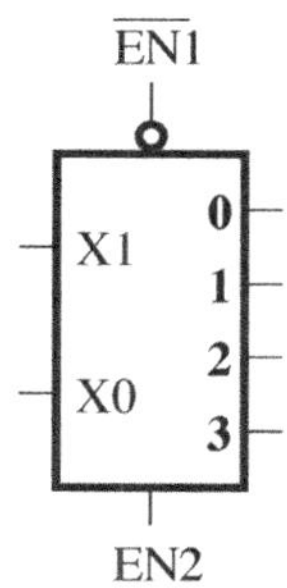

Figure 7.37. *Decoder*

$\overline{EN1}$	$EN2$	X1	X0	0	1	2	3
x	0	x	x	0	0	0	0
1	x	x	x	0	0	0	0
0	1	0	0	1	0	0	0
0	1	0	1	0	1	0	0
0	1	1	0	0	0	1	0
0	1	1	1	0	0	0	1

Table 7.3. *Truth table*

EXERCISE 7.2.– BCD-to-7-segment decoder.

A number with four bits A, B, C, D (where is the least significant bit) is applied to the decoder inputs, yielding the signals a, b, c, d, e, f and g, which can be used to control a seven-segment display (see Figure 7.38) generating numbers from 0 to 9.

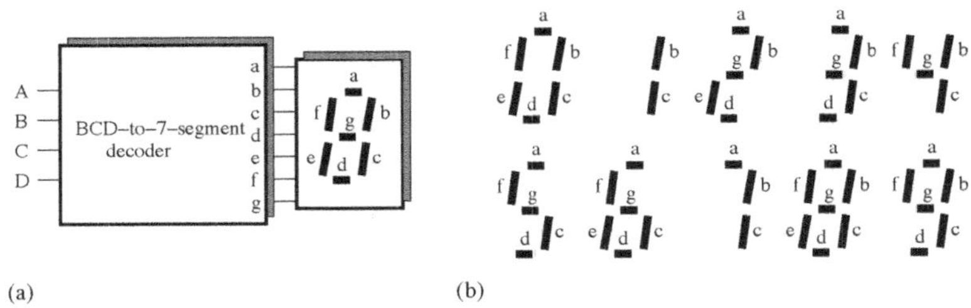

Figure 7.38. *a) BCD-to-7-segment decoder; b) display of numbers from 0 to 9*

Implement the BCD-to-7-segment decoder using a PLA circuit.

We will assume that the input combinations 1010, 1011, 1100, 1101 and 1111 are unused.

EXERCISE 7.3.– Full adder.

A full adder can be implemented as shown in Figure 7.39, where the bits to be added are A_i and B_i, C_i denotes the carry-in, the sum corresponds to S_i and the carry-out is given by C_{i+1}.

By appropriately decomposing the logic equations for a full adder, implement the logic circuit i using logic gates.

EXERCISE 7.4.– Implementation of a 9-dot display decoder.

A 9-dot display can be used to implement a circuit for lighting animation. The decoder inputs for the 9-dot display consist of 4-bit numbers, A, B, C and D, that determine the state of the display (see Figure 7.40). Each of the decoder's nine outputs ($\overline{a}$, $\overline{b}$, $\overline{c}$, $\overline{d}$, $\overline{e}$, $\overline{f}$, $\overline{g}$, $\overline{h}$ and $\overline{i}$) is active-low and is connected to a corresponding dot of the display.

The different states of the display associated with the inputs $ABCD$ are illustrated in Figure 7.41. No dot of the display is enabled for inputs associated with 0, while inputs related to 9 are used to activate all dots of the display:

– Draw up the truth table of the 9-dot display decoder.

– Simplify the logic expressions $\overline{a}$, $\overline{b}$, $\overline{c}$, $\overline{d}$, $\overline{e}$, $\overline{f}$, $\overline{g}$, $\overline{h}$ and $\overline{i}$ (consider, if necessary, the cases where the functions are independent and dependent).

– Implement the decoder using a PLA.

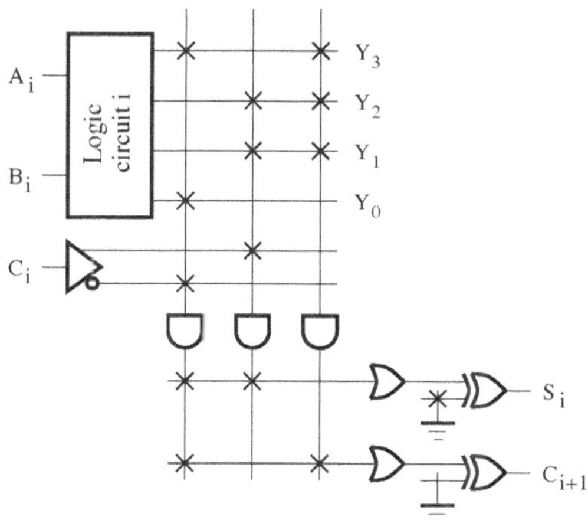

Figure 7.39. *Logic circuit of a full adder*

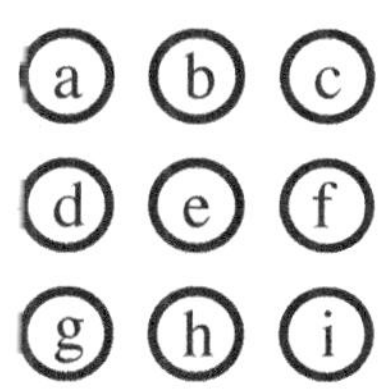

Figure 7.40. *Representation of the 9-dot display*

EXERCISE 7.5.– Multiplexer circuit.

The multiplexer circuit shown in Figure 7.42 has two data inputs (A and B), two outputs (F and G) and two control inputs (S_1 and S_0).

– Draw up its truth table based on A and B.

– Determine the role of this circuit.

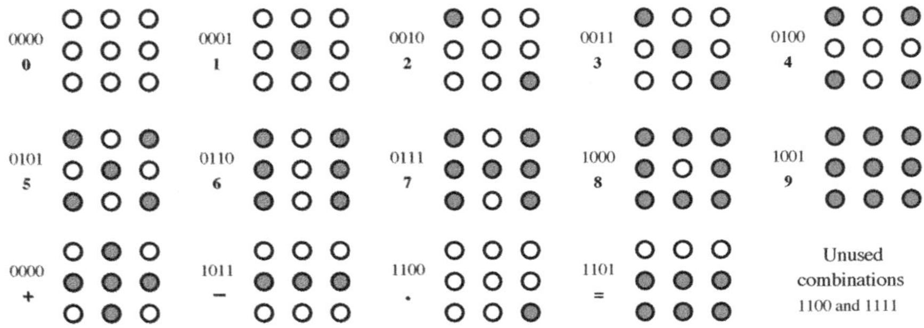

Figure 7.41. *Operation of the 9-dot display*

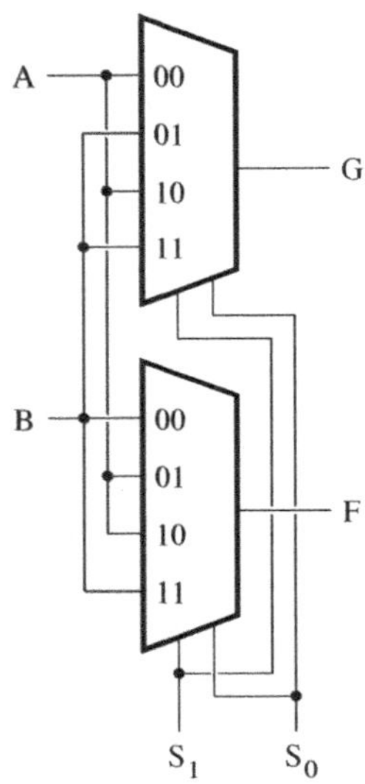

Figure 7.42. *Circuit using two 4 : 1 multiplexers*

EXERCISE 7.6.– Circuit for a logic element.

The circuit shown in Figure 7.43 is a logic element that can be used to implement any 3-variable boolean function.

Complete the column F of the function table that is shown in Table 7.44.

Determine the combination of the variables A, B, S_2, S_1 and S_0, that are required for the implementation of the function $F = X \cdot Y \oplus Y \cdot Z \oplus Y \cdot Z$.

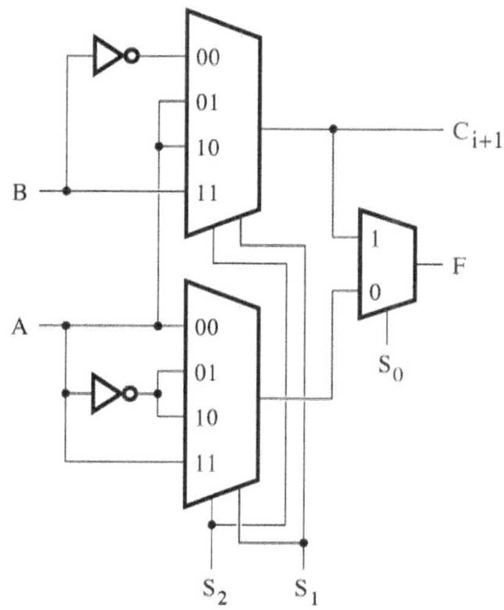

Figure 7.43. *Logic circuit of a logic element*

A	B	S_2	S_1	S_0	F	C_{i+1}
X	Y	X	Z	Z		–
0	X	Y	Z	1		–
0	X	$\overline{Y}$	$\overline{Y}$	Z		–
0	X	$\overline{Y}$	Z	$\overline{Y}$		–
0	0	X	Y	Z		–
0	Y	0	X	Z		–
X	1	Y	Z	1		–
X	Y	1	1	Z		–
Z	Z	1	$\overline{X}$	$\overline{Y}$		–
X	1	Y	Z	0	$X \oplus Y \oplus Z$	$X \cdot Y + X \cdot Z + Y \cdot Z$

Figure 7.44. *Function table (F)*

Specify the choice of variables A, B, S_2, S_1 and that can be used to implement a full adder, whose inputs are X and Y and whose carry-in is C_i.

EXERCISE 7.7.– Implementation of a 5-variable function based on 4-input LUTs.

Consider the 5-variable Boolean function that can be written as follows:

$$F(A,B,C,D,E) = [A{\cdot}B{\cdot}\overline{D}{\cdot}\overline{E}+(\overline{A}+E)D+(A\oplus B)E]\odot[B+(A\oplus C)] \quad [7.38]$$

Implement this function using two or three LUTs.

EXERCISE 7.8.– Implementation of an adder and a comparator using 4-input LUTs.

A 4-input LUT can be implemented by combining two 3-input LUTs and a 2 : 1 multiplexer, as shown in Figure 7.45.

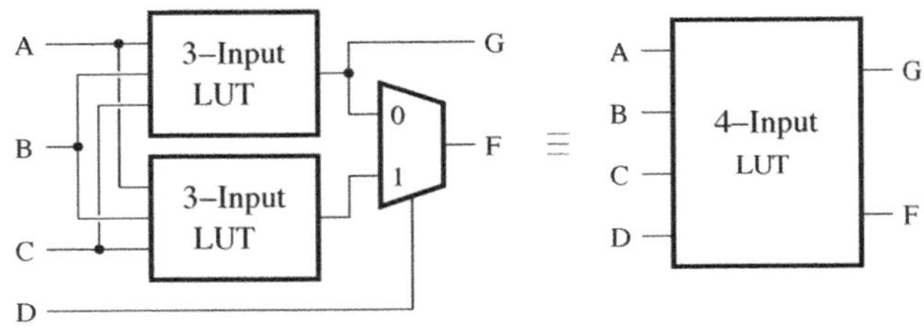

Figure 7.45. *Four-input LUT*

Determine the logic expressions for G_1, F_1, G_0 and F_0, considering that the circuit of Figure 7.46 plays the role of an adder whose inputs are the numbers $A = A_1A_0$ and $B = B_1B_0$, and the carry-in C_i, and whose outputs are the sum $S = S_1S_0$ and the carry-out C_2.

Determine the logic expressions for F_{01} and F_{23} so that the circuit of Figure 7.47 may operate as a comparator setting the $S_{A=B}$ to 1 when the two binary numbers, $A = A_3A_2A_1A_0$ and $B = B_3B_2B_1B_0$, are equal.

EXERCISE 7.9.– Logic element circuit.

Consider a programmable circuit based on a logic element that is composed of an XOR gate, a D flip-flop and a 2 : 1 multiplexer, as shown in Figure 7.48:

– complete the truth table given in Table 7.4 for all the possible values of the control signals S_1 and S_0. The input signal of the flip-flop is represented by D and the output of the multiplexer is denoted by F;

– using three-state buffers, modify the circuit shown in Figure 7.48 so that the pin X can be used as an input or output depending on the state of the control signal S_2;

– modify the resulting circuit by using an AND gate to allow for the synchronous reset of the D flip-flop via the control signal S_3.

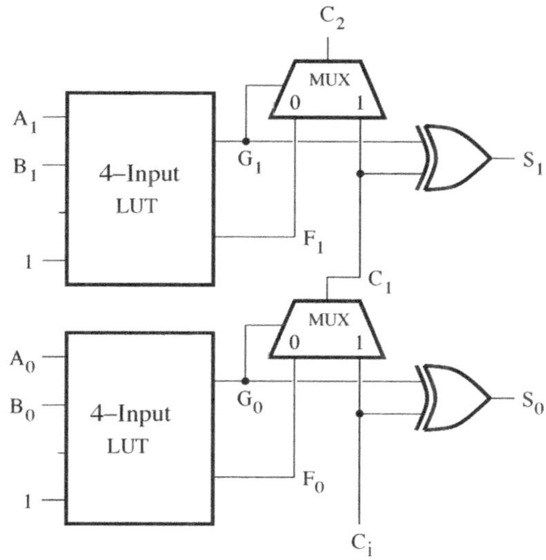

Figure 7.46. *Adder*

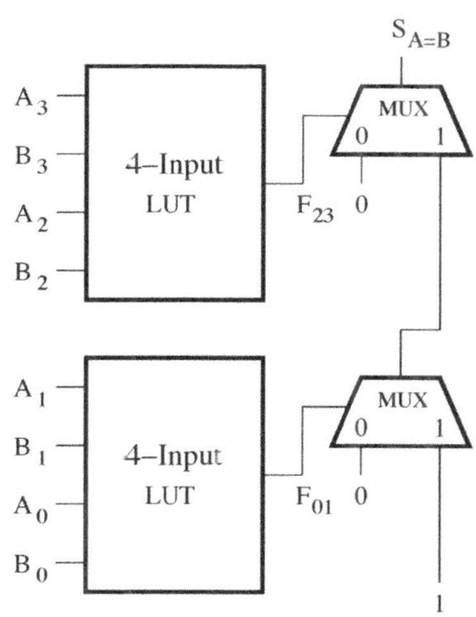

Figure 7.47. *Comparator*

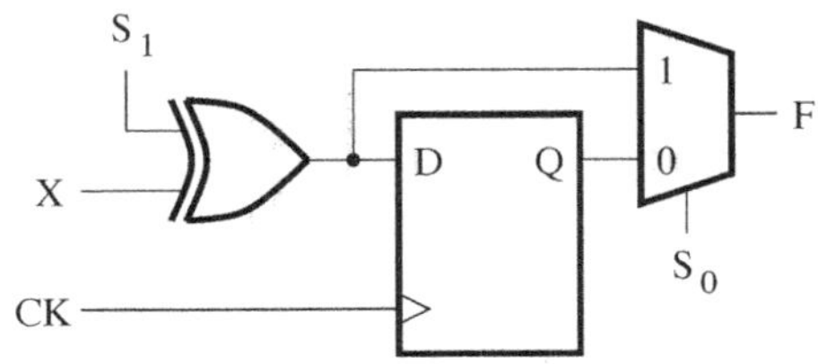

Figure 7.48. *Logic element circuit*

S_1 S_0	D F

Table 7.4. *Truth table to be completed*

EXERCISE 7.10.– Timing diagram of a logic element for a programmable circuit.

Consider a programmable circuit based on the logic element of Figure 7.49a that consists of logic gates (AND, OR, XOR), a 2 : 1 multiplexer and a D flip-flop.

Complete the timing diagram shown in Figure 7.49(b).

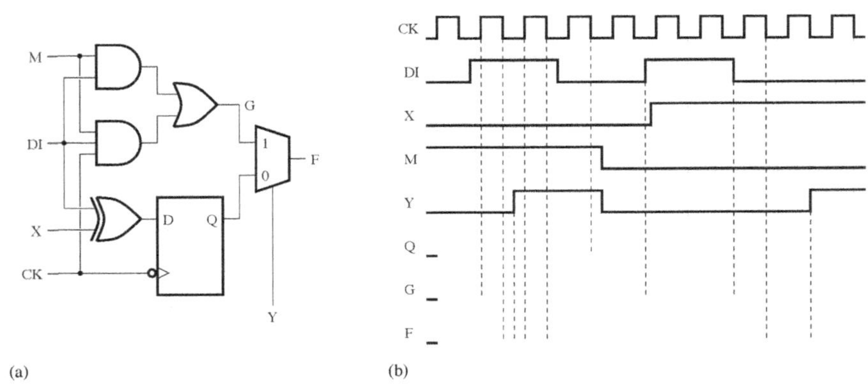

Figure 7.49. *a) Logic circuit; b) timing diagram*

EXERCISE 7.11.– Logic function implementations using an LUT.

The circuit shown in Figure 7.50 constitutes a logic element in an FPGA, whose functional resource usage is to be optimized. It is composed of two 2-input LUTs, two 2 : 1 multiplexers and logic gates (inverter and OR).

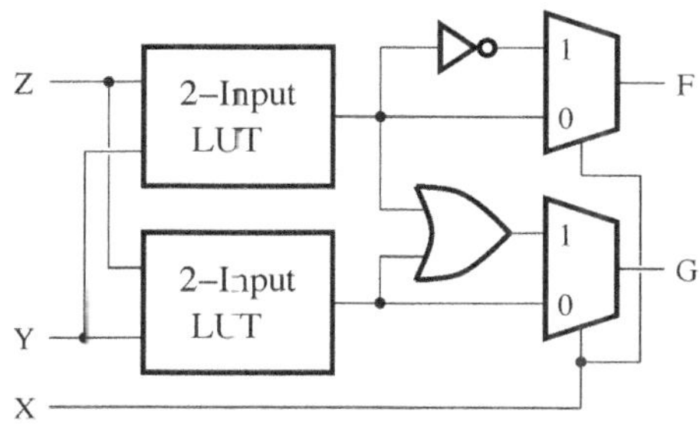

Figure 7.50. *Logic circuit*

Show that this circuit can be used to implement each of the following functions (see the truth tables shown in Tables 7.5 and 7.6):

– a full adder (S and C_0);

– a comparator with an output indicating equality, E_0.

C_i	A	B	S	C_0
0	0	0	0	0
0	0	1	1	0
0	1	0	1	0
0	1	0	0	1
1	0	0	1	0
1	0	1	0	1
1	1	0	0	1
1	1	0	1	1

Table 7.5. *Truth table*

E_i	A	B	E_O
0	0	0	0
0	0	1	0
0	1	0	0
0	1	0	0
1	0	0	1
1	0	1	0
1	1	0	0
1	1	0	1

Table 7.6. *Truth table*

7.7. Solutions

SOLUTION 7.1.– Implementation of the logic functions F and G.

The logic functions F and G can be written as follows:

$$F(A,B,C,D) = A\cdot\overline{C} + B\cdot\overline{C}\cdot D$$

$$= \sum m(5,8,9,12,13) \quad [7.39]$$

$$G(A,B,C,D) = A\cdot\overline{B}\cdot\overline{C} + \overline{B}\cdot D + \overline{A}\cdot\overline{C}\cdot D$$

$$= \sum m(1,3,5,8,9,11) \quad [7.40]$$

Figure 7.51(a) depicts a possible implementation of the functions F and G based on decoders and OR gates. When the variables C and D are chosen to activate the decoders, the combination $C\cdot\overline{D}$ is not required.

Using a PROM, the functions F and G can be implemented as shown in Figure 7.51(b).

To implement the functions F and G using a PLA, it is useful to simplify these two functions while bringing out their common terms. Thus, the Karnaugh maps shown in Figure 7.52 can be used to obtain expressions of the form:

$$F(A,B,C,D) = A\cdot\overline{C} + \overline{A}\cdot B\cdot\overline{C}\cdot D \quad [7.41]$$

$$G(A,B,C,D) = A\cdot\overline{B}\cdot\overline{C} + \overline{B}\cdot D + \overline{A}\cdot B\cdot\overline{C}\cdot D \quad [7.42]$$

The PLA that can be used to realize the functions F and G is shown in Figure 7.53.

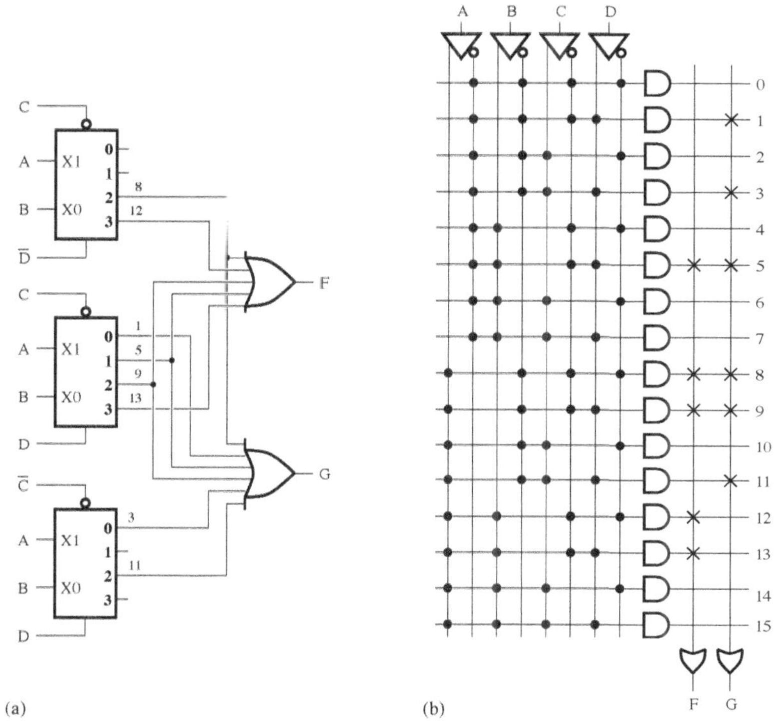

Figure 7.51. *Implementation of F and G using a) decoders and b) a PROM*

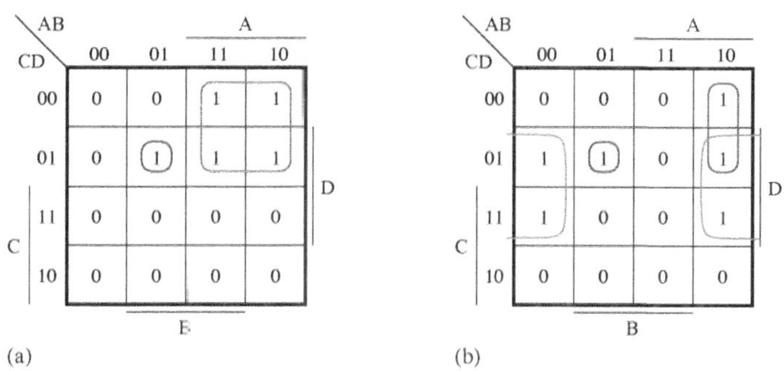

Figure 7.52. *Karnaugh maps: a) F; b) G. For a color version of this figure, see www.iste.co.uk/ndjountche/electronics2.zip*

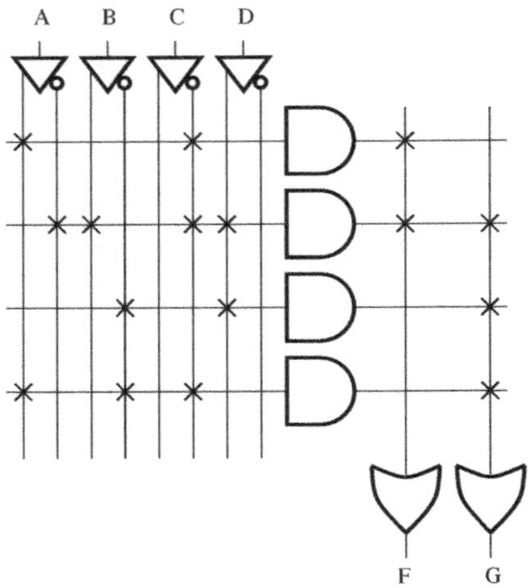

Figure 7.53. *PLA for the implementation of F and G*

SOLUTION 7.2.– BCD-to-7-segment decoder.

The truth table and Karnaugh maps for each of the outputs can be used to obtain the following logic equations:

$$a = \sum m(0,2,3,5,6,7,8,9)$$
$$= A + C + B \cdot D + \overline{B} \cdot \overline{D} \quad [7.43]$$

$$b = \sum m(0,1,2,3,4,7,8,9)$$
$$= \overline{B} + C \cdot D + \overline{C} \cdot \overline{D} \quad [7.44]$$

$$c = \sum m(0,1,3,4,5,6,7,8,9)$$
$$= B + \overline{C} + D \quad [7.45]$$

$$d = \sum m(0,2,3,5,6,8,9)$$
$$= A + \overline{B} \cdot C + \overline{B} \cdot \overline{D} + C \cdot \overline{D} + B \cdot \overline{C} \cdot D \quad [7.46]$$

$$e = \sum m(0,2,6,8)$$
$$= \overline{B} \cdot \overline{D} + C \cdot \overline{D} \quad [7.47]$$

$$f = \sum m(0,4,5,6,8,9)$$

$$= A + B \cdot \overline{C} + B \cdot \overline{D} + \overline{C} \cdot \overline{D} \qquad [7.48]$$

and

$$g = \sum m(2, 3, 4, 5, 6, 8, 9)$$
$$= A + B \cdot \overline{C} + \overline{B} \cdot C + C \cdot \overline{D} \qquad [7.49]$$

Because the size (or number of horizontal lines) of the PLA that can be used to implement logic functions depends on the number of different product terms, the above equations must be written in other forms to highlight the common terms. This is shown in Figures 7.54–7.60.

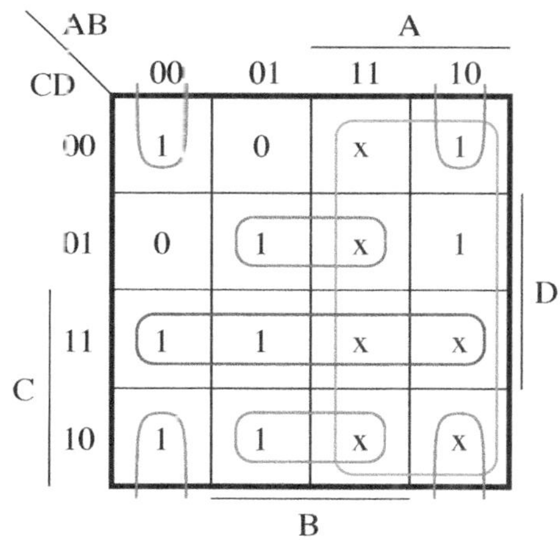

Figure 7.54. *Signal a:* $a = A + \overline{B} \cdot \overline{D} + C \cdot D + B \cdot C \cdot \overline{D} + B \cdot \overline{C} \cdot D$. *For a color version of this figure, see www.iste.co.uk/ndjountche/electronics2.zip*

Comparing these two solutions we can see that the number of different product terms has been reduced from 15 to 9.

In the case of the PLA, the best implementation is based on Boolean equations obtained by using a multiple-output logic minimization approach. This consists of minimizing a set of logic functions while maximizing the sharing of the product terms.

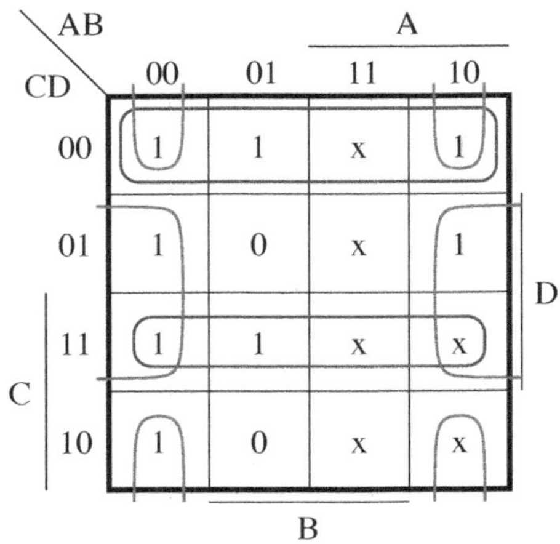

Figure 7.55. *Signal b:*
$b = \overline{B} \cdot D + \overline{B} \cdot \overline{D} + C \cdot D + \overline{C} \cdot \overline{D}$. *For a color version of this figure, see www.iste.co.uk/ndjountche/electronics2.zip*

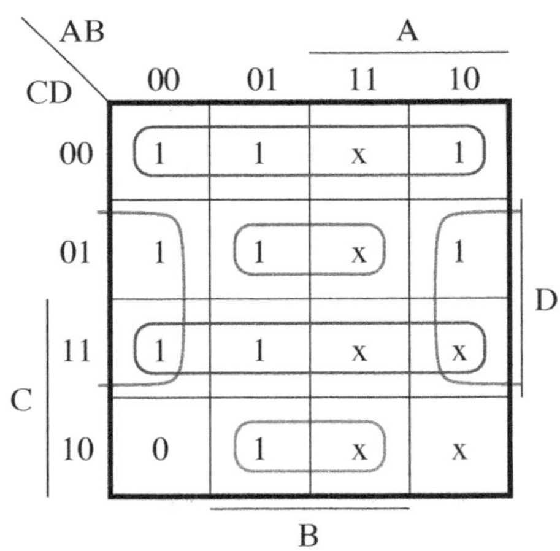

Figure 7.56. *Signal c:*
$c = \overline{B} \cdot D + C \cdot D + \overline{C} \cdot \overline{D} + B \cdot C \cdot \overline{D} + B \cdot \overline{C} \cdot D$. *For a color version of this figure, see www.iste.co.uk/ndjountche/electronics2.zip*

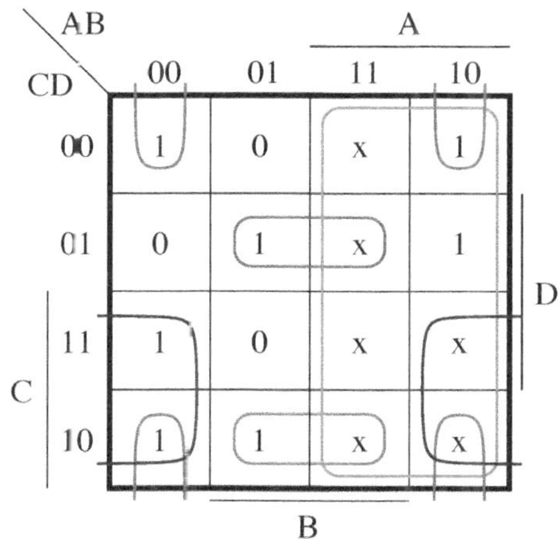

Figure 7.57. *Signal d:*
$d = A + \overline{B} \cdot C + \overline{B} \cdot \overline{D} + B \cdot C \cdot \overline{D} + B \cdot \overline{C} \cdot D$. *For a color version of this figure, see www.iste.co.uk/ndjountche/electronics2.zip*

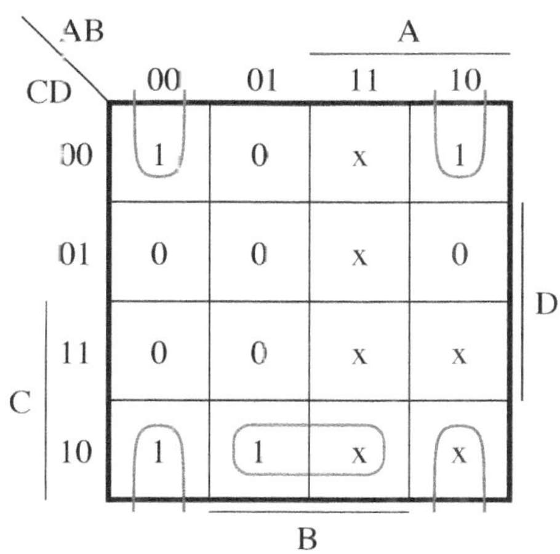

Figure 7.58. *Signal e:*
$e = \overline{B} \cdot \overline{D} + B \cdot C \cdot \overline{D}$. *For a color version of this figure, see www.iste.co.uk/ndjountche/electronics2.zip*

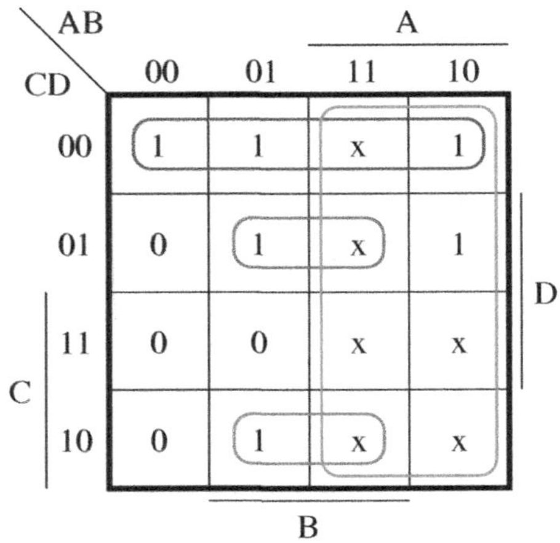

Figure 7.59. *Signal f:*
$f = A + \overline{C} \cdot \overline{D} + B \cdot \overline{C} \cdot D + B \cdot C \cdot \overline{D}$. *For a color version of this figure, see www.iste.co.uk/ndjountche/electronics2.zip*

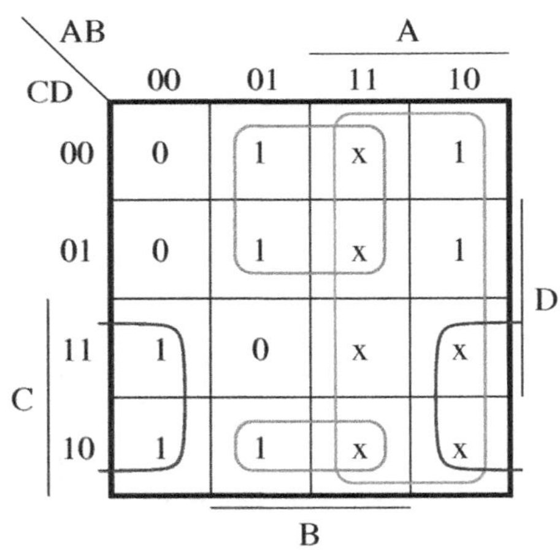

Figure 7.60. *Signal g:*
$g = A + B \cdot \overline{C} + \overline{B} \cdot C + B \cdot C \cdot \overline{D}$. *For a color version of this figure, see www.iste.co.uk/ndjountche/electronics2.zip*

Figure 7.61 depicts the logic circuit for the BCD-to-7-segment decoder.

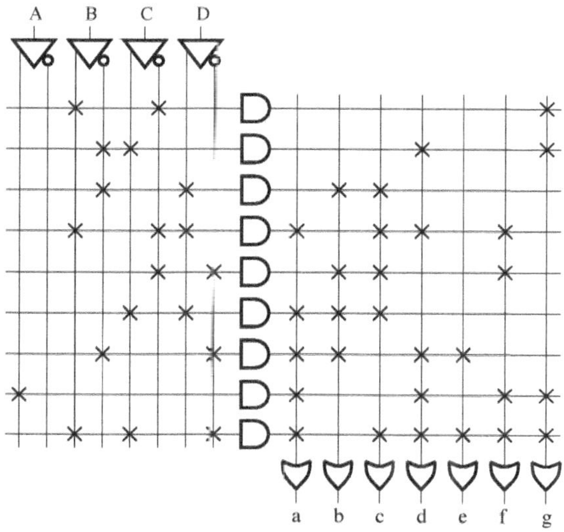

Figure 7.61. *Implementation of the BCD-to-7-segment decoder using a PLA*

SOLUTION 7.3.– Full adder.

A full adder is a circuit with three inputs: the first number being A_i, the second B_i and a carry-in C_i. It provides the sum S_i and the carry-out, C_{i+1}, that can be expressed as follows:

$$S_i = (A_i \cdot \overline{B_i} + \overline{A_i} \cdot B_i)\overline{C_i} + (A_i \cdot B_i + \overline{A_i} \cdot \overline{B_i})C_i \quad [7.50]$$

and:

$$C_{i+1} = A_i \cdot B_i + (A_i + B_i)C_i = A_i \cdot B_i + (\overline{A_i} \cdot B_i + A_i \cdot \overline{B_i})C_i \quad [7.51]$$

The sum S_i can be put into the form:

$$S_i = (\overline{A_i} + \overline{B_i})(A_i + B_i)\overline{C_i} + (\overline{A_i} + B_i)(A_i + \overline{B_i})C_i \quad [7.52]$$

The complement of the carry-out, C_{i+1}, is given by:

$$\overline{C_{i+1}} = \overline{A_i \cdot B_i + (A_i + B_i)C_i} \quad [7.53]$$

$$= (\overline{A_i} + \overline{B_i})(\overline{A_i} + \overline{C_i})(\overline{B_i} + \overline{C_i}) \quad [7.54]$$

$$= \overline{A_i} \cdot \overline{B_i} + (\overline{A_i} + \overline{B_i})\overline{C_i} \quad [7.55]$$

But it can also be decomposed as follows:

$$\overline{C_{i+1}} = \overline{A_i \cdot B_i + (\overline{A_i} \cdot B_i + A_i \cdot \overline{B_i})C_i} \quad [7.56]$$

$$= (\overline{A_i} + \overline{B_i})(\overline{A_i} + B_i)(A_i + \overline{B_i})$$

$$+ (\overline{A_i} + \overline{B_i})(1 + \overline{A_i} + A_i + \overline{B_i} + B_i)\overline{C_i} \quad [7.57]$$

$$= (\overline{A_i} + \overline{B_i})(\overline{A_i} + B_i)(A_i + \overline{B_i})$$

$$+ (\overline{A_i} + \overline{B_i} + \overline{A_i} \cdot \overline{B_i} + A_i \cdot \overline{B_i} + \overline{A_i} \cdot B_i)\overline{C_i} \quad [7.58]$$

Because:

$$\overline{A_i} \cdot \overline{B_i} = (\overline{A_i} + \overline{B_i})(\overline{A_i} + B_i)(A_i + \overline{B_i}) \quad [7.59]$$

$$\overline{A_i} = \overline{A_i}(B_i + \overline{B_i}) = \overline{A_i} \cdot B_i + \overline{A_i} \cdot \overline{B_i} \quad [7.60]$$

and:

$$\overline{B_i} = (A_i + \overline{A_i})\overline{B_i} = A_i \cdot \overline{B_i} + \overline{A_i} \cdot \overline{B_i} \quad [7.61]$$

we obtain:

$$\overline{C_{i+1}} = (\overline{A_i} + \overline{B_i})(\overline{A_i} + B_i)(A_i + \overline{B_i}) + (\overline{A_i} \cdot \overline{B_i} + A_i \cdot \overline{B_i} + \overline{A_i} \cdot B_i)\overline{C_i} \quad [7.62]$$

$$= (\overline{A_i} + \overline{B_i})(\overline{A_i} + B_i)(A_i + \overline{B_i})(1 + \overline{C_i}) + (A_i \cdot \overline{B_i} + \overline{A_i} \cdot B_i)\overline{C_i} \quad [7.63]$$

$$= (\overline{A_i} + \overline{B_i})(\overline{A_i} + B_i)(A_i + \overline{B_i}) + (\overline{A_i} + \overline{B_i})(A_i + B_i)\overline{C_i} \quad [7.64]$$

Figure 7.62 depicts the logic circuit of the decoder, which can generate *maxterms* that appear in the decomposition of S_i and C_{i+1}.

SOLUTION 7.4.– Decoder of a 9-dot display for a lighting animation circuit.

The truth table of the 9-dot display decoder can be drawn up as shown in Table 7.7. The simplification of the equation for each of the outputs using Karnaugh maps is shown in Figures 7.63–7.70.

It must be noted that the expression for $\overline{h}$ presents the advantage of comprising only preexisting terms but it is different from the minimal form that contains the term $B \cdot \overline{D}$ instead of the term $A \cdot B \cdot \overline{D}$.

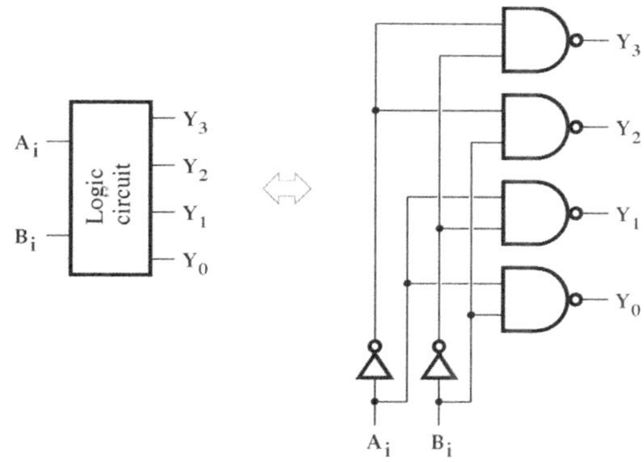

Figure 7.62. *Logic circuit of the decoder*

Symbol	A	B	C	D	$\overline{a}$	$\overline{b}$	$\overline{c}$	$\overline{d}$	$\overline{e}$	$\overline{f}$	$\overline{g}$	$\overline{h}$	$\overline{i}$
0	0	0	0	1	1	1	1	1	1	1	1	1	1
1	0	0	0	1	1	1	1	1	0	1	1	1	1
2	0	0	1	0	0	1	1	1	1	1	1	1	0
3	0	0	1	1	0	1	1	1	0	1	1	1	0
4	0	1	0	0	0	1	0	1	1	1	0	1	0
5	0	1	0	1	0	1	0	1	0	1	0	1	0
6	0	1	1	0	0	1	0	0	1	0	0	1	0
7	0	1	1	1	0	1	0	0	0	0	0	1	0
8	1	0	0	0	0	0	0	0	1	0	0	0	0
9	1	0	0	1	0	0	0	0	0	0	0	0	0
+	1	0	1	0	1	0	1	0	0	0	1	0	1
−	1	0	1	1	1	1	1	0	0	0	1	1	1
.	1	1	0	0	1	1	1	1	1	1	1	1	0
=	1	1	0	1	1	1	1	0	0	0	0	0	0
	1	1	1	0	x	x	x	x	x	x	x	x	x
	1	1	1	1	x	x	x	x	x	x	x	x	x

Table 7.7. *Truth table of the 9-dot display decoder*

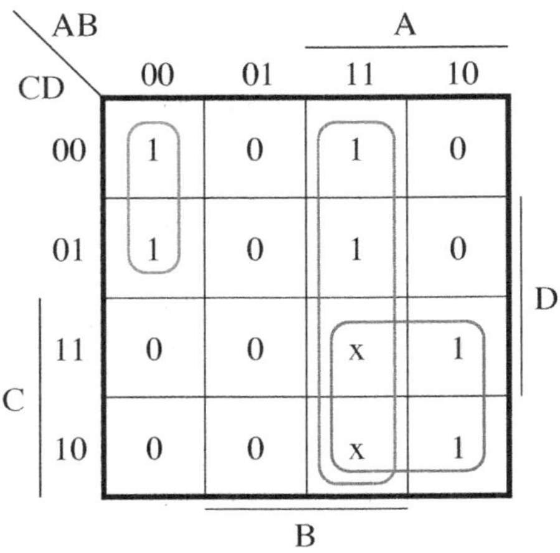

Figure 7.63. *Signal* $\overline{a}$*:*
$\overline{a} = A \cdot B + A \cdot C + \overline{A} \cdot \overline{B} \cdot \overline{C}$. *For a color version of this figure, see www.iste.co.uk/ndjountche/electronics2.zip*

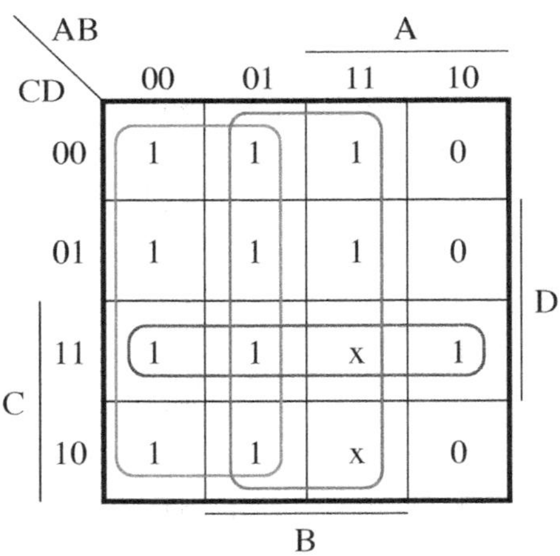

Figure 7.64. *Signal* $\overline{b}$*:*
$\overline{b} = \overline{A} + B + C \cdot D$. *For a color version of this figure, see www.iste.co.uk/ndjountche/electronics2.zip*

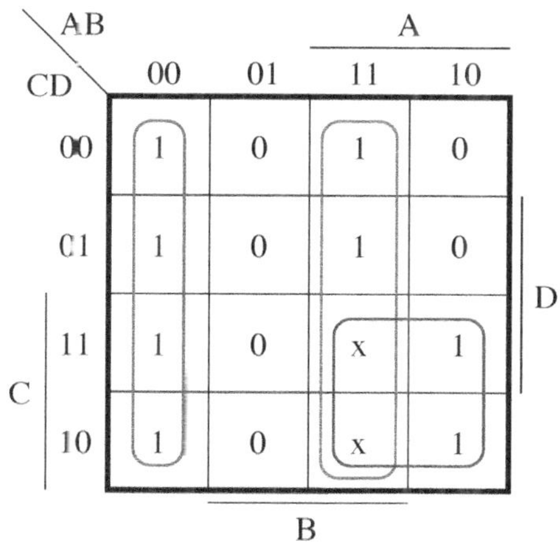

Figure 7.65. *Signal* $\overline{c}$*:*
$\overline{c} = A \cdot B + A \cdot C + \overline{A} \cdot \overline{B}$. *For a color version of this figure, see www.iste.co.uk/ndjountche/electronics2.zip*

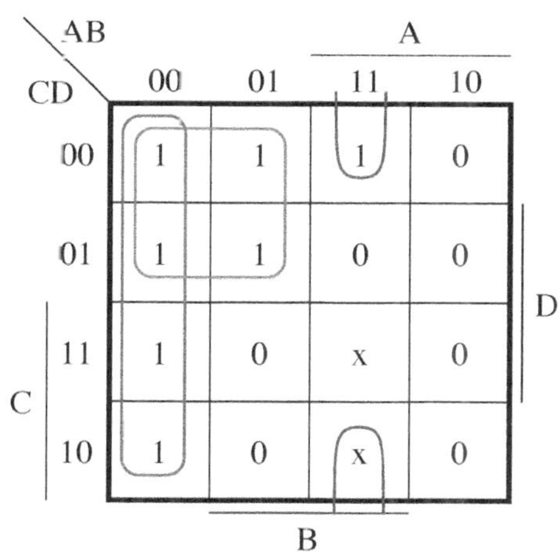

Figure 7.66. *Signals* $\overline{d}$ *and* $\overline{f}$*:*
$\overline{d} = \overline{f} = \overline{A} \cdot \overline{B} + \overline{A} \cdot \overline{C} + A \cdot B \cdot \overline{D}$. *For a color version of this figure, see www.iste.co.uk/ndjountche/electronics2.zip*

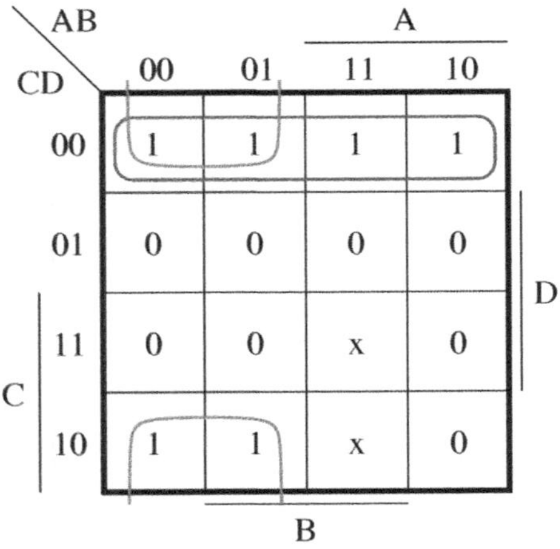

Figure 7.67. *Signal $\overline{e}$: $\overline{e} = \overline{A} \cdot \overline{D} + \overline{C} \cdot \overline{D}$. For a color version of this figure, see www.iste.co.uk/ndjountche/electronics2.zip*

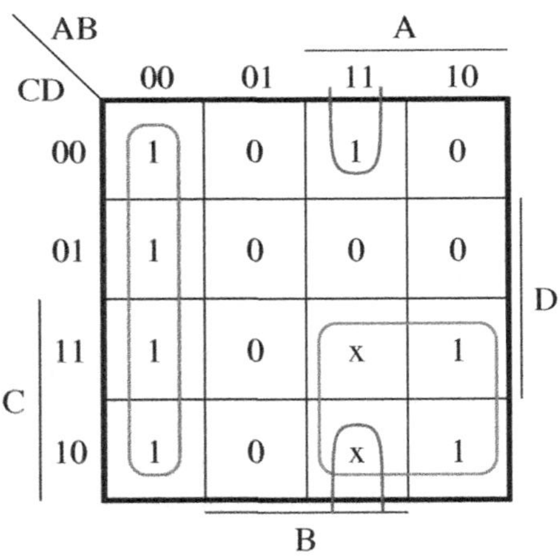

Figure 7.68. *Signal $\overline{g}$: $\overline{g} = \overline{A} \cdot \overline{B} + A \cdot C + A \cdot B \cdot \overline{D}$. For a color version of this figure, see www.iste.co.uk/ndjountche/electronics2.zip*

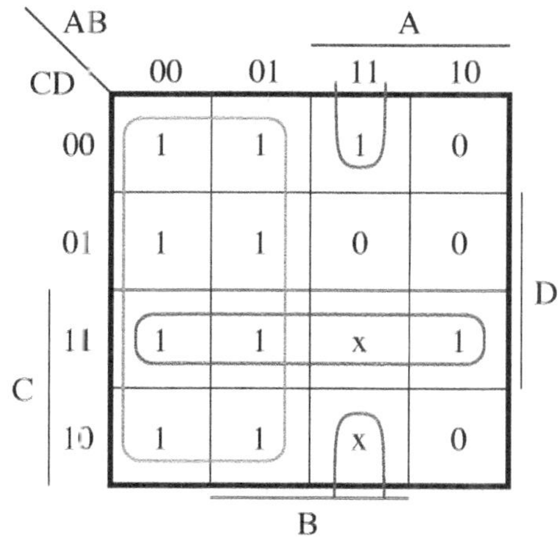

Figure 7.69. *Signal* $\overline{h}$*:*
$\overline{h} = \overline{A} + C \cdot D + A \cdot B \cdot \overline{D}$*. For a color version of this figure, see www.iste.co.uk/ndjountche/electronics2.zip*

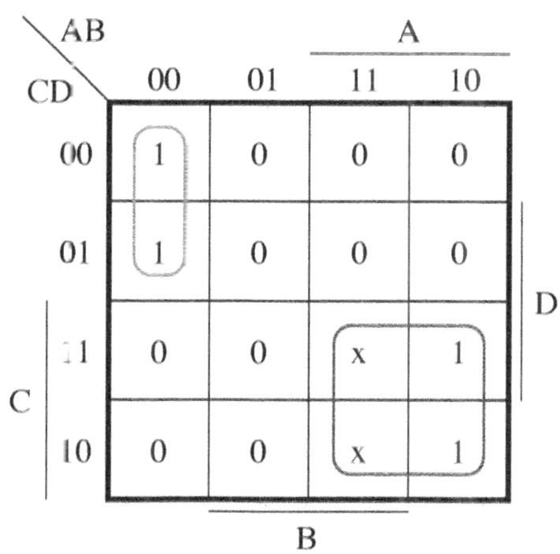

Figure 7.70. *Signal* $\overline{i}$*:*
$\overline{i} = A \cdot C + \overline{A} \cdot \overline{B} \cdot \overline{C}$*. For a color version of this figure, see www.iste.co.uk/ndjountche/electronics2.zip*

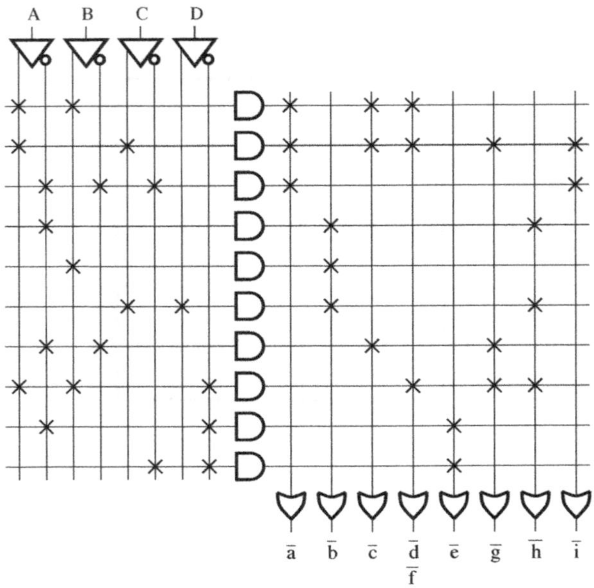

Figure 7.71. *Logic circuit of the 9-dot display decoder*

Figure 7.71 shows the 9-dot display decoder implemented using a PLA.

SOLUTION 7.5.– Multiplexer circuit.

Analyzing the proposed logic circuit, we obtain the following equations:

$$F = A \cdot \overline{S_1} \cdot \overline{S_0} + B \cdot \overline{S_1} \cdot S_0 + A \cdot S_1 \cdot \overline{S_0} + B \cdot S_1 \cdot S_0 \qquad [7.65]$$

and

$$G = B \cdot \overline{S_1} \cdot \overline{S_0} + A \cdot \overline{S_1} \cdot S_0 + A \cdot S_1 \cdot \overline{S_0} + B \cdot S_1 \cdot S_0 \qquad [7.66]$$

The truth table can then be drawn up as shown in Table 7.8.

This is a cross-bar switch, the symbol for which is given in Figure 7.72.

SOLUTION 7.6.– Logic element circuit.

The analysis of the logic element provides the following equation:

$$F = [A \cdot \overline{S_2} \cdot \overline{S_1} + \overline{A}(S_2 \cdot \overline{S_1} + \overline{S_2} \cdot S_1) + A \cdot S_2 \cdot S_1]\overline{S_0}$$
$$+ [\overline{B} \cdot \overline{S_2} \cdot \overline{S_1} + A(S_2 \cdot \overline{S_1} + \overline{S_2} \cdot S_1) + B \cdot S_2 \cdot S_1]S_0 \qquad [7.67]$$

S_1	S_0	F	G
0	0	A	B
0	1	B	A
1	0	A	A
1	1	B	B

Table 7.8. *Truth table of the cross-bar switch*

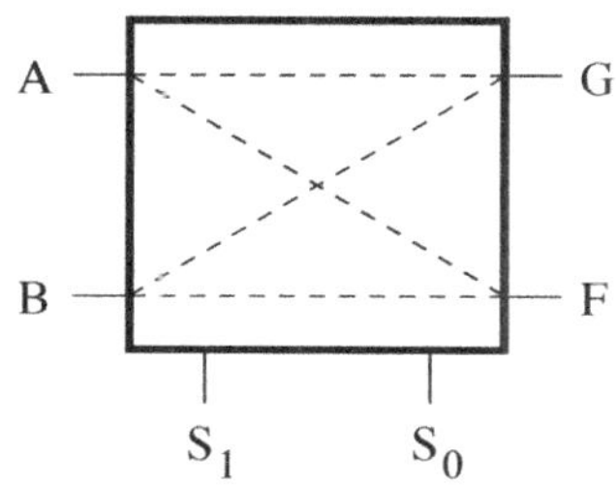

Figure 7.72. *Symbol of the cross-bar switch*

A	B	S_2	S_1	S_0	F	C_{i+1}
X	Y	X	Z	Z	$X \cdot Y \cdot Z$	–
0	X	Y	Z	1	$X \cdot Y \cdot Z + \overline{X} \cdot \overline{Y} \cdot \overline{Z}$	–
0	X	$\overline{Y}$	$\overline{Y}$	Z	$(X \oplus Y)Z$	–
0	X	$\overline{Y}$	Z	$\overline{Y}$	$(X + Y)Z$	–
0	0	X	Y	Z	$X \cdot \overline{Y} \cdot \overline{Z} + \overline{X} \cdot Y \cdot \overline{Z} + \overline{X} \cdot \overline{Y} \cdot Z$	–
0	Y	0	X	Z	$X \cdot \overline{Z} + \overline{X} \cdot \overline{Y} \cdot Z$	–
X	1	Y	Z	1	$X \cdot Y + Y \cdot Z + X \cdot Z$	–
X	Y	1	1	Z	$X \cdot \overline{Z} + Y \cdot Z$	–
Z	Z	1	$\overline{X}$	$\overline{Y}$	$(X \cdot Y) \oplus Z$	–
X	1	Y	Z	0	$X \oplus Y \oplus Z$	$X \cdot Y + Y \cdot Z + X \cdot Z$

Table 7.9. *Function table (F)*

Table 7.9 shows the function table of the logic element.

For the combination of the variables $A = X$, $B = 1$, $S_2 = Y$, $S_1 = Z$ and $S_0 = 1$, we have:

$$F = X \cdot Y + Y \cdot Z + X \cdot Z = X \cdot Y \oplus Y \cdot Z \oplus X \cdot Z \quad [7.68]$$

The choice of variables that can be used to implement a full adder is given below:

$$A = X, B = 1, S_2 = Y, S_1 = C_i \text{ and } S_0 = 0.$$

That is:

$$F = S = X \oplus Y \oplus C_i \qquad \text{and} \qquad C_{i+1} = X \cdot Y + X \cdot C_i + Y \cdot C_i \quad [7.69]$$

SOLUTION 7.7.– Implementation of a 5-variable function using a 4-input LUT.

The Boolean function to be implemented is given by:

$$F(A,B,C,D,E) = [A \cdot B \cdot \overline{D} \cdot \overline{E} + (\overline{A} + E)D + (A \oplus B)E] \odot [B + (A \oplus C)] \quad [7.70]$$

In this form, it can be implemented using two LUTs, as shown in Figure 7.73(a). The LUT 1 implements the function $X = [A \cdot B \cdot \overline{D} \cdot \overline{E} + (\overline{A} + E)D + (A \oplus B)E]$, and the LUT 2 implements the function $X \odot [B + (A \oplus C)]$.

The function F can also be decomposed according to Shannon's expansion theorem. Thus:

$$F(A,B,C,D,E) = \overline{A} \cdot F(0,B,C,D,E) + A \cdot F(1,B,C,D,E) \quad [7.71]$$

where:

$$F(0,B,C,D,E) = (D + B \cdot E) \odot (B + C) \quad [7.72]$$

and:

$$F(1,B,C,D,E) = (B \cdot \overline{D} \cdot \overline{E} + D \cdot E + \overline{B} \cdot E) \odot (B + \overline{C}) \quad [7.73]$$

The circuit obtained in this case is illustrated in Figure 7.73(b). The functions $F(0,B,C,D,E)$ and $F(1,B,C,D,E)$ are implemented by LUT 1 and LUT 2, while the LUT 3 is configured as a 2 : 1 multiplexer whose select signal is A.

SOLUTION 7.8.– Implementation of an adder and comparator using 4-input LUTs.

A full adder is characterized by the following Boolean equations:

$$S_i = (A_i \oplus B_i) \oplus C_i \quad [7.74]$$

$$C_{i+1} = (A_i \oplus B_i)C_i + A_i \cdot B_i \quad [7.75]$$

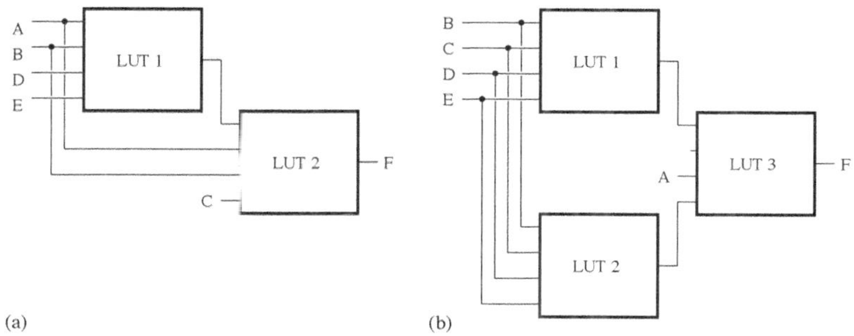

Figure 7.73. *Circuits for the implementation of F*

We have:

$$S_1 = G_1 \oplus C_1 \quad [7.76]$$

$$S_0 = G_0 \oplus C_i \quad [7.77]$$

$$C_2 = C_1 \cdot G_1 + F_1 \cdot \overline{G_1}$$

and:

$$C_1 = C_i \cdot G_0 + F_0 \cdot \overline{G_0} \quad [7.78]$$

By comparison, we can deduce the following logic expressions:

$$G_1 = A_1 \oplus B_1 \quad [7.79]$$

$$F_1 = A_1 \cdot B_1 \quad [7.80]$$

$$G_0 = A_0 \oplus B_0 \quad [7.81]$$

and:

$$F_0 = A_0 \cdot B_0 \quad [7.82]$$

For a 4-bit comparator, the logic equation for the output $S_{A=B}$ is written as:

$$S_{A=B} = \overline{(A_3 \oplus B_3)}\ \overline{(A_2 \oplus B_2)}\ \overline{(A_1 \oplus B_1)}\ \overline{(A_0 \oplus B_0)} \quad [7.83]$$

Because:

$$S_{A=B} = F_{23} \cdot F_{01} \quad [7.84]$$

it follows that:

$$F_{01} = \overline{(A_1 \oplus B_1)}\;\overline{(A_0 \oplus B_0)} \qquad [7.85]$$

and:

$$F_{23} = \overline{(A_3 \oplus B_3)}\;\overline{(A_2 \oplus B_2)} \qquad [7.86]$$

SOLUTION 7.9.– Logic element circuit.

By analyzing the circuit for the logic element in Figure 7.74, we have:

$$F = D \cdot S_0 + Q \cdot \overline{S_0} \qquad [7.87]$$

and

$$Q^+ = D = X \oplus S_1 \qquad [7.88]$$

where Q^+ denotes the next state of the flip-flop output. Table 7.10 shows the truth table of the logic element.

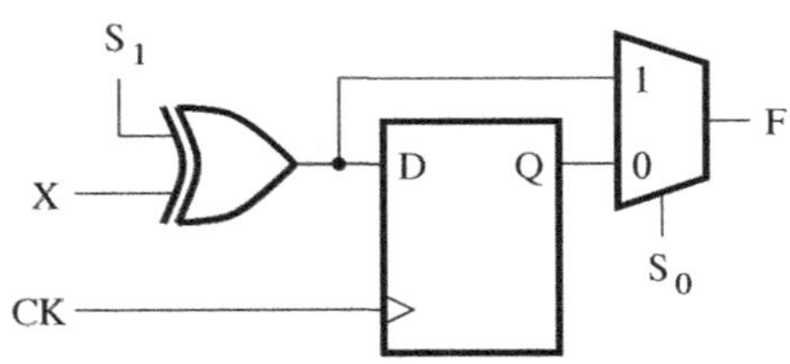

Figure 7.74. *Logic element circuit*

S_1	S_0	D	F
0	0	X	Q
0	1	X	D
1	0	$\overline{X}$	Q
1	1	$\overline{X}$	D

Table 7.10. *Truth table*

The logic element circuit with an I/O pin is shown in Figure 7.75.

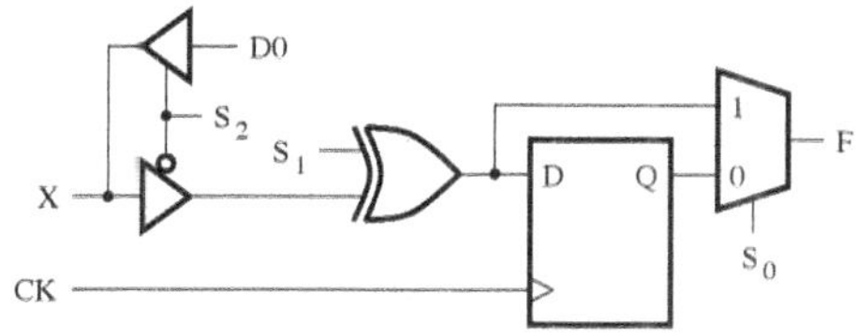

Figure 7.75. *Circuit of a logic element with an I/O pin*

The logic element circuit with a synchronous reset input for the flip-flop is shown in Figure 7.76.

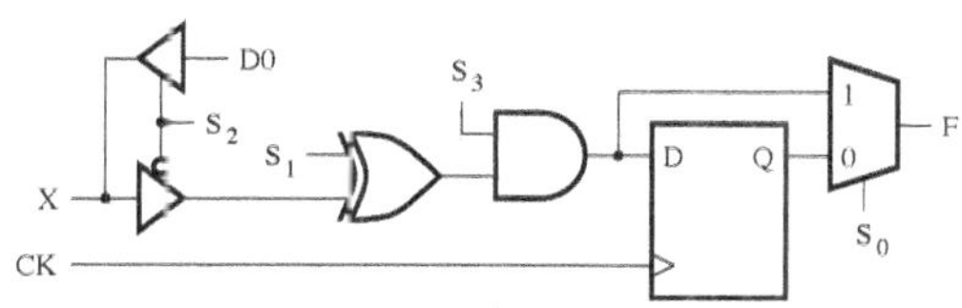

Figure 7.76. *Circuit of a logic element with a synchronous reset input*

SOLUTION 7.10.– Timing diagram of the logic element for a programmable circuit.

An analysis of the logic element circuit yields the following equations:

$$F = G \cdot Y + Q \cdot \overline{Y} \quad [7.89]$$

where

$$G = M \cdot DI + M \cdot DI \cdot \overline{CK} = M \cdot DI(1 + CK) = M \cdot DI \quad [7.90]$$

and the next state of the flip-flop is defined by:

$$Q^{+} = D = X \oplus DI \quad [7.91]$$

The timing diagram of the logic element shown in Figure 7.77(a) is represented in Figure 7.77(b).

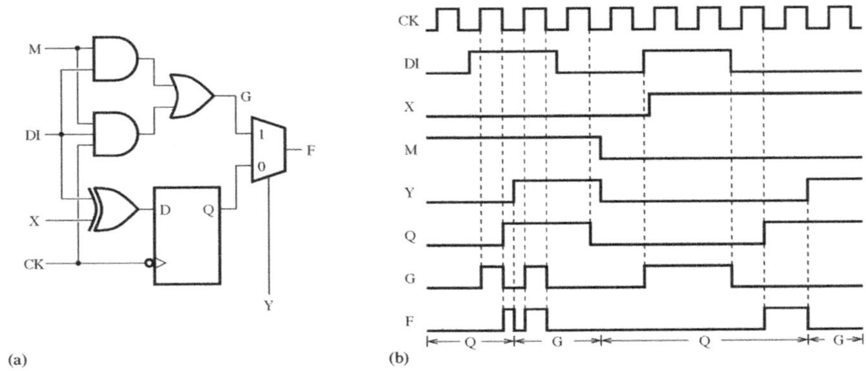

Figure 7.77. *a) Logic circuit; b) timing diagram*

SOLUTION 7.11.– Implementation of logic functions using an LUT.

Shannon's theorem can be used to decompose the sum, S, as follows:

$$S = \overline{C_i} \cdot S(0, A, B) + C_i \cdot S(1, A, B) \qquad [7.92]$$

Because, based on the truth table shown in Table 7.11, the different values of $S(0, A, B)$ are 0110, and the values of $S(1, A, B)$ are 1001, it follows that $1001 = \overline{0}\,\overline{1}\,\overline{1}\,\overline{0}$, or similarly $S(1, A, B) = \overline{S(0, A, B)}$. Thus:

$$S = \overline{C_i} \cdot S(0, A, B) + C_i \cdot \overline{S(0, A, B)} \qquad [7.93]$$

C_i	A	B	S	C_O
0	0	0	0	0
0	0	1	1	0
0	1	0	1	0
0	1	0	0	1
1	0	0	1	0
1	0	1	0	1
1	1	0	0	1
1	1	0	1	1

Table 7.11. *Truth table of the adder*

For the carry-out, C_O, we have:

$$C_O = \overline{C_i} \cdot C_O(0, A, B) + C_i \cdot C_O(1, A, B) \quad [7.94]$$

The truth table gives the representation 0001 for $C_O(0, A, B)$, 0111 for $C_O(1, A, B)$ and 0110 for $S(0, A, B)$. This can be exploited to write $C_O(1, A, B) = S(0, A, B) + C_O(0, A, B)$, and finally:

$$C_O = \overline{C_i} \cdot C_O(0, A, B) + C_i[S(0, A, B) + C_O(0, A, B)] \quad [7.95]$$

The logic equation for the output E_O of the comparator can take the form:

$$E_O = \overline{C_i} \cdot E_O(0, A, B) + C_i \cdot E_O(1, A, B) \quad [7.96]$$

Because, according to the truth table shown in Table 7.12, $E_O(0, A, B)$ and $E_O(1, A, B)$ take the values 0000 and 1001, respectively, it can be established that $E_O(1, A, B) = E_O(0, A, B) + E_O(1, A, B)$. Hence:

$$E_O = \overline{C_i} \cdot E_O(0, A, B) + C_i[E_O(0, A, B) + E_O(1, A, B)] \quad [7.97]$$

E_i	A	B	E_O
0	0	0	0
0	0	1	0
0	1	0	0
0	1	0	0
1	0	0	1
1	0	1	0
1	1	0	0
1	1	0	1

Table 7.12. *Truth table of the comparator*

The input and output specifications of the logic element are given in Table 7.13 for each of the functions to be implemented.

Z	Y	X	F	G	Function
A	B	C_i	S	C_0	Adder
A	B	C_i	-	E_0	Comparator

Table 7.13. *Input and output specifications of the logic element*

Appendix

Digital Integrated Circuits and Software

A1.1. Combinational circuit

- 74HC00 Quad 2-Input NAND Gate
- 74HC02 Quad 2-Input Nor Gate
- 74HC04 Hex Inverter
- 74HC86 Quad 2-input XOR Gate
- 74HC138 3 of 8 Decoder
- 74HC139 Dual 2 of 4 Decoder
- 74HC157 Quad 2 to 1 Multiplexer non-inverting
- 74HC153 Dual 4 to 1 Data Selector/Multiplexer
- 74HC151 8 to 1 Data Selector/Multiplexer
- 74HC139 Dual 2 of 4 Decoder
- 74HC147 10-Decimal to 4-BCD Priority Encoder
- 74HC148 8 to 3 Octal Priority Encoder
- 74HC154 4 of 16 Line Decoder/Demultiplexer
- 74LS47 BCD-7 Segment Decoder/Driver
- 74HC42 BCD/Decimal Decoder
- 74HC283 4 Bit Binary Full Adder with Fast Carry
- 74HC583 4 Bit BCD Full Adder with Fast Carry
- 74HC182 4 Bit Lookahead Carry Generator

– 74HC181 4 Bit ALU/Function Generator

– 74HC384 8 by 1 Bit Twos-Complement Multiplier

– 74HC85 4 Bit Magnitude Comparator

– 74HC365 Hex Buffer W/Common Enable

A1.2. Sequential circuit

– 74LS279 RS Latch

– 74HC75 Quad Bistable Transparent Latch

– 74HC74 Dual D-Flip-Flop with Preset and Clear

– 74HC109 Dual JK Flip-Flop with Preset and Clear

– 74HC194 Four Bit Bidirectional Universal Shift Register

– 74HC670 Four by Four Register File

– 74HC93 Four Bit Binary Ripple Counter

– 74HC161 Four Bit Synchronous Counter

– 74HC699 Binary Up/Down Counter, Synchronous Clear Latch

A1.3. Memory and programmable circuit

– 74HC189 64 (16 × 4) Bit RAM Inverting

– CY6264 8 K × 8 Static RAM

– Altera: Stratix-10 FPGA; Cyclone V FPGA; Max V CPLD

– Xilinx: Virtex-7 FPGA; Spartan-6 FPGA; CoolRunner-II CPLD

A1.4. Computer-aided design software

– Quartus Prime Design Software: *www.altera.com*

– ISE Design Suite; Vivado Design Suite: *www.xilinx.com*

Bibliography

[BRO 08] BROWN S., VRANESIC Z., *Fundamentals of digital logic with VHDL design*, 3rd ed., McGraw-Hill Education, New York City, NY, 2008.

[CLE 00] CLEMENTS A., *The principles of computer hardware*, 3rd ed., Oxford University Press, Oxford, UK, 2000.

[COM 95] COMER D.J., *Digital logic and state machine design*, 3rd ed., Oxford University Press, New York City, NY, 1995.

[DUE 01] DUECK R.K., *Digital design with CPLD applications and VHDL*, Delmar Thomson Learning, Albany, NY, 2001.

[GIV 03] GIVONE D., *Digital principles and design*, McGraw-Hill, New York City, NY, 2003.

[HAY 93] HAYES J.P., *Introduction to digital logic design*, Addison-Wesley Publishing Company, Boston, MA, 1993.

[HAY 98] HAYES J.P., *Computer architecture and organization*, McGraw-Hill, New York City, NY, 1998.

[KAT 05] KATZ R.H., BORRIELO G., *Contemporary logic design*, 2nd ed., Prentice Hall, Upper Saddle River, NJ, 2005.

[MAN 01] MANO M.M., *Digital design*, 3rd ed., Prentice Hall, Upper Saddle River, NJ, 2001.

[MAR 10] MARCOVITZ A.B., *Introduction to logic design*, 3rd ed., McGraw-Hill Education, New York City, NY, 2010.

[NDJ 11] NDJOUNTCHE T., *CMOS analog integrated circuits: high-speed and power-efficient design*, CRC Press, Boca Raton, FL, 2011.

[ROT 04] ROTH JR. C.H., *Fundamental of logic design*, 5th ed., Brooks/Cole – Thomson Learning, Belmont, CA, 2004.

[SAN 02] SANDIGE R.S., *Digital design essentials*, Prentice Hall, Upper Saddle River, NJ, 2002.

[TIN 00] TINDER R.F., *Engineering digital design*, Academic Press, San Diego, CA, 2000.

[TOC 03] TOCCI R.J., AMBROSIO F.J., *Microprocessors and microcomputers*, 6th ed., Prentice Hall, Upper Saddle River, NJ, 2003.

[WAK 00] WAKERLY J.F., *Digital design principles and practices*, 3rd ed., Prentice Hall, Upper Saddle River, NJ, 2000.

[WIL 98] WILKINSON B., *The essence of digital design*, Prentice Hall Europe, Hemel Hempstead, UK, 1998.

[YAR 97] YARBROUGH J.M., *Digital logic – Applications and design*, West Publishing Company, St. Paul, MN, 1997.

Index

N, O, P

R

S

T, W

Other titles from

iSTE

in

Electronics Engineering

2015

DURAFFOURG Laurent, ARCAMONE Julien
Nanoelectromechanical Systems

2014

APPRIOU Alain
Uncertainty Theories and Multisensor Data Fusion

CONSONNI Vincent, FEUILLET Guy
Wide Band Gap Semiconductor Nanowires 1: Low-Dimensionality Effects and Growth
Wide Band Gap Semiconductor Nanowires 2: Heterostructures and Optoelectronic Devices

GAUTIER Jean-Luc
Design of Microwave Active Devices

LACAZE Pierre Camille, LACROIX Jean-Christophe
Non-volatile Memories

TEMPLIER François
OLED Microdisplays: Technology and Applications

THOMAS Jean-Hugh, YAAKOUBI Nourdin
New Sensors and Processing Chain

2013

COSTA François, GAUTIER Cyrille, LABOURE Eric, REVOL Bertrand
Electromagnetic Compatibility in Power Electronics

KORDON Fabrice, HUGUES Jérôme, CANALS Agusti, DOHET Alain
Embedded Systems: Analysis and Modeling with SysML, UML and AADL

LE TIEC Yannick
Chemistry in Microelectronics

2012

BECHERRAWY Tamer
Electromagnetism: Maxwell Equations, Wave Propagation and Emission

LALAUZE René
Chemical Sensors and Biosensors

LE MENN Marc
Instrumentation and Metrology in Oceanography

SAGUET Pierre
Numerical Analysis in Electromagnetics: The TLM Method

2011

ALGANI Catherine, RUMELHARD Christian, BILLABERT Anne-Laure
Microwaves Photonic Links: Components and Circuits

BAUDRANT Annie
Silicon Technologies: Ion Implantation and Thermal Treatment

BESNIER Philippe, DÉMOULIN Bernard
Electromagnetic Reverberation Chambers

DEFAY Emmanuel
Integration of Ferroelectric and Piezoelectric Thin Films: Concepts ans Applications for Microsystems

DEFAY Emmanuel
Ferroelectric Dielectrics Integrated on Silicon

LANDIS Stefan
Nano-lithography

2010

LANDIS Stefan
Lithography

PIETTE Bernard
VHF / UHF Filters and Multicouplers

2009

DE SALVO Barbara
Silicon Non-volatile Memories / Paths of Innovation

DECOSTER Didier, HARARI Joseph
Optoelectronic Sensors

FABRY Pierre, FOULETIER Jacques
Chemical and Biological Microsensors / Applications in Fluid Media

GAUTIER Jacques
Physics and Operation of Silicon Devices in Integrated Circuits

MOLITON André
Solid-State Physics for Electronics

PERRET Robert
Power Electronics Semiconductor Devices

SAGUET Pierre
Passive RF Integrated Circuits

2008

CHARRUAU Stéphane
Electromagnetism and Interconnections

2007

RIPKA Pavel, TIPEK Alois
Modern Sensors Handbook

Printed and bound by CPI Group (UK) Ltd, Croydon, CR0 4YY

13/09/2026

14961054-0002